[redacted]

The Major Histocompatibility Complex Region of Domestic Animal Species

CRC Series in

Comparative Immunology

Edited by

Rodney R. Dietert

New and Forthcoming Titles

Experimental Immunotoxicology
Ralph J. Smialowicz and Michael P. Holsapple

The Major Histocompatibility Complex Region of Domestic Animal Species
Lawrence B. Schook and Susan J. Lamont

The Physiology of Immunity
James A. Marsh and Marion D. Kendall

The Major Histocompatibility Complex Region of Domestic Animal Species

Edited by
Lawrence B. Schook
Susan J. Lamont

CRC Press
Boca Raton New York London Tokyo

Publisher:	Bob Stern
Project Editor:	Debbie Didier
Marketing Manager:	Susie Carlisle
Direct Marketing Manager:	Becky McEldowney
Cover Design:	Denise Craig
Manufacturing:	Sheri Schwartz

Library of Congress Cataloging-in-Publication Data

The major histocompatibility complex region of domestic animal species
/ edited by Lawrence B. Schook, Susan J. Lamont.
p. cm. -- (Comparative immunology)
Includes bibliographical references and index.
ISBN 0-8493-8032-4
1. Major histocompatibility complex. 2. Veterinary immunogenetics. I. Schook, Lawrence B. II. Lamont, Susan J., 1953- . III. Series: CRC series in comparative immunology.
SF757.234.M34 1996
636.089′6079--dc20 96-12590
CIP

Direct all inquiries to CRC Press, Inc., 2000 Corporate Blvd., N.W., Boca Raton, Florida 33431.

International Standard Book Number 0-8493-8032-4
Library of Congress Card Number 96-12590
Printed in the United States of America 1 2 3 4 5 6 7 8 9 0
Printed on acid-free paper

Preface

Recent years have witnessed an explosion in information regarding the major histocompatibility complex (Mhc) of domestic animal species. There are many factors contributing to this exponential increase in knowledge. New research technologies, especially those related to molecular genetic analysis, allow researchers to explore the genetic structure of the Mhc at a level of detail heretofore impossible. A "trickle-down" effect of the tremendous resources invested in the human genome project has enhanced the availability of technical approaches applicable of study of the Mhc of domestic animal species. Moreover, as the conservation of structure and function of the biologically essential genes of the Mhc region becomes increasingly evident, studies can efficiently focus on synteny groups and functionally homologous genes by utilizing information from better-characterized species. The Mhc is one of the few identified gene systems in domestic animals that is associated with quantitative traits such as disease resistance, immune response, growth, and reproduction. As knowledge continues to accumulate, we move closer toward the reality of genetic enhancement of animal welfare and production efficiency via integrated management of the Mhc. Thus, the time was right to publish this comprehensive text on the fundamental and practical knowledge now available on the Mhc of domestic animal species. We envision this book to be of use not only to researchers in this specific field, but also to university and postgraduate students, veterinarians, comparative immunologists and geneticists, and animal breeders.

Our gratitude is extended to many people: to Rod Dietert, who invited us to edit this book and recognized the importance and timeliness of the topic; to Harvey Kane, Bob Stern, Debbie Didier, and Carol Messing from CRC Press for their help with the editorial process. Our thanks go to all the chapter authors who prepared such excellent contributions for this book, and to the unsung heroes who play such essential support roles in the research laboratories and offices of all the authors. Special thanks are due to Ann Shuey and Gretchen Triplett. Most of all, our appreciation is extended to our families who, once again, did without us as we attended to this contribution to the scientific community.

Susan J. Lamont
Lawrence B. Schook

The Editors

Lawrence B. Schook, Ph.D., is Professor and Chair of Veterinary PathoBiology in the College of Veterinary Medicine and Director of the Food Animal Biotechnology Center at the University of Minnesota, St. Paul.

Dr. Schook received his B.A. degree from Albion College, Albion, Michigan, in 1972. He obtained his M.S. and Ph.D. degrees in 1975 and 1978, respectively, from the Department of Immunology and Microbiology, Wayne State School of Medicine, Detroit, Michigan. After performing postdoctoral work at the Institute of Clinical Immunology, Berne, Switzerland, and the Departments of Microbiology and Surgery, University of Michigan Medical School, he was appointed an Assistant Professor of Microbiology and Immunology at the Medical College of Virginia in 1982 and an Associate Professor of Molecular Immunology in the Department of Animal Sciences at the University of Illinois at Urbana-Champaign in 1986, and Professor in 1990. In 1993, Dr. Schook was appointed Professor and Chair of Veterinary PathoBiology at the University of Minnesota. During 1992-1993, Professor Schook was a Visiting Professor, Faculty of Medicine, University of Lausanne, Switzerland, in the Ludwig Institute for Cancer Research.

In 1990, while at the University of Illinois, Professor Schook was selected as a University Scholar and the recipient of the H.H. Mitchell Award for Excellence in Graduate Training and Research. In 1992, he received the Paul A. Funk Award for Outstanding and Meritorius Research Achievement in Agriculture. He has served on numerous national and international committees and is the former Chair of the USDA National Animal Genome Research Program and served as the 1992 USDA National Research Initiatives Panel Manager for Animal Molecular Genetics.

He is a member of the American Society of Animal Science, American Society of Human Genetics, American Association of Immunology, American Society for Microbiology, International Society of Animal Genetics, and Society of Toxicology. He received fellowships from the National Cancer Institute, Fogerty/Swiss National Fund, and the Elsa Pardee Foundation.

Dr. Schook is the recipient of research grants from the National Institutes of Health, the National Science Foundation, the USDA Research Inititatives Program, and private industry. He has published more than 100 papers and edited 5 books and is the Editor-in-Chief of *Animal Biotechnology*. His current research interests include identifying genes associated with cell-mediated immunity and developing the pig as a biomedical model.

Susan J. Lamont, Ph.D., is Professor of Animal Science in the College of Agriculture at Iowa State University.

Dr. Lamont received her B.A. degree from Trinity Christian College, Palos Heights, Illinois in 1975. She obtained her Ph.D. degree in 1980 from the University of Illinois School of Basic Medical Sciences, Chicago. After doing postdoctoral work at the University of Massachusetts, Amherst, she was appointed an Assistant Professor of Animal Science at Iowa State University in 1983. She became Associate Professor of Animal Science in 1987 and Professor in 1991. She became an Allied Professor in Microbiology, Immunology, and Preventive Medicine in 1991.

Dr. Lamont is a member of the American Association of Immunologists, the Conference of Research Workers in Animal Disease, the International Society of Anim. Genetics, the Poultry Science Association, the World's Poultry Science Association, and the honorary societies, Sigma Xi and Gamma Sigma Delta. She served as Secretary of the International Society of Animal Genetics and as a Director of the U.S. Branch of the World's Poultry Association, both from 1992 to 1996.

Dr. Lamont has been the recipient of research grants from the U.S. Department of Agriculture, the Binational Agriculture Research and Developmental Foundation, and private industry. She has presented 40 invited lectures at national and international meetings. She has published more than 50 research papers and has been author or co-author of eight book chapters. Her current major research interests relate to molecular genetics and immunogenetics of poultry.

Contributors

Leif Andersson
Department of Animal Breeding and Genetics
Swedish University of Agricultural Sciences
Uppsala, Sweden

Douglas F. Antczak
James A. Baker Institute for Animal Health
New York State College of Veterinary Medicine
Cornell, University
Ithaca, New York

Ernest Bailey
Gluck Equine Research Center
University of Kentucky
Lexington, Kentucky

Keith Ballingall
Moredun Research Institute
Edinburgh, United Kingdom

Marite Bradshaw
Department of Veterinary PathoBiology
College of Veterinary Medicine
University of Minnesota
St. Paul, Minnesota

Johannes Buitkamp
Versuchsstation Thalhausen
Technical University Munich
Kranzberg, Germany

Allan M. Crawford
AgResearch Molecular Biology Unit
Department of Biochemistry and
Centre for Gene Research
Otago University
New Zealand

Bernadette M. Dutia
Department of Veterinary Pathology
University of Edinburgh
Edinburgh, United Kingdom

Joerg Thomas Epplen
Molekulare Humangenetik
Ruhr-Universität Bochum
Bochum, Germany

Elaine D. Ferguson
Department of Veterinary Pathology
University of Edinburgh
Edinburgh, United Kingdom

Heinz Gerber
Klinik für Nutztiere und Pferde
University of Berne
Berne, Switzerland

Unni Grimholt
Division of Genetics
Department of Morphology, Genetics, and Aquatic Biology
Norwegian College of Veterinary Medicine
Oslo, Norway

David Groth
School of Biomedical Sciences
Curtin University
Perth, Western Australia

John Hopkins
Department of Veterinary Pathology
University of Edinburgh
Edinburgh, United Kingdom

Irma Joosten
Institute of Infectious Disease and Immunology
Faculty of Veterinary Medicine
University of Utrecht
Utrecht, The Netherlands

James F. Kaufman
Basel Institute for Immunology
Basel, Switzerland

Susan J. Lamont
Department of Animal Science
Iowa State University
Ames, Iowa

Sandor Lazary
Division of Immunogenetics
Institute of Animal Breeding
University of Berne
Berne, Switzerland

Jong-Keuk Lee
Department of Veterinary PathoBiology
College of Veterinary Medicine
University of Minnesota
St. Paul, Minnesota

Harris A. Lewin
Laboratory of Immunogenetics
Department of Animal Sciences
University of Illinois
Urbana, Illinois

Øystein Lie
Division of Genetics
Department of Morphology, Genetics, and Aquatic Biology
Norwegian College of Veterinary Medicine
Oslo, Norway

Joan K. Lunney
Immunology and Disease Resistance Lab
LPSI, ARS, USDA
Beltsville, Maryland

Jill Maddox
Centre for Animal Biotechnology
University of Melbourne
Melbourne, Australia

Eliane Marti
Klinik für Nutztiere und Pferde
University of Berne
Berne, Switzerland

Susan M. Rhind
Department of Veterinary Pathology
University of Edinburgh
Edinburgh, United Kingdom

Gabriela Obexer-Ruff
Division of Immunogenetics
Institute for Animal Breeding
Veterinary School
University of Berne
Berne, Switzerland

Mark S. Rutherford
Department of Veterinary PathoBiology
College of Veterinary Medicine
University of Minnesota
St. Paul, Minnesota

David Sargan
Department of Clinical Veterinary Medicine
University of Cambridge
Cambridge, United Kingdom

Ulla M. Sarmiento
Department of Pathology
Amgen Inc.
Thousand Oaks, California

Lawrence B. Schook
Department of Veterinary PathoBiology
College of Veterinary Medicine
University of Minnesota
St. Paul, Minnesota

Franz-Werner Schwaiger
Max Planck Institut für Psychatrie
Abt. Neuromorphologie
Martinsried, Germany

Yun-Chao Shia
Department of Veterinary PathoBiology
College of Veterinary Medicine
University of Minnesota
St. Paul, Minnesota

Nopporn Sittisombut
Department of Microbiology
Faculty of Medicine
Chiang Mai University
Chiang Mai, Thailand

Peter A. Swarbrick
Ag Research
Dunedin, New Zealand

Gabor Szalai
Division of Immunogenetics
Institute of Animal Breeding
University of Berne
Berne, Switzerland

John Wetherall
School of Biomedical Sciences
Curtin University
Perth, West Australia

Harry Wright
Moredun Research Institute
Edinburgh, United Kingdom

Contents

CHAPTER 1

Major Histocompatibility Complex Evolution

Leif Andersson

CONTENTS

I. INTRODUCTION

Mhc class I and class II genes encode membrane-bound molecules whose function is to bind and present peptides to T cells of the immune system. Thereby they play a key role in the initiation of an immune response.[1,2] The Mhc class I molecules present endogenously derived peptides to cytotoxic T cells while Mhc class II molecules present exogenously derived peptides to T helper cells. The Mhc class I and class II molecules are structurally strikingly similar, and they are both expressed on the surface of antigen-presenting cells as heterodimeric molecules. Class I molecules are composed of a light chain, β_2 microglobulin (β_2m), and a heavy α chain.

0-8493-8032-4/96/$0.00+$.50

The class II molecule is similarly composed of an α and a β chain. The Mhc subunits are all encoded in the Mhc region except β_2m which is encoded by a gene on a separate chromosome in all mammals investigated.

The Mhc is one of the most well-studied gene regions among vertebrate species. A major impetus for this strong interest is that some Mhc genes and gene products exhibit a remarkably high level of genetic polymorphism. Moreover, the Mhc is particularly interesting since it is one of the few polymorphic systems where it has been possible to firmly establish a functional significance for the observed genetic variation.[1] Thus, the Mhc has been intensively studied in the fields of immunology, genetics, and evolutionary biology. The available data on Mhc structure and function strongly imply that the Mhc diversity has evolved as a response to the selection pressure on the vertebrate immune system to recognize and eliminate invading parasites and microorganisms. Genetic diversity of Mhc molecules is generated by two mechanisms, gene duplications and allelic polymorphism. Both peptide-presenting class I and class II molecules are encoded by multiple loci. The presence of two or more linked loci, which encode distinct molecules, are denoted "fixed heterozygosity", implying that even homozygotes possess multiple forms of antigen-presenting molecules.

This review of the evolution of Mhc genes and their polymorphism will focus primarily on the Mhc in mammalian species which are by far the most well studied. The Mhc of the chicken will be covered in detail in Chapter 3. As an introduction to the subject, the organization of the Mhc region in humans and mouse will be briefly reviewed.

II. ORGANIZATION OF THE Mhc REGION IN HUMANS AND MICE

A. General Organization

The Mhc in humans, denoted HLA for human leukocyte antigens, and in mice, denoted H-2 for histocompatibility locus 2, reside on chromosome 6 and chromosome 17, respectively. The genomic organization of the HLA region is schematically depicted in Figure 1. The HLA spans about 4 million base pairs and thus constitutes about 0.1% of the human genome. The Mhc region is divided into three major subregions: the class I region harboring the peptide-presenting class I molecules, the class II region encoding the peptide-presenting class II molecules, and the class III region containing various genes which are not structurally related to the class I and class II genes. The overall organization of the Mhc region is fairly well conserved among distantly related mammals such as humans, mice, and cattle. However, there are differences between species. For instance, the mouse class I region is divided into two parts as the H-2K locus is found centromeric to the class II region (i.e., to the left in Figure 1). Moreover, the number of certain genes varies between as well as within species. Polymorphisms in the number of genes have been well documented for murine class I genes,[3] for cattle class II DQ genes,[4] and for human DRB, C4, and CYP21 genes.[5]

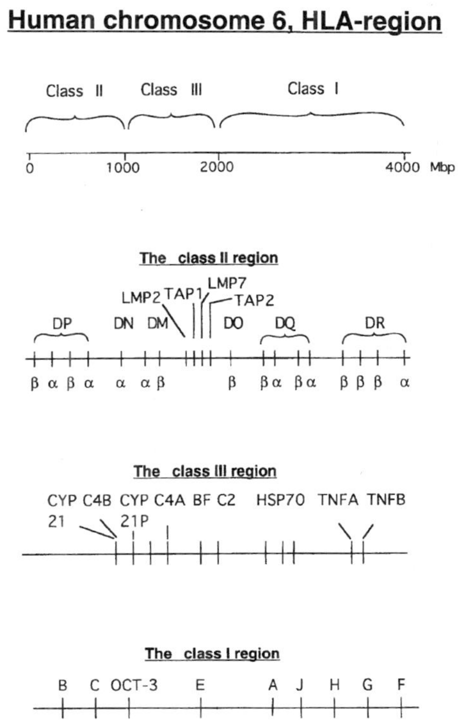

Figure 1 The HLA region of humans (based on Campbell and Trowsdale[5]). Not all genes identified in the HLA region are shown. Mbp, million base pairs.

B. The Class I Region

Molecular characterization of the class I regions has revealed the presence of many more class I genes than was expected on the basis of previous immunogenetic and immunological studies.[6] In the mouse there are about 30 class I genes and in humans about 20. However, there are only three genes in each species which encodes the classical class I genes which are expressed on a majority of tissues and which play a major role for peptide presentation. These are the HLA-A, -B, and -C loci in man and H-2D, -K, and -L in the mouse. These loci are also highly polymorphic. Then there is a set of nonclassical class I genes, including HLA-E, -F, and -G in humans and H-2Qa and Tla in the mouse. The nonclassical genes have an expression pattern distinct from the classical genes and show less polymorphism. In addition there are a number of nonexpressed pseudogenes or class I gene fragments in each species. The class I region also harbors a gene for the transcription factor OCT-3 and several other novel genes which are not related to the class I Mhc genes.[5]

C. The Class II Region

There are five subclasses of mammalian class II genes. These are the DQ, DR, DP, DOB/DNA, and DM in humans which are homologous to the mouse H-2A, -E,

-P, -O/N, and -M loci. Each subclass contains at least one α and one β chain gene except the DP homolog in the mouse which is represented by a single β pseudogene denoted H-2Pb. The DQ, DR, and DP loci in humans and the H-2A and -E loci in mice encode the classical class II molecules which are highly polymorphic and play a major role for peptide presentation. The fourth subclass HLA-DNA/DOB, corresponding to H-2N/O, may be denoted as nonclassical class II genes. These are not highly polymorphic and have a distinct expression pattern compared with the classical genes. The DNA and DOB genes have been identified by gene cloning but no expression of protein products has yet been observed, and consequently it has not been possible to assess whether the DNA and DOB gene products form a functional heterodimer. However, it has been shown that the murine homologs Na and Ob encode a class II heterodimer with an unusual tissue distribution.[7]

The most recently described subclass of class II genes is HLA-DM[8] corresponding to H-2M.[9] This is a highly divergent member of the Mhc class II family and is assumed to have diverged early from the other class II genes (see below). In fact, the sequences of these genes are almost as closely related to class I genes as they are to class II genes. It has recently been shown that the DM heterodimer plays an essential role for class II peptide presentation.[10,11] Cell lines that lack DM expression also show a defect in antigen presentation which can be rescued by gene transfection of functional DM genes. However, the precise role of the DM heterodimer for class II presentation is not yet clear.

Similar to the class I region, the class II region contains a number of genes that are not structurally related to the Mhc class II genes.[5,12] The most interesting genes from an immunological point of view is a set of genes involved in antigen processing. These include TAP1 and TAP2 encoding a peptide transporter that transports peptides into the endoplasmic reticulum for class I presentation. Two other genes, LMP2 and LMP7, encode subunits of a large cytoplasmic protein complex called the proteasome. This complex is involved in the degradation of intracellular proteins into peptides and may play a role for Mhc class I presentation.

D. The Class III Region

The chromosome segment in between the Mhc class I and class II gene regions is denoted the class III region. More than 30 class III genes have been detected,[5] but none of these is structurally related to the Mhc class I and class II genes. The class III region harbors the genes for three complement components (BF, C2, and C4), for 21 hydroxylase (CYP21), for tumor necrosis factor α and β (TNFA and TNFB, respectively), and for heat shock protein 70 (HSP70). In addition a number of novel genes, whose function remains to be resolved, has been identified.

E. Is the Organization of Genes in the Mhc Functionally Significant?

The finding that the linkage relationships in the Mhc complex is well conserved among distantly related mammals has stimulated proposals on the significance of the existing organization. A functional significance has been proposed for the linkage of Mhc class I and class II genes, the location of the complement component genes

C2, C4, and BF (which have a function in the immune system) in the Mhc, and more recently the Mhc linkage of the LMP2, LMP7, TAP1, and TAP2 genes which are involved in the processing and transport of peptides in the cell.

Linkage is advantageous if there is interaction between loci so that certain haplotype combinations are more favorable than random combinations of alleles. Selection will then cause linkage disequilibrium and favor reduced recombination in the chromosome segment. It is very likely that this is the case for the close linkage of class I loci as well as of class II loci. It is quite obvious that certain combinations of alleles provide a better peptide-presenting repertoire than a random combination of alleles. This argument is particularly strong for the linkage of class II α and β chain genes encoding polymorphic subunits. There are in fact experimental data in the mouse showing that haplotype-matched subunits are functionally superior compared with random combinations of subunits.[13] The close linkage of class I and class II genes may be advantageous since class I and class II molecules are both involved in thymic T-cell selection and thereby shape the T-cell repertoire of an individual.[6]

The Mhc linkage of the genes for transporters associated with antigen processing (TAP1 and TAP2) and subunits of a proteasome complex (LMP2 and LMP7) has also been suggested to be functionally significant.[12] This idea gains support from the finding that genetic polymorphism in the rat TAP-2 gene influences the repertoire of peptides presented by class I molecules.[14] However, LMP7 (and most likely also LMP2, TAP1, and TAP2) maps to the class IIb region in cattle (C. J. Davies, personal communication) which is separated by a recombination distance of 17 cM from the classical class I and class II genes.[15] This shows that close linkage of the TAP/LMP genes and the Mhc genes encoding the classical peptide-presenting molecules is not obligatory in mammals.

The Mhc complex spans a physical distance of about 4 million base pairs, and it should be noted that the maintenance of linkage within a chromosome segment of this size between distantly related mammals is the rule rather than the exception.[16] Thus, the location of the complement component genes as well as many other "non-Mhc genes" in the Mhc region may well be coincidental and no selection mechanism needs to be invoked to explain the maintenance of Mhc organization among mammals.

III. THE EVOLUTIONARY ORIGIN OF Mhc MOLECULES

The evolution of Mhc molecules is most likely closely associated with the evolution of the vertebrate immune system. The presence of Mhc molecules has been documented in all vertebrate classes except agnathan fishes.[17] Exon shuffling appears to be an important mechanism in the evolution of proteins. It is believed that all existing proteins originate from a rather restricted number of exons encoding specific protein structures.[18] Exon shuffling means that new protein molecules are formed by combining protein domains from previously existing proteins. This appears also to be the case for the Mhc, and this phenomenon has been referred to as the "composite origin of Mhc molecules."[17] The Mhc molecules are composed of three distinct types of protein domains, a peptide binding domain, an immunoglobulin-like domain, and a membrane-spanning domain as illustrated in Figure 2.

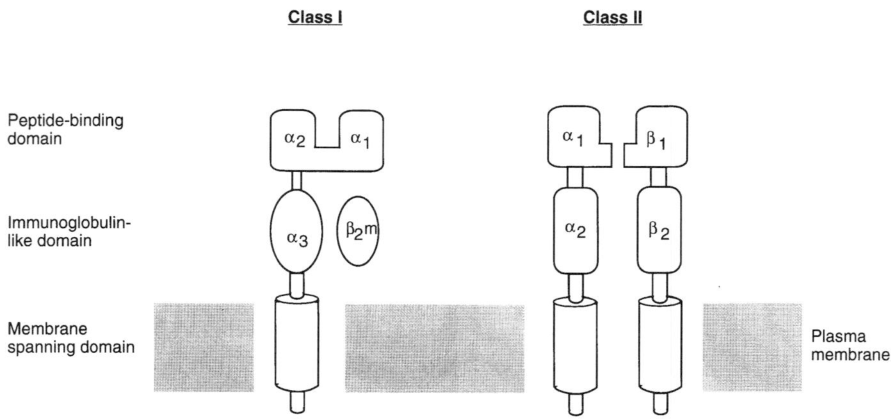

Figure 2 Schematic presentation of structural domains in Mhc class I and class II molecules.

It is well established that the class I α3, β_2m, class II α2, and class II β2 domains all belong to the immunoglobulin superfamily.[19] The origin of the peptide-binding domain is more uncertain. It shows structural similarities to the immunoglobulin family, the interleukin-8 family as well as the heat shock proteins, but none of these similarities are sufficiently strong to provide evidence of homology.[17] The membrane-spanning domain is composed of a stretch of hydrophobic amino acids, and its evolutionary origin can hardly be traced.

It has been proposed[6] that the most parsimonious model for the evolution of Mhc molecules is a one where the primordial Mhc molecule was a class II-like molecule which probably formed a homodimer at the cell surface. Tandem duplication of this ancestral gene followed by gene diversification may have given rise to an α and a β chain gene forming a heterodimeric antigen-presenting class II-like molecule. Finally, class I and β_2m genes may have arisen by unequal crossing over between two class II genes forming one gene with two exons encoding the peptide-binding domain and the other (β_2m) lacking a peptide-binding domain. This model is supported by a recent phylogenetic analyses of Mhc-related sequences.[20]

IV. EVOLUTION BY GENE DUPLICATION

It is clear that gene duplications and subsequent diversification have played a major role in the evolution of Mhc diversity. In most species, if not all, there are several expressed nonallelic forms of both class I and class II molecules. However, it is worth noticing that there seems to be an upper limit for the number of expressed classical class I and class II genes; classical loci encode the predominantly expressed and often highly polymorphic peptide presentation molecules. Most mammals possess two to three class I genes as well as two to three class II genes of this type. A possible explanation for this is that an increasing number of expressed genes has a positive effect on the diversity of peptides which can be presented by Mhc molecules but a negative effect on the T-cell repertoire due to the elimination of self-reactive T cells.[21]

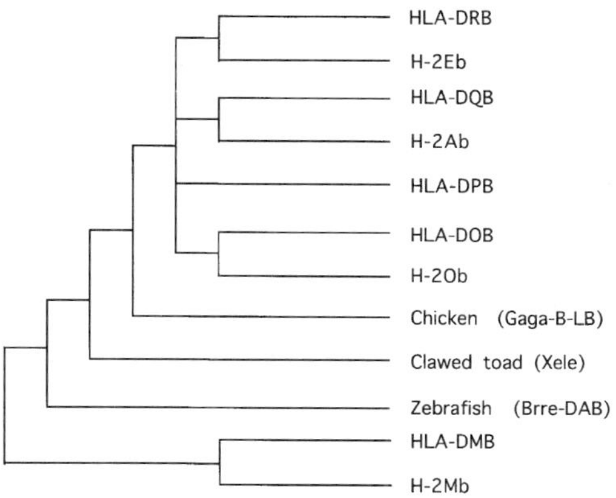

Figure 3 Schematic phylogenetic tree of Mhc class II genes in different vertebrate species. (Modified from Rask et al.,[51] Schneider et al.,[52] and Ono et al.[53])

A schematic phylogenetic tree of vertebrate class II β chain gene sequences is depicted in Figure 3; a similar tree is formed by the corresponding α chain sequences. Eutherian mammals exhibit five different isotypic forms of class II molecules denoted DR, DQ, DP, DO/DN, and DM. It is clear that the split between DM and other class II sequences provides the deepest root in the class II phylogenetic tree. The DM genes obviously diverged from the other types of class II sequences prior to the split between fishes and other vertebrate classes. The phylogenetic tree also shows that the DR, DQ, DP, and DO genes must have diverged from a common ancestral form subsequent to the split between birds and mammals; the tree has been drawn with a common root for these four types of class II genes since their phylogenetic relationship has not yet been resolved with confidence. The organization of the class II region in placental mammals is fairly well conserved since DQ, DR, and most likely also DO/DN are expressed in a majority of species whereas the expression of DP apparently has been lost in several species including cattle and mice. Moreover, DRA, DOB, and DNA are maintained as single copy genes whereas the number of DQA, DQB, and DRB genes varies within as well as between species. Thus, the expansion and contraction of the class II region is an ongoing process.

That the orthologous relationships of class II genes are obvious provides compelling evidence that these genes have evolved largely independent of each other (i.e., exchange of sequence information has not played a major role) during the evolution of eutherian mammals. However, an exception to this rule has been observed as strong indications for the occurrence of concerted evolution in a segment of the first domain exon of polymorphic class II B loci have been found.[22,23] Concerted evolution is a common phenomenon among multigene families[24] and implies that duplicated genes do not evolve independently of each other but exchange sequence information by unequal crossing-over or gene conversion. The segment in question encodes the α-helical region of the peptide binding site. In this region there is only weak locus divergence between polymorphic class II B loci (i.e., DQB, DRB, and DPB) in sharp contrast to the marked locus divergence in the segment encoding the β strand region of the peptide binding site as well as in class II A genes. The reason for the different evolutionary histories in these two subregions is not clear but it was proposed that the interaction between nonallelic class II molecules and a common ligand may cause a selective constraint on the divergence of the segment encoding the α-helical region in class II β chains.[23]

The phylogeny of mammalian class I genes is strikingly different from the one of class II genes in the respect that it has not been possible to establish orthologous relationships between class I genes in distantly related species. This is illustrated in Figure 4 which shows that a given class I gene is more closely related to other class I genes in the same species than to any class I gene in the other species. Thus, the human HLA-A, -B, and -C loci have evolved from an ancestral form well after the divergence from the rodent lineage. That multiple expressed class I genes are present in all species studied implies that multiple class I genes were present also in a mammalian ancestor. The lack of orthologous relationships between distantly related mammals can be explained by concerted evolution or by a fairly rapid turnover of class I genes during mammalian evolution. Concerted evolution has apparently played a more prominent role in the rodent lineage[25] than in primates[26] or artiodactyls.[27] The cause for the difference in evolutionary histories of class I and class II genes remains to be answered but it is possible that the class I loci in the mammalian ancestor had not diverged to such an extent to prohibit subsequent concerted evolution (see Reference 24).

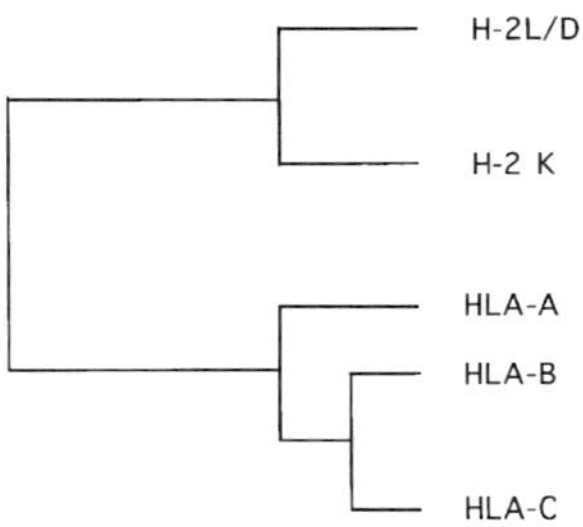

Figure 4 Schematic phylogenetic tree of human and mouse Mhc class I sequences. (Based on Lawlor et al.[6])

The number of class I genes varies extensively between mammalian species from a low number of 6 genes in the pig up to about 60 in the rat.[28] This implies a plasticity in the evolution of the class I region involving expansions and contractions of the gene region. Thus, it is no surprise that some species exhibit a polymorphism in the number of class I genes in the genome.

V. MONO- AND POLYMORPHIC Mhc GENES

Although extensive polymorphism is a hallmark of the Mhc system, it is important to notice that not all Mhc genes are highly polymorphic and there are also a few species, such as the cheetah[29] and the European beaver,[30] with no or very restricted Mhc polymorphism. The "nonpolymorphic" genes may not be entirely monomorphic but they do not show the unique polymorphism exhibited by the "polymorphic" Mhc genes (see below). The extensive polymorphism of certain genes is due to natural selection-promoting genetic diversity of Mhc molecules and therefore it is the predominantly expressed classical peptide-presenting molecules which are highly polymorphic i.e., the HLA-A, -B, and -C (class I) and HLA-DQ, -DR, and -DP (class II) molecules in humans and their homologs in other species with the exception that DP genes are not present or not expressed in several species (see above). However, there is a paradox regarding the DR molecule since the DRA gene (encoding the α chain) is nonpolymorphic while the DRB gene (encoding the β chain) is highly polymorphic. The monomorphism of the DRA gene appears widespread and calls for an explanation. One possible explanation could be that the lack of polymorphism in the DR α chain ensures the binding of some important peptide motifs to the DR molecule.

The "nonpolymorphic" nature of some Mhc genes may occur because they are not expressed and thus not exposed to a selection pressure promoting genetic diversity. Alternatively, they may be expressed but have a distinct function and exposed to a distinct selection pressure compared with the classical peptide-presenting molecules. This seems to be the case for the DM and DN/DO class II molecules and possible for some of the nonpolymorphic class I molecules.

VI. EVOLUTION OF Mhc POLYMORPHISM

Some Mhc genes exhibit a unique level of genetic polymorphism compared with other protein-encoding genes. The Mhc polymorphism shows several characteristic features. First, a large number of alleles occur at each locus. An example of this is illustrated in Figure 5 showing the cattle DRB3 locus; more than 30 alleles have been found at this locus. Second, allelic gene products differ by a remarkably large number of amino acid substitutions (Figure 5). Third, Mhc polymorphisms show a trans-species mode of evolution.[31,32] This implies that allelic lineages persist for long evolutionary periods (millions of years) and that their origin often predates speciation events. The long persistence time allows the accumulation of multiple substitutions between allelic lineages and provides an explanation for the characteristic large genetic distances between Mhc alleles. Fourth, the polymorphism occurs predominantly at functionally important sites, i.e., residues involved in peptide binding.[33,34] Consequently, the majority of highly polymorphic codons are located in exon 2 and 3 of class I genes encoding the α1 and α2 domains and in exon 2 of class II α and β chain genes encoding the α1 and β1 domains (see Figure 2). Fifth, linkage disequilibrium is frequently observed among class I loci as well as among class II loci and may even be found between class I and class II loci.

There is compelling evidence that the Mhc polymorphism is maintained by some form of balancing selection. First, the Mhc allele frequency distribution deviates significantly from the one expected for selectively neutral alleles.[35] The characteristic feature of Mhc polymorphism is an even allele frequency distribution in which no allele predominates in the population. Second, there is a significant excess of nonsynonymous substitutions (nucleotide substitutions causing amino acid substitutions)

```
          10         20         30         40         50         60         70        80         90
BoDRB3    +  + +               + +       ++                    +    +     +  + +  +   +    +   +
  4A      EY STSECHFFNG TERVRFLDRY FYNGEEYVRF DSDWGEFQAV TELGRPDAKY WNSQKDILER ERAAVDTYCR HNYGGVESFTV
  1A      -- -K-------- ---------- YT----T--- -------R-- -----Q--E- ------F--E K--E--RV-- -----M-----
  1B      Q- HKG------- -----L---H ------F--- ----D--R-- ------A-EQ ------F--Q K--E--RV-- -----------
  2A      -- ---------- ---------- -H----F--- ------YR-- --------E- ---- E---- A--------- -----------
  3       -- CK-------- -------E-S ------F--- ------YR-- ---------- ------L--Q K--N------ ----VG-----
  5       -- HK-------- ---L-Y---- ---------- ------YR-- ---------- -----E---- K--N------ ----V------
  6       -- CKR------- -----L---C -H----F--- -------R-- -----RV-EH L----E---- K--E---V-- ----VG-----
  7A      -- CKR------- ---------C -H----F--- -------R-- -----RV-EQ ------F--- R--E--RV-- ----V------
  8A      -- A--------- ---------- -H----L--- -------R-- ------S-VH L-----F--D ---S--T--- ----V------
  9A      -- -K-------- -------E-S ------N--- ------YR-- --------E- -----E---- K--N--RV-- ----VG-----
  10      -- -K-------- ---------- -H-------- ------YR-- ----QRV-E- C-----F--- A--------- ----VG-----
  11      Q- HKG------- -----L---H ---------- ----D--R-- ------S-E- ------F--- R--E---V-- ----V------
  12      -- TKK------- -------N-- -H----F--- ------YR-- --------E- -----E---- A--------- ----VG-----
  13A     -- LK-------- -------E-- ---------- ------YR-- ---------- ------L--- K--N------ ----V------
```

Figure 5 Alignment of the amino acid sequence of the β1 domain in 14 cattle DRB3 alleles (based on data from Sigurdardóttir et al.[40]). + indicates positions in the peptide binding site.[34] A dash indicates identity to the DRB3*4A allele. A blank in position 65 of DRB3*2A indicates the presence of a deleted codon in this allele.

over synonymous substitutions (nucleotide substitutions *not* causing amino acid substitutions) at the peptide-binding site of both class I and class II genes.[36,37] Third, the long-term persistence time of Mhc alleles (millions of years) by far exceeds the one expected for selectively neutral alleles.[38] Two of these features, a too-even allele-frequency distribution and a significant excess of nonsynonymous substitutions, have also been found for cattle DQ and DR loci.[39-41]

The cause of selection favoring Mhc polymorphism has been much disputed over the years. Although not yet formally proven available data strongly suggest that the cause is related to the host defense against infectious diseases. This is based on the fact that the only well-documented function of Mhc molecules is in the immune system. Moreover, there is solid evidence that allelic differences of Mhc molecules significantly influence the recognition of foreign peptides by the immune system.[1,2] The hypothesis of pathogen-driven selection promoting Mhc polymorphism is thus quite well founded, but it has been difficult to verify it by experimental or empirical data. Although there is a long list of well-documented Mhc disease associations,[42] few concern infectious diseases. However, there are two prominent examples of Mhc associations to life-threatening infectious diseases, namely, the association between certain Mhc alleles and protection to malaria in man[43] and Marek's disease in chicken.[44,45] A reason why it has been so difficult to firmly establish associations between Mhc polymorphism and infectious diseases is of course related to the problems of studying the genetic basis of resistance to infectious diseases. It is difficult to control the previous and current exposure to the pathogen, and the outcome of the infection will depend on the host's health status and a number of environmental as well as non-Mhc genetic factors. It has been proposed that the selection pressure favoring Mhc polymorphism may be quite small and thus would need very large sample sizes to be recognized.[32]

The type of balancing selection promoting Mhc polymorphism is either over-dominant selection or frequency-dependent selection or a combination of the two. Overdominant selection occurs when the heterozygote(s) are more fit than the homozygotes. The classical example of overdominant selection is the sickle-cell mutation in the human hemoglobin β chain which in the heterozygous condition is associated with resistance to malaria but in the homozygous condition causes sickle-cell anemia. The selective advantage of being heterozygous at Mhc loci may be a broader immune response repertoire due to the expression of two allelic forms of peptide-presenting molecules compared with a single one in homozygotes.[46] Frequency-dependent selection implies that the fitness of an allele decreases when its frequency in the population increases. A classical example of operating frequency-dependent selection is self-incompatibility systems in plants which have evolved to avoid inbreeding. The scenario for the immune system is that the Mhc genotype influences the set of foreign peptides which can be recognized by the immune system. This will cause a selection pressure favoring those pathogen genotypes which stimulate a poor immune response. As a consequence of this coevolution, pathogens tend to adopt to the immunological environment set by the more common alleles causing a selective advantage for individuals carrying rare Mhc alleles.[47] There is not yet any compelling experimental or empirical evidence supporting one or the other of these mechanisms.

The accumulation of point mutations is the basic mechanism generating Mhc diversity. However, there has been a long ongoing debate concerning if and to which extent intra- and intergenic recombination events contribute to the generation of Mhc diversity. Intra- and intergenic recombination involve exchanges of short sequence motifs between allelic or nonallelic genes (Figure 6). These types of events resemble the non-reciprocal transfer of sequence information denoted gene conversion which is well documented in yeast and other lower eukaryotes. The occurrence of gene conversion-like events in the evolution of Mhc polymorphism has been suggested by the characteristic patchwork pattern of Mhc alleles, i.e., the frequent sharing of sequence motifs among alleles. For instance, cattle DRB3 alleles 5, 6, 9A, and 13A shares the motif KRAN at amino acid residues 71 to 74 but at other positions these alleles share sequence motifs with other alleles (Figure 5). There are strong indications that intragenic recombination in fact contributes to the generation of Mhc diversity by shuffling existing sequence motifs into new allelic forms.[26,48] The significance of this mechanism has been underlined by the observation of new recombinant class I alleles in tribes of South American Indians.[49,50] It is clear that also *inter*genic recombination occurs at class I loci[6] and to a lesser extent also at class II loci[23] but this mechanism plays a relatively minor role in the generation of Mhc polymorphism.

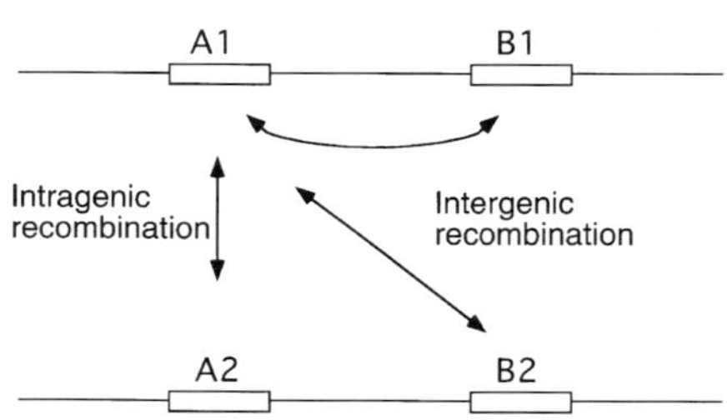

Figure 6 Illustration of the possible exchange of sequence information within (intragenic recombination) and between loci (intergenic recombination).

REFERENCES

1. Klein, J., *Natural History of the Major Histocompatibility Complex,* John Wiley & Sons, New York, 1986.
2. Cresswell, P., Assembly, transport and function of Mhc class II molecules, *Annu. Rev. Immunol.,* 12, 259, 1994.
3. Flavell, R. A., Allen, H., Burkly, L. C., Sherman, D. H., Waneck, G. L., and Widera, G., Molecular biology of the H-2 histocompatibility complex, *Science,* 233, 437, 1986.
4. Andersson, L. and Rask, L., Characterization of the Mhc class II region in cattle. The number of DQ genes varies between haplotypes, *Immunogenetics,* 27, 110, 1988.
5. Campbell, R. D. and Trowsdale, J., Map of the human Mhc, *Immunol. Today,* 14, 349, 1993.
6. Lawlor, D. A., Zemmour, J., Ennis, P. D., and Parham, P., Evolution of class-I Mhc genes and proteins: from natural selection to thymic selection, *Annu. Rev. Immunol.,* 8, 23, 1990.

7. Karlsson, L., Surh, C. D., Sprent, J., and Peterson, P. A., A novel class II Mhc molecule with unusual tissue distribution, *Nature,* 351, 485, 1991.
8. Kelly, A. P., Monaco, J. J., Cho, S., and Trowsdale, J., A new human HLA class II-related locus, DM, *Nature,* 353, 571, 1991.
9. Cho, S., Attaya, M., and Monaco, J. J., New class II-like genes in the murine Mhc, *Nature,* 353, 573, 1991.
10. Fling, S. P., Arp, B., and Pious, D., HLA-DMA and HLA-DMB genes are both required for Mhc class II/peptide complex formation in antigen-presenting cells, *Nature,* 368, 554, 1994.
11. Morris, P., Shaman, J., Attaya, M., Amaya, M., Goodman, S., Bergman, C., Monaco, J. J., and Mellins, E., An essential role for HLA-DM in antigen presentation by class II major histocompatibility molecules, *Nature,* 368, 551, 1994.
12. Trowsdale, J. and Powis, S. H., The Mhc: relationship between linkage and function, *Curr. Opin. Genet. Dev.,* 2, 492, 1992.
13. Germain, R. N., Bentley, D. M., and Quill, H., Influence of allelic polymorphism on the assembly and surface expression of class II (Ia) Mhc molecules, *Cell,* 43, 233, 1985.
14. Powis, C. J., Deverson, E. V., Coadwell, W. J., Ciruela, A., Huskisson, N. S., Smith, H., Butcher, G. W., and Howard, J. C., Effect of polymorphism of an Mhc-linked transporter on the peptides assembled in a class I molecule, *Nature,* 357, 211, 1992.
15. Andersson, L., Lundén, A., Sigurdardóttir, S., Davies, C. J., and Rask, L., Linkage relationships in the bovine Mhc region. High recombination frequency between class II subregions, *Immunogenetics,* 27, 273, 1988.
16. O'Brien, S. J., Womack, J. E., Lyons, L. A., Moore, K. J., Jenkins, N. A., and Copeland, N. G., Anchored reference loci for comparative genome mapping in mammals, *Nature Genetics,* 3, 103, 1993.
17. Klein, J. and O'hUigin, C., Composite origin of major histocompatibility complex genes, *Curr. Opin. Genet. Dev.,* 3, 923, 1993.
18. Pathy, L., Exons — original building blocks of proteins?, *BioEssays,* 13, 187, 1991.
19. Williams, A. F. and Barclay, A. N., The immunoglobulin superfamily-domains for cell surface recognition, *Annu. Rev. Immunol.,* 6, 381, 1988.
20. Hughes, A. L. and Nei, M., Evolutionary relationships of major histocompatibility complex genes, *Immunogenetics,* 37, 337, 1993.
21. Kappler, J. W., Roehm, N., and Marrack, P., T cell tolerance by clonal elimination in the thymus, *Cell,* 49, 273, 1987.
22. Andersson, L., Gustafsson, K., Jonsson, A.-K., and Rask, L., Concerted evolution in a segment of the first domain exon of polymorphic Mhc class II β loci, *Immunogenetics,* 33, 235, 1991.
23. Sigurdardóttir, S., Borsch, C., Gustafsson, K., and Andersson, L., Exon encoding the antigen-binding site of Mhc class II β chains is divided into two subregions with different evolutionary histories, *J. Immunol.,* 148, 968, 1992.
24. Walsh, J. B., Sequence-dependent gene conversion: can duplicated genes diverge fast enough to escape conversion, *Genetics,* 117, 543, 1987.
25. Rada, C., Lorenzi, R., Powis, S. J., van den Bogaerde, J., and Parham, P., Concerted evolution of class I genes in the major histocompatibility complex of murine rodents, *Proc. Natl. Acad. Sci. U.S.A.,* 87, 2167, 1990.
26. Parham, P., Lawlor, D. A., Lomen, C. E., and Ennis, P. D., Diversity and diversification of HLA-A, B, C alleles, *J. Immunol.,* 142, 3937, 1989.
27. Garber, T. L., Hughes, A. L., Letvin, N. I., Templeton, J. W., and Watkins, D. I., Sequence and evolution of cattle Mhc class I cDNAs: concerted evolution has not taken place in cattle, *Immunogenetics,* 38, 11, 1993.

28. Kindt, T. J. and Singer, D., Class I major histocompatibility genes in vertebrate species: what is the common denominator?, *Immunol. Res.*, 6, 57, 1987.
29. O'Brien, S. J., Roelke, M. E., Marker, L., Newman, A., Winkler, C. A., Meltzer, D., Cooly, L., Evermann, J. F., Bush, M., and Wildt, D. E., Genetic basis for species vulnerability in the cheetah, *Science,* 227, 1428, 1985.
30. Ellegren, H., Hartman, G., Johansson, M., and Andersson, L., Major histocompatibility complex monomorphism and low levels of DNA fingerprinting variability in a reintroduced and rapidly expanding population of beavers, *Proc. Natl. Acad. Sci. U.S.A.,* 90, 8150, 1993.
31. Klein, J., Generations of diversity at Mhc loci: implications for T-cell receptor repertoires, in *Immunology 80,* Fougerau, M. and Dausset, J., Eds., Academic Press, London, 1980, 239.
32. Klein, J., Satta, Y., O'hUigin, C., and Takahata, N., The molecular descent of the major histocompatibility complex, *Annu. Rev. Immunol.,* 11, 269, 1993.
33. Bjorkman, P. J., Saper, M. A., Samraoui, B., Bennett, W. S., Strominger, J. L., and Wiley, D. C., Structure of the human class I histocompatibility antigen HLA-A2, *Nature,* 335, 167, 1988.
34. Brown, J. H., Jardetzky, T., Saper, M. A., Samraoui, B., Bjorkman, P. J., and Wiley, D. C., A hypothetical model of the foreign antigen-binding site of class II histocompatibility molecules, *Science,* 332, 845, 1988.
35. Hedrick, P.W. and Thomson, G., Evidence for balancing selection at HLA, *Genetics,* 104, 449, 1983.
36. Hughes, A. L. and Nei, M., Pattern of nucleotide substitutions at major histocompatibility complex class I loci reveals overdominant selection, *Nature,* 335, 167, 1988.
37. Hughes, A. L. and Nei, M., Nucleotide substitutions at major histocompatibility complex class II loci: evidence for overdominant selection, *Proc. Natl. Acad. Sci. U.S.A.,* 86, 958, 1989.
38. Takahata, N. and Nei, M., Allelic genealogy under overdominant and frequency-dependent selection and polymorphism of major histocompatibility complex loci, *Genetics,* 124, 967, 1990.
39. Sigurdardóttir, S., Lundén, A., and Andersson, L., Restriction fragment length polymorphism of DQ and DR class II genes of the bovine Mhc, *Anim. Genet.,* 1988.
40. Sigurdardóttir, S., Borsch, C., Gustafsson, K., and Andersson, L., Cloning and sequence analysis of 14 DRB alleles of the bovine Mhc by using the polymerase chain reaction, *Anim. Genet.,* 22, 199, 1991.
41. Sigurdardóttir, S., Borsch, C., Gustafsson, K., and Andersson, L., Gene duplications and sequence polymorphism of bovine class II DQB genes, *Immunogenetics,* 35, 205, 1992.
42. Kostyu, D. D., The HLA gene complex and genetic susceptibility to disease, *Curr. Opin. Genet. Dev.,* 1, 40, 1991.
43. Hill, A. V. S., Allsop, C. E. M., Kwiatkowski, D., Anstey, N. M., Twumasi, P., Rowe, P. A., Bennett, S., McMichael, A. J., and Greenwood, B. M., Common West African HLA antigenes are associated with protection from severe malaria, *Nature,* 352, 595, 1991.
44. Pazderka, F., Longenecker, B. M., Law, G. R. J., Stone, H. A., and Ruth, R. F., Histocompatibility of chicken populations selected for resistance to Marek's disease, *Immunogenetics,* 2, 93, 1975.
45. Briles, W. E., Stone, H. A., and Cole, R. K., Marek's disease: effects of B histocompatibility alloalleles in resistant and susceptible chicken lines, *Science,* 195, 193, 1977.
46. Doherty, P. C. and Zinkernagel, R. M., Enhanced immunological surveillance in mice heterozygous at the *H-2* gene complex, *Nature,* 256, 50, 1975.

47. Bodmer, W., Evolutionary significance of the *HL-A* system, *Nature,* 237, 139.
48. Wakeland, E. K., Boheme, S., and She, J. X., The generation and maintenance of *Mhc* class II gene polymorphism in rodents, *Immunol. Rev.,* 113, 207.
49. Belich, M. P., Madrigal, J. A., Hildebrand, W. H., Zemmour, J., Williams, R. C., Luz, R., Petz-Erler, M. L., and Parham, P., Unusual *HLA-B* alleles in two tribes of Brazilian Indians, *Nature,* 357, 326, 1992.
50. Watkins, D. I., McAdam, S. N., Liu, X., Strang, C. R., Milford, E. L., Levine, C. G., Garber, T. L., Dogon, A. L., Lord, C. I., Ghim, S. H., Troup, G. M., Hughes, A. L., and Letvin, N. L., New recombinant *HLA-B* alleles in a tribe of South American Amerindians indicate rapid evolution of Mhc class I loci, *Nature,* 357, 329, 1992.
51. Rask, L., Andersson, L., Gustafsson, K., and Jonsson, A.-K., Parsimony analysis of mammalian class II histocompatibility genes, *Immunol. Rev.,* 113, 187, 1990.
52. Schneider, S., Vincek, V., Tichy, H., Figueroa, F., and Klein, J., Mhc class II genes of a marsupial, the red-necked wallaby (*Macropus rufogriseus*): identification of new gene families, *Mol. Biol. Evol.,* 8, 753, 1991.
53. Ono, H., Klein, D., Vincek, V., Figueroa, F., O'hUigin, C., Tichy, H., and Klein, J., Major histocompatibility complex class II genes of zebrafish, *Proc. Natl. Acad. Sci. U.S.A.,* 89, 11886, 1992.

CHAPTER 2

The Major Histocompatibility Complex of Fish: Genetic Structure and Function of the Mhc of Teleost Species

Øystein Lie and Unni Grimholt

CONTENTS

0-8493-8032-4/96/$0.00+$.50

I. INTRODUCTION

A. Driving Forces Behind Fish Mhc Research

Fish major histocompatibility complex (Mhc) research attracts wide attention for a series of good reasons, such as obtaining new insight into evolution of immunity through a number of questions which can be addressed only by studying lower vertebrates. Among these, fish occupy a critical position since typical features of higher vertebrate Mhc presence, such as acute graft rejection, vary dramatically between species and fade when moving downward through the lineages. Teleostei display acute graft rejection capacity, whereas this reaction is slow or chronic in cartilaginous and jawless fishes. Questions that may be relevant to ask are: when did the ancestral Mhc molecule emerge? Is it a chordate specific system (550 to 600 million years) or does it originate in the boundary between invertebrates and vertebrates (650 million years)? If invertebrates and vertebrates have common Mhc like ancestor molecule(s), why are not Ig-superfamily similar molecules found in invertebrates (although they may be present according to recent findings in *Staphylococcus Aureus,* see below)? Has the Ig-superfamily diverged and evolved under different selective pressure to undertake totally different biological functions in the two phyla, or have these molecules disappeared in invertebrates? Where is the defense molecule link between vertebrates and invertebrates? Last, but not the least, how has Mhc polymorphism evolved (convergent vs. trans-species, neutralistic vs. selectionistic, or by combination) and, since the debate on neutralism vs. positive selection now is almost closing in favor of the latter, which are the selective forces: pathogen driven, reproductive selection, inbreeding depression, or combinations? [1,2]

A second major driving force in fish Mhc research is the belief in the applied and biomedical usefulness of dissecting its anatomy, polymorphism, and function. Fish may serve as an efficient model system for understanding basic immunodevelopmental and immunopathological mechanisms (autoimmunity, infections, neoplasms) for the benefit of medical science. Moreover, extensive knowledge of fish Mhc may be commercially utilized in aquaculture, in vaccine development, in selective breeding, or by using molecular typing to survey structure and dynamics in wild populations and to monitor and ensure the integrity of the Mhc diversity in fish stocks subjected to intensive selective breeding.

B. Attempts Aiming at Identifying or Providing Evidence for a Fish Mhc

Over the last 2 decades a considerable number of functional studies embracing tests like mixed lymphocyte reactivity (MLR), *in vitro* assays with alloantisera, allorecognition by graft rejection, *in vitro* and *in vivo* antibody responsiveness, and biochemical analyzes in many fish species have revealed typical features attributable to Mhc, thus strongly indicating the existence of functional Mhc molecules in teleostei. In spite of this substantial volume of functional studies, we are not aware of any work to date employing *in vitro* cell-mediated lymphocytotoxicity (CML) tests to strengthen the likelihood of functional class I existence in fish. In parallel, but mostly over the last decade, a series of attempts have been carried out with interspecific putative cross-hybridizing/reactive probes and antibodies from higher vertebrates aiming at obtaining molecular evidence of a fish Mhc with no particular success. A breakthrough came with a series of studies employing the polymerase chain reaction (PCR) and sequence information on evolutionary conserved motifs in molecular cloning (see below).

II. CURRENT STATUS OF FISH Mhc FINDINGS

A. Studies of Products and Function

As indicated above, the Mhc product and function-oriented studies in fish can be grouped into several categories: immune responsiveness, allograft rejection, serology, cell-mediated mechanisms, and biochemical studies (see Table 1).

1. Immune Responsiveness

Any species, which can be shown to display an immune response regulation feature embracing properties such as self non-self recognition, thymus dependence, allogeneic associated restriction, cooperation of cells homologous to T-cells, B-cells, and antigen-presenting cells and major gene immune response patterns when using simple uniform antigens, is very likely to possess governing molecules homologous to those of higher vertebrate Mhc.

Several studies in fish provide data to meet most of the above requirements. Thus, cell populations attributable to T-cells, B-cells, and macrophages have been identified by structural and functional studies.[3-5] Furthermore, studies on *in vitro* responsiveness to thymus-dependent hapten-carrier complexes have demonstrated clearcut cell cooperation with T-cells, B-cells, and macrophages involved.[6,7] Vallejo et al.[8] provided data on immune restriction by revealing only strong MLRs when no detectable *in vitro* antibody response was produced.[8] Moreover, a recent *in vivo* antibody response study in gynogenetic carps has revealed response patterns with major gene effects.[9]

2. Grafting

Allograft recognition and rejection, and the kinetics of it, are a classical test system which together with serology were employed in both mice (mid thirties) and man (late fifties) to define the major histocompatibility complex. Also in several teleost fish species such as the common carp (*Cyprinus carpio*),[10] goldfish *(Carrassius auratus)*,[11] poecilids (*Poecilia formosa*), and platyfish (*Xiphophorus couchianus*),[12] a series of transplantation studies have been carried out reflecting the effect of major and minor loci. The studies have been performed with a diverse array of aims such as the elucidation of the genetics of transplantation, selection of target animals for alloserum production, or the study of developmental aspects of cellular immunity. Like the powerful tool with inbred or at best congenic mice, employment of naturally occurring or artificially generated gynogenetic fish has in recent years established well-funded evidence for the existence of a few, e.g., one or two, major histocompatibility loci[10,13] (see also review by Stet and Egberts[14]).

3. Serology

To define a putative Mhc in a new species with serological tools requires certain prerequisites. With a potentially vast number of polymorphic gene products of both Mhc and non-Mhc origin capable at eliciting alloantisera, this challenge is obvious. Strategies employing inbred or gynogenetic lines and family material together with analysis of cosegregation of several Mhc-associated features are thus preferable. Such an approach was used by Du Pasquier et al.[15,16] to define an Mhc in *Xenopus laevis*. Kaastrup et al.[17] employed an analog strategy in fish with convincing data on cosegregation of serologically determined specificities with graft rejection kinetics in gynogenetic families of the common carp (*Cyprinus carpio*).

4. Mixed Lymphocyte Reaction (MLR)

MLR has been observed in several jawless and jawed fish species, some examples of which are lamprey (*Lampetra fluviatilis*),[18] hagfish (*Eptatretus stoutii*),[19] Atlantic salmon *(Salmo salar)*,[20,21] rainbow trout *(Oncorhyncus mykiss)*,[22] channel catfish *(Ictalurus punctatus)*,[23] and the common carp *(Cyprinus carpio)*[24,25] which likely reflects Mhc class II involvement (see also reviews by Du Pasquier[1] and Stet and Egberts[14]). A one-way matrix of MLR analysis in one single carp family resulting from a cross between two individuals of different geographical origin showed stimulation indices consistent with one locus and three alleles (one shared between parents).[26] However, there are a number of obstacles to the interpretation of MLR responses in the context of detailed immunogentic dissection. Such are the variability in height and kinetics of both allogeneic and autologous responses, overall repeatability of the test, and surprising results in salmonids (e.g., rainbow trout). Thus, Kaastrup et al.[27] indicated in a two-way MLR matrix in full sibs from two rainbow trout families that each of the two progenies displayed more genotypes than predicted, the reason for which could be the pseudotetraploid status of salmonids.

5. Biochemistry

Up to the time of the successful molecular cloning of both genomic and expressed Mhc in fish, a quite extensive series of experiments has been carried out with cross-reactive xenoantibodies, alloantisera, and two-dimensional (2-D) electrophoresis aiming at isolating products with expected characteristics. Several of these studies indicate the presence of both classes of Mhc, either directly or indirectly (e.g., via β_2-microglobulin as demonstrated in goldfish and carp[28,29] and in cod [*Gadus morhua*]),[30] in many teleost species but with no firm evidence.[31,32] Also 2-D gelelectrophoresis has revealed Mhc class I-like signals (40 to 44 kD), with the MW range possibly corresponding to the different alleles of the K locus present in the cell lysate of carp material.[14] Products in both rainbow trout and Atlantic salmon with size and pI typical for both Mhc classes, have also been identified employing 2-D electrophoresis of PBL lysates, labeled during preincubation with ^{35}S (Jones et al., personal communication).

B. Molecular Studies

The first successful isolation of Mhc molecules in fish came with Hashimoto et al.[33] They took advantage of the polymerase chain reaction (PCR) method and used primers based on conserved motifs flanking the invariant cysteins of the $\beta 2$ and $\alpha 3$ domains in man, mouse, and chicken Mhc sequences. The resultant genomic common carp (*Cyprinus carpio L.*) fragment of approximately 190 bp contained fragments resembling both Mhc class I α and class II β molecules. These two classes of fragments were then employed as probes to isolate genomic Mhc clones which were shown to contain convincing similarity to those of higher vertebrates (Figures 1 and 2). This approach, using PCR with primers based on conserved regions, has subsequently resulted in the successful isolation of both genomic and cDNA Mhc molecules from other lower vertebrates such as amphibians, reptiles, various teleostei, and cartilaginous fishes.

Full-length Mhc class I-α chain cDNA molecules have been described in Atlantic salmon (*Salmo salar*),[34] the common carp (*Cyprinus carpio*, van Erp et al., personal communication), and zebrafish (*Brachydanio rerio*).[35] Although promising homology exists between current fish Mhc class I genes and the classical counterparts in higher vertebrates with regard to sequence similarity (see Figure 1), intron/exon organization and gene organization (see below),[36] major requirements for being defined as classical class I molecules,[37] (i.e., high contents of polymorphism; strong, wide, and fairly uniform pattern of tissue expression; main function to present endogenous peptides to cytotoxic T-cells; and location within a typical Mhc-clustered region) are not yet fulfilled for any fish Mhc class I molecule.

Expressed β_2-microglobulin molecules, required for a functional Mhc class I protein, have been identified in zebrafish (*Brachydanio rerio*),[38] as well as carp (*Cyprinus carpio*) and tilapia (*Oreochromis niloticus L.*),[39] thus strengthening the hypothesis that the identified fish Mhc class I α cDNA molecules are translated into functional proteins.

Full-length Mhc class II β chain cDNA molecules have been identified in Atlantic salmon (*Salmo salar*),[40] rainbow trout (*Oncorhynchus mykiss*),[41] common carp (*Cyprinus carpio*)[42] (van Erp et al., personal communication), zebrafish (*Brachydanio rerio*),[43] an African great lake cichlid (*Cyphotilapia frontosa*),[44] and striped bass (*Morone saxatilis*)[45] (Figure 2).

The first indication of a functional teleost Mhc came in rainbow trout with a partial class II β-chain cDNA which was also used in Northern analysis to reveal an interesting expression polymorphism.[46] A more recent study solely concentrating on Mhc class II β chain RNA expression in carp (*Cyprinus carpio*) revealed high expression in thymus; intermediate in peripheral blood, spleen, head kidney, and gut; and no detectable expression in muscle and erythrocytes.[47] A similar study in

Mhc CLASS I alpha chain

```
                                                                                                              CHO
ALPHA 1           1                   20                    40                    60                   80    |
Atantic salmon    VTHALKYFYTA SSEVPNFPEFVVVGVVDGVQMVHYDSN  SQRAVPKQDWVNKAADPQYWERNTGIFKGSQQTFKANIDIAKQRFNQS  G
Common carp       K-Y-R-MF-VMLKAD TL-V-SA--ES-HI-IS--SIEEQ        V-MRENLTEDD-DQAPEGPP-TRDWYLDL-K-    LS-CTESS
Frog              RS-T-HFHI-LA-API-GLSQYAIIVYM----YGR-N-DI  RR  AQFYSASLNPLSV-LDMQ-KFAQTFEVWQRHRLNFLMGVF-TTN -
Chicken           EL-T-R-IQ--MTDPG-GQ-W--T--Y---ELF---N-TA  R-Y--RTE-IAAK--Q---DGQ-Q-GQ-NE-IDRE-LG-LQR-Y--T  -
Mouse             GP-S-R--V--V-RPGLGE-R-IA--Y--DT-F-RF--DADNP-FE-RAP-M EQEG-E---EQ-QRA-SDE-W-RVSLRT-QRYY---K -
HLA-A             GS-SMR----SV-RPGRGE-R-IA--Y--DT-F-RF--DAA---ME-RAP-I EQEG-E--DGE-RKV-AHS-THRVDLGTLRGYY---EA

ALPHA 2                       100                   120                   140                   160                   180
Atlantic salmon   GV  HVNQWMYGCEWD   DEAGVTEGFEQWGYDGEDFIAFDLKTKSWIAPTHRSVITKLKWDSDTAQNEHRKNYYTQICIEWLKKYVDYGKSTLMRT
Common carp       E  LY-L-RLI---LEKLP-GTVNLTPVDEY-FG---CV--NSD-LQGSDKSPNVKE-EI-R-  HQVKLQA   FLKN-LD-ISTFNN        TKK
Frog              NGHI--Y-RKSA--LH   -DGT IG-YQDIAF--KE-LV--KERAVY-TS-QEA-MVSHL-NKRYDSTDS -MFLEID--KHM-M-LP-ISND-EKK
Chicken           -S  -TV------DIL   EGGP IR-YY-MA---R--T---KG-MTFT-AVPEA-P--R--E EESEP-RW---LEET-V---RR--E---AE-G-R
Mouse             -S  -TF-R-F--DVG   SDWRLLR-YH-FA---R-Y--LNEDL-T-T-ADTAAL--RR--E QAGDA-YYRA-LEGE-V---RR-LEL-NE--L--
HLA-A             -S  -TV-R----DVG   SDWRFLR-YH-YA---K-Y--LKEDLR--T-ADMAAQT--H--E AAHVA-QLRA-LEGT-V---RR-LEN--E--Q--

ALPHA 3                            200                  220                   240                   260
Atlantic salmon   VPPSVSLLQKTPSS    P  VTCHATGFYPSGVMVSWQKDGQDHHEDVEYGETLQNDDGTFQKSSHLTVTPEEWKNNKYQCVVQVTGVKEDFIKVLT
Common carp       NS-D-HVFARRAPDDHSKLVLI-LV-----RDIEMN IRINRINI-SQISSGIRP---ES--MR-SVKIDRNH  RGS-D-H-IHSSLTKPVT -EW
Shark              K---NITTHRLHRGKD-LLLI--VN------INAT-LHN-GTIQQE-LSSRI-P-T-----TTLQIS---QS  RDT-T-Q-EHSSSTDKLT ATW
Frog              ---N-KISSSESE-GIK  LH-WVYR---RD-E-K-I-N-T-EIYSE-SA-I-P-P---Y-IRVSVE-----  GAT-S-H-DHSSLENPLVVPFV
Chicken           ER-E-RV WGKEADGILT  LS-R-H----RPIV---L---AVRGQ-AHS-GIVP-G---YHTWVTIDAQ-GD  GD----R-EHASLPQPGL YSW
Mouse             DS-KAHVTYHPR-QVDVT  LR-W-L----ADITLT--LN-E-LTQ-M-LV--RPAG------WAAVV-PLGK  EQN-T-H-HHK-LP-PLT LRW
HLA-A             DA-KTHMTHHAV-DHEAT  LR-W-LS---AEITLT--R--E-QTQ-T-LV--RPAG------WAAVV-PSGQ  EQR-T-H--HE-LPKPLT LRW

CP/ TM / CYT                  280                  300                   320
Atlantic salmon    ESEIKTNWNEPNIVLIIVVVVALLLLVVAVVVGSSLEEEEQERLCSGQHFRH
Frog               PFVPVESNKKN        ILYIIIPVGSGLL-LLIVLGVFIY-KVPSGLQSVSTEDGD
Chicken            -PP        QPNLVPIVAGVAV AIVAI-IM--  VGFIIYR-H A-KKGKGYNIAPGSNPAI
Mouse              KLP        P STVSNTVIIAV-VV-GA-I-T-AVVAFVMKM-RNT-GKGVNYALAPGS     QTSDLSLPDGKVMVHDPHSLA
HLA-A              -PS        SQPTIPIVGIIAG-V-FG --IT-AVVAAVMWR-KS- DRKGGSYSQAASSDSA QGSDVSLTACKV
```

Figure 1 Dots represent identities with numbering according to *HLA-A2*. Sequence references are as follows: Atlantic salmon (*Salmo salar*, p30) class I;[34] common carp (*Cyprinus carpio*, TLa1α-1) class I;[33] shark (*Triakis scyllia*);[55] frog (*Xenopus laevis*);[75] chicken (*B-F*);[64] mouse (*H-2K^d*);[76] and man (*HLA-A2.4α*).[77]

Mhc class II beta chain

BETA 1 DOMAIN

```
                 1                   20                  40                  60                  80
Atlantic salmon  GYFEQVVRQCRYSSKDLQGIEFIDSYVFNKAEYIRFNSTVGKFVGYTELGVKNAEAWNSDAAVLAVERGELERYCKHNADLHYSTILDKT
Cichlid          -FMNYM-SR-DFN-TE-KD-Q--R--YY-MM-VV--D-S--EY----DF---Q-KYF--QP-N--AM-AQKGT--Q--IGID--N--SKS
Zebrafish        --YQYTMLE-I--TS-YSDMVLLE-GS---VVDVQY------Y-----Q--IF-RNF-KNQ-Y-QQRKA-V-SF-R---QISD-AVR--A
Common carp      H-HYGY-QML-  HASIS-NV--TF-VNY--F--L-Y---EK-V-------E-W--DY-   MI-LAKG-LPVQQ-RQLRM-TDPFAALT
Frog             D-VF-WKS--  -YRNGTDNVRYLEG-AN-QE--AY-D-D--EYKAKNDF-EVQ-KY---QKDL-DQK-AVVDTV-RY-YQ-YKPYTV-RK
Chicken          FFQWSATVE-  HFLNGTERVR-LVRH-Y-RQQ-VH-D-D--L--AD-V--EPS-KLF--QPD--EKN-AAV-ML-NY-YEIVAPLT-QRR
Mouse            HFVV-FKGE-  -YTNGT-R-RLVTR-IY-RE--V-YD-D--EYRAV---RPD--Y---QPEI-ERT-A-VDTA-R--YEGPETSTSLRRL
HLA-DRB          DFVY-FKGM-  -FTNGTERVRLVSRSIY-RE-VV--D-D--E-RAV-L--LPA--Y---QKDI-ERK-AAVD-V-R--YQ-ELR-T-QRR
```

BETA 2 DOMAIN

```
                      100                 120                 140                 160                 180
Atlantic salmon  VEPHVRLSSVAPPSGRHPAMLMCSAYDFYPKPIRVTWLRDGREVKSDVTSTEELANGDWYYQIHSHLEYTPRSGEKISCMVEHISLTEPMVYHW
Rainbow trout                   ---------------Q-------------------------------------K------------------M---
Cichlid          A--Y---H-ST----N-Q---V--V------Y-K-S-----EK-SA-------M-D----------------------V---A--K--LKTD-
Zebrafish        -K-K-TIQ--MQAE-K-----L-D--E----K-KMS----DKV-T------I-M---N-------------K-----Q-V---A-S-Q-ITKE-
Common carp      -K-E-IIR-DREAK-NEK-V-V-----------KL--M--DK--TT---------D------------F-KP------V---A-SHK--I---
Frog             SQ-D-KIVNTKTLDLE-ENLIT-IVDG-F-PM-K----KN-I-EREQ---SAL-K----TFE--VM--T-IKH-DTFT-Q---S--QQ-VSVN-
Chicken          --K--IFALQSG-LPQTDR-A-YVTG---PE-E-K-FQN-Q-ETER-V--DVICN---T--VLVV--IS--H-DSYV-Q---T--QQ-ITQR-
Mouse            EQ-N-AI-LSRTEALN-HNT-V--VT----AK-K-R-F-N-QEETVG-S--QLIR----TF-VLVM--M--HQ--VYT-H---P--KS-ITVE-
HLA-DRB          ---T-TI-PSRTEALN-HNL-V--VT----AQ-K-R-F-NDQ-ETAG-V--PLIR----TF--LVM--M--QR-DVYT-H---P--QS-ITVE-
```

```
                  CP              TM                     CYT
                                      200                    220
Atlantic salmon  DPS  LPEAERN   KIAIGASGLVLGAILALAGLIY  YKKKSSG VL
Rainbow trout    ---  -------   ------------T---------  ------- --
Cichlid          ---SSMP-S---   -L-------I--L--S---F--  --R-AR-RI-VPTN
Zebrafish        N-H  IS-SD--   -F----------I-I-I-----  -----T-RI-VP N
Frog             E-D  VS-SA--   -MLT-IV-F---S-FIIV--VV  -LRSKKTMTRFSVVQNENLMS
Chicken          DEP  PGDVSRS   -LLM-VG-F---LVYLAL-IFF  FLCSKKGCPDPTSPGILN
HLA-DRB          -RA  QSESAQS   -MLS-IG-F---L-FLGL---I  HHRSQKGRPVAAAPGMLN
```

Figure 2 Dashes indicate identities with numbering according to *HLA-DR*β. Sequence references are as follows: Atlantic salmon (*Salmo salar,* c157);[41] zebrafish (*Brachydanio rerio,* Brre-DAB1*01);[43] cichlid (*Cyphotilapia frontosa,* cyfr-T-141c);[44] carp (*Cyprinus carpio,* Cyca-TLa);[33] rainbow trout (*Oncorhynchus mykiss,* Onmy-OM#55);[46] chicken (p14);[78] frog (*Xenopus laevis,* F3);[79] mouse (Aβ^d);[80] and man, *HLA-DR*.[81]

Atlantic salmon (*Salmo salar*) also shows typical RNA expression patterns for the Mhc class II β molecule with a high expression in gills, head kidney, and spleen and low or missing expression in brain and muscle (Koppang et al., this lab).

Full-length Mhc class II α chain cDNA molecules have been identified in Atlantic salmon (*Salmo salar*)(Grimholt et al., unpublished), the common carp (*Cyprinus carpio*) (van Erp et al., personal communicaton), zebrafish (*Brachydanio rerio*),[48] striped bass (*Morone saxatilis*),[49] and the nurse shark *(Ginglymostoma cirratum)*[50] (Figure 3).

Partial cDNAs and genomic Mhc fragments are now being identified in a rapidly growing number of different fish species (for review see also Dixon et al.[36]).

Recently, molecules resembling the major Mhc collaborating molecule, the T-cell receptor, have also been identified in fish. Thus, α and β chain-resembling cDNAs have been identified in rainbow trout (*Oncorhynchus mykiss*),[51] pufferfish (*Spheroides nephelus*),[52] and horned shark (*Heterodontus francisci*),[53] further strengthening the assumption that fish have a functional Mhc.

Shark belong to the elasmobranch lineage which split off from the bony fishes more than 400 million years ago. Until recently there were no indications of a functional Mhc system in elasmobranchs, among the lowest vertebrates studied so far, since typical Mhc reactions such as acute allograft rejection have not been detected.[54] The identification of a partial Mhc class I cDNA fragment in shark (*Triakis scyllia*)[55] and a full-length class II α molecules in a nurse shark (*Ginglymostoma cirratum*)[50] was therefore a surprise. Fragments resembling Mhc class I were later found in a coelacanth (*Latimeria chalumnae*),[56] which is the sole survivor of another phylogenetic lineage that was founded more than 400 million years ago. Thus, the Mhc genes and probably also Mhc class I and class II genes already existed 400 million years ago. An Mhc class II-resembling molecule has even been identified in *Staphylococcus aureus*,[57] but this may be the result of a competent bacterium adopting a naked Mhc molecule from the surroundings and integrating it into its own genome. If, however, *S. aureus* does contain a functional Mhc class II analog, the Mhc is extremely old.

C. Differences and Similarities Across Species

The fish Mhc class I, class II β, and class II α molecules have many sequence similarities to those of higher vertebrates, probably reflecting both functional and structural requirements (Figures 1 to 3). An intradomain disulphide bridge found in all class I α2 and class II β1 domains (except possibly the nonclassical MIC molecule), but which is not present in the class I α1 and class II α1 domains, is also found in all fish sequences. A potential glycosylation site is found between α1 and α2 domains (*HLA-A2* position 86) in all class I molecules (except CD1 and MIC molecules) and in all class II molecules in roughly the same position *(HLA-DRα* position 78). Another potential glycosylation site is found between β1 strands 1 and 2 (*HLA-DRβ* position 19) for all class II molecules except for those in some bony fish. There are several highly conserved salt bridges in comparable positions in peptide binding regions (PBR) of all classical and nonclassical class I except CD1 (H3-D29, H93-D119) and in all class II α1 domains including *HLA-DM* but not

Mhc class II alpha chain

```
ALPHA 1 DOMAIN
                  1                      20                    40                      60                    80
Atlantic salmon   HKVLHIDLV     ITGCSDSDGLDMYGLDGEEMWYADFNKQEGVVALPPFADPFTFP GFYEQAVGNQGVCKGNLAKCIKAYKNPEEKI
Striped bass      ADD--E-FH     -G-------E--IA-----V-V---INKK--EPQ-S-I---NYQE-V-----A--QI-RQ--EVER----DLPLTL
Zebrafish         AQAE-R-VD     FF----TEKEYLQ-F----LYHS--IRKV---TA-D----MSY- ----NS-AQME---QD--TD----NS---QL
Common carp       TQ-VNR-VQ     FV----TEREFLI-F----L-H---IRK----TV-D----IG-- ----TG-ALME---Q---LN--V--PTD-QL
Shark             GK  YLYDFTQVYFVQQRSPEKHFDVME--D-IF-M--NLKKE-ARI-E--HLYMQGGEA  AISA-IAIV-N--KVDMNLSAGTP-PK
HLA-DQA           IVAD-VASYGVNLY QSYGPS-QYTHEF--D-QF-V-LGRK-T-WC--VLRQF R-DPQ-  AL T-IA-T-H--NIL--RSNSTAATN
HLA-DPA           IKAD-VSTYA AFV QTHRPT-EF-FEF-ED--F-V-LD-K-T-WH-EE-GQA-S-EAQG  GL A-IAILNN--NTL-QRSNHTQATN
HLA-DRA           IKEE-VIIQA EFY LNP-QS-EF-FDF--D-IFHV-MA-K-T-WR-EE-GRFAS-EAQG  AL A-IA-D-A--EIMT-RSNYTPITN

ALPHA 2 DOMAIN
                             100                     120                     140                    160
Atlantic salmon DPPHSS IYPRDDVDLGVENTLICHVSGFFPAPVRVRWTRNNQNLTEGVRFSTPYPNADFTLNQFSSLPFTPEEGDIYGCTVEHKGLAEPL TRIW
Striped bass    -R-S-PM-----N-E--EK-------T--Y----K-Y--K-GK-V---TSINV----K-GSFT-T-R-E-I-QL--M-S-S-K-LS-KD-- T-F-
Zebrafish       ---VT- --SE-E-V-DER-------T----P--N-S--K--DIV--EIS--QYRR-S-G-F-M--A-K---A-----S---N-RSIQGQPN-KT-
Common carp     A--DA- V-SEG--V---Q-------T-L--P--N-S--K---IV--D-SL-QYRRKN-G-F-I----K---A-----S---Y--A-ESRFI-KT-
Shark           VR-EV- V-SE-L-EW-QL-----FAD--Y-PHITMK-R---EPM-D-DNITEF-IKE---FRR--Y-SIV-SP--M-S-H---SS-QD-VT VF-
HLA-DQA         EV-EVT VFSKSP-T--QP-----L-DNI--PV-NIT-LS-GHSV----SETSFLSKS-HSFFKI-Y-T-L-SADE--D-K---W--D---L KH-
HLA-DPA         ---EVT VF-KEP-E--QP------IDK---PVLN-T-LC-GELV----AE-LFL-RT-YSFHK-HY-T-V-SAE-F-D-R---W--DQ--L KH-
HLA-DRA         V--EVT VLTNSP-E-REP-V---FIDK-T-PV-N-T-L--GKPV-T--SETVFL-RE-HLFRK-HY---L-STE-V-D-R---W--DE--L KH-

                CP           TM                            CYT
                180                     200                    220
Atlantic salmon EPEVIQP   SVGP  DVFCGVGLTLGLLGVAAGTFFLI  KGNQCN
Striped bass    DV-KPE-   -I--  A----L---V-------------  ---E-S
Zebrafish       -VD-EL-   ----  A-------V--------------  ---N--
Common carp     -VD-AV-   G---  A-------S--------------  -L-N--
Shark           DQG-PEEKSGH-    TII-AL-----IISAVV-IIL--  -E RQRLQAQQHG I
HLA-DQA         ---IPA-MSELTE   T-V-AL--SV--V-IVV--V-I-  R-LRSVGASRHQGPL
HLA-DPA         -AQEPIQMPETTE   T-L-AL--V---V-IIV--VLI-  -SLRSGHDPRAQGTL
HLA-DRA         -FDAPS-LPETTE   N-V-AL---V--V-III--I-I-  --VRKSNAAERRGPL
```

Figure 3 Dashes indicate identities with numbering according to *HLA-DQα*. Sequence references are as follows: salmon (*Salmo salar,* Grimholt et al., unpublished); striped bass (*Morone caxatilis,* Mosa-A-S5);[49] zebrafish (*Brachydanio rerio,* Brre-1.3.4);[48] common carp (*Cyprinus carpio,* Cyca-D[10A]A), van Erp et al., personal communication; shark (*Ginglymostoma cirratum,* pSαB-1);[50] and man, *HLA-DQA*),[82] *HLA-DPA,*[83] and *HLA-DRA.*[84]

shark (H5-D27 in *HLA-DRα*). A salt bridge between the two helices (E56-R/K170) is the only other salt bridge that is well conserved (except in chicken). The exact function of these conserved features is uncertain, but they may be important for the peptide loading mechanism.

The immunoglobulin-like constant domains are the most conserved domains between different vertebrate groups (roughly 20 to 40%) probably due to structural constraint as well as conservation of sequences participating in interactions with the α1 chain itself or CD8. Once the protein structures of fish Mhc molecules have been resolved, individual residues important for interdomain contact will be easier to identify. Assignment of residues that are important in the structure of the PBRs is even harder and awaits the unravelling of protein structures.

The eight residues located at the ends of the binding site that bind the N- and C-termini of the peptide in mammalian class I molecules are highly conserved in phylogeny with the exception of R84 in nonmammalian vertebrates instead of the mammalian Y84 counterpart. Similarly, four of the eight highly conserved residues of class II that bind mainchain atoms of the peptide (Nα62, Nα69, Nβ82, Wβ61) are conserved in all class II sequences from nonmammalian vertebrates except two bony fish sequences. Also other regions of the Mhc contain residues that are highly conserved through phylogeny such as several of the residues in the transmembrane region of Mhc class II molecules (for review, see Kaufman et al.[58]).

D. Studies on Polymorphism and Genetic/Genomic Organization

Several groups have reported considerable fish Mhc variation for class I,[35,56,60] class II β,[42-44,59] and class II α,[48,49] but the total number of fish Mhc loci is still unclear. RFLP studies have been performed using probes from the conserved class I α3 and class II β2 domain regions to identify the number of genes in some fish species. In the common carp,[61] a total of 9 to 12 Mhc class I and 3 to 5 Mhc class II strongly hybridizing fragments were identified. The number of *Mhc-Cyca* class I genes is comparable to some species such as rabbit,[62] swine,[63] and chicken,[64] but considerably lower than other vertebrates such as mouse[65] and man[66] which have between 17 and 36 Mhc class I genes. The number of strongly hybridizing *Mhc-Cyca* class II fragments, on the other hand, is comparable to the estimates for mouse,[65] man,[67] chicken,[64] and frog.[68] In rainbow trout,[46] hybridization with a class II β2 probe in a family material resulted in two strongly hybridizing fragments with an allelic segregation pattern suggesting one locus and two alleles. In Atlantic salmon, preliminary pulse-field data indicate a minimum of five Mhc class I loci clustered in a genomic region of approximately 1 Mb[69] (Lundin et al., this lab). Since the number of loci and their sequence similarities are uncertain, defining alleles is difficult and polymorphism (i.e., defining the frequency of alleles at a given locus in a population) has therefore not been established.

Data on the organization of fish Mhc genes is still very limited although several studies at both gene and product level (see above) have shown the existence of multiple genes in many species. The fact that the two classes appear to be split in zebrafish and Atlantic salmon (see below) and that amphibian as well as avian Mhc genes do not display the same degree of discrete regions as mammals indicate that we need to be open to more surprises in fish. The currently rapid accumulation of cloned sequences in many fish species is an excellent basis for solving the gene organization question through segregation analysis and walking strategies. Most data on gene structure (sequence: see above and exon/intron organization), though still scarce, is consistent with those of higher vertebrates (for reviews, see Trowsdale[70] and Dixon et al.[36]).

So far the linkage group (LG) assignment of the Mhc class I and class II genes is only available in zebrafish where they surprisingly are located on different chromosomes, LG 19 and LG 8, respectively (due to extensive gynogenetically based centromere mapping, LGs and chromosomes are synonymes in zebrafish). Class I (*Brre-UAA*) is syntenic with "no tail" and rxr genes as in mouse, reflecting a conserved synteny on one side of the Mhc[71] (Postlethwait and Figueroa, personal communication). Also in Atlantic salmon, a linkage analysis using satellite markers residing in the *Mhc-Sasa* class I and class II α genes shows no linkage, indicating that the two classes may reside on different chromosomes also in this species (Slettan et al., this lab). Among higher vertebrates, only chicken and frogs may have Mhc-related genes mapping to more than one chromosome (see also review by Trowsdale[70] and for an overview, see Table 1).

III. FUTURE PROSPECTS

A. Further Unraveling of Structure and Function of Fish Mhc

Forming the largest group of vertebrates with some 20,000 living species, allocated in an interesting evolutionary position, quite a few of which with high commercial and food supply value, teleost fishes are, not surprisingly, currently subjected to comprehensive immunological research with Mhc occupying an expanding position. This is clearly reflected in most recent fish immunological reviews and also in the present chapter, depicting a steep increase in fish Mhc papers after the breakthrough in 1990 by Hashimoto et al.[33] As emphasized, fundamental immunodevelopmental/pathological and evolutionary issues together with applied potentials of disease resistance and population surveillance are major driving forces behind this "explosive" development in international fish Mhc research.

Together with these needs and driving forces, current development and refinement of molecular genetic techniques open up extraordinary new possibilities to carry out prior impossible projects in volume and depth.

B. Evolutionary Speculations and Implications of Fish Mhc Research

As indicated in the introductory paragraph, teleost fish Mhc research may contribute to revealing the evolutionary origin of the Mhc, finding the link between vertebrate and invertebrate defense molecules (if existing), and answering how and when the Mhc gene cluster with its polymorphism has evolved. A number of hypotheses have been proposed to account for the evolutionary origin of the two classes of Mhc genes of vertebrates. All of the Mhc proteins show homology to one another

Table 1 Data Reflecting Current Status of Fish Mhc Research

Phylogenic group	Species	Product/Function		Molecular Studies				Variation and Genetic Organization		Map Data
					Mhc	Mhc class II		Mhc		Mhc class I/
		MLR/GR/IR	SD/BD	β2-μ-globulin	class I	α	β	class I	class II αβ	class II
	Atlantic salmon (*Salmo salar*)	MLR	(BD)		+	+	+	+	+	(+/+)
	Rainbow trout (*Oncorhynchus mykiss*)	MLR	(BD)				+		+	
	Common Carp (*Cyprinus carpio*)	MLR, GR, IR	SD, (BD)	+	+	+	+	+	+	
	Zebrafish (*Brachydanio rerio*)			+	+	+	+		+	+/+
Gnathostomata	Tilapia (*Oreochromis niloticus*)			+						
	Channel Catfish (*Ictalurus punctatus*)	MLR, IR								
	Ciclides (*Cyphotilapia frontosa*)						+		+	
(chondrichtyes, sarcopterygii)	Goldfish (*Carassius auratus*)	GR	(BD)							
	Striped bass (*Morone saxatilis*)					+	+		+	
	Coelacanth (*Latimeria chalumnae*)				(+)					
	Shark (*Triakis scyllia, Ginglomystoma cirratum*)	(+)			(+)	+				
Agnatha	Lamprey (*Lampetra fluviatilis*)	MLR								
	Hagfish (*Eptatretus stoutii*)	MLR								

Abbreviations: MLR, mixed lymphocyte reaction; GR, graft rejection; IR, immune response; SD, serologically defined; BD, biochemically defined. Brackets indicate uncertain determination. See text for references.

in their conserved domains,[72,73] which are clearly homologous to the C-domains of the immunoglobulin (Ig) gene superfamily.[74] The relationship between the peptide binding (PB) domains is, however, still subjected to controversial opinion. Recent data, however, support the hypothesis that the Mhc class I genes originated from a Mhc class II $\beta1/\beta2$ predecessor molecule which acquired a Mhc class II$\alpha1$ domain, leaving the class II $\alpha2$ domain to evolve into $\beta2$-microglobulin (Grimholt et al., this lab). Regardless of speculations and unsolved questions, the identification of both Mhc class I and class II genes in fish has stated that these genes must have split at least 400 million years ago. Other issues, open to debate are: how did the ancestral Mhc molecule evolve to become a complex? Since the phenomenon allorecognition is present in both vertebrates and invertebrates, does it stem from a common progenitor organism where defense against pathogens is more or less equivalent with defense against members of the same species? Are Mhc molecules major elements in reproduction compatibility/incompatibility and through this role maintaining its diversity by obstructing inbreeding? How important is the Mhc complex for nonimmunological functions? How meaningful is its polymorphism and polygenism? Fish Mhc research will contribute to give more complete answers to such questions.

C. Applied Aspects

The potentials of such aspects are dependent upon two major features of Mhc, for which there are very few or no data available in fish so far: extent of polymorphism and linkage or linkage disequilibria with disease. A polymorphic fish Mhc would be a potentially useful tool in itself by adding to existing informative genetic systems in monitoring the dynamics and structure of wild stocks. Linkage to disease would in turn create new tools in selective breeding for improved resistance to the greatest problem of international aquaculture: infectious diseases. The latter aspect would also contribute to elucidate major forces and dynamics behind short- and long-term evolvement of Mhc diversity. Several fish species, not least the salmonids, experiencing dramatical and recurrent shifts in environmental conditions during life cycle and long distance migration, would be excellent experimental animals for such studies. Thus, the knowledge accumulated through such investigations would contribute to resolve and give answer to an array of questions and challenges ranging from the evolution of Mhc diversity to genetic and environmental measures in our management of both wild and cultured fish species.

REFERENCES

1. Du Pasquier, L., Evolution of the immune system, in *Fundamental Immunology*, William, E. P., Ed., Raven Press Ltd., New York, 1989, chap. 7.
2. Potts, W. K. and Wakeland, E. K., Evolution of Mhc genetic diversity: a tale of incest, pestilence and sexual preference, *Trends Genetics*, 9, 408, 1993.
3. DeLuca, D., Wilson, M., and Warr, G. W., Lymphocyte heterogeneity in the trout, *Salmo gairdneri*, defined with monoclonal antibodies to IgM, *Euro. J. Immunol.*, 13, 546, 1983.

4. Secombes, C. J., Van Groningen, J. J. M., and Egberts, E., Separation of lymphocyte subpopulations in carp *Cyprinus carpio* L. by monoclonal antibodies: immunohistochemical studies, *Immunology,* 48, 165, 1983.
5. Miller, N. W., Bly, J. E., Van Ginkel, F., Ellsaesser, C. F., and Clem, L. W., Phylogeny of lymphocyte heterogeneity: identification and separation of functionally distinct subpopulations of channel catfish lymphocytes with monoclonal antibodies, *Dev. Comp. Immunol.,* 11, 739, 1987.
6. Clem, L. W., Sizemore, R. C., Ellsaesser, C. F., and Miller, N. M., Monocytes as accessory cells in fish immune responses, *Dev. Comp. Immunol.,* 9, 803, 1985.
7. Miller, N. W., Sizemore, R. C., and Clem, L. W., Phylogeny of lymphocyte heterogeneity: the cellular requirements for *in vitro* antibody responses of channel catfish leukocytes, *J. Immunol.,* 134, 2884, 1985.
8. Vallejo, A. N., Miller V. M., Jørgensen, T., and Clem, W., Phylogeny of immune recognition: antigen processing/presentation in channel catfish immune responses to hemocyanins, *Cell. Immunol.,* 130, 364, 1990.
9. Wiegertjes, G. F., Stet, R. J. M., and Van Muiswinkel, W. B., Divergent selection for antibody production in common carp (*Cyprinus carpio L.*) using gynogenesis, *Anim. Genet.,* 25, 251, 1994.
10. Komen, J., Van den Dobbelsteen, P. J. M., Slierendrecht, W. J., and Van Muiswinkel, W. B., Skin grating in gynogenetic common carp (*Cyprinus carpio L.*), *Transplantation,* 49, 788, 1990.
11. Nakanishi, T., Histocompatibility analysis intetraploid induced from clonal triploid crucian carp and in gynogenetic diploid goldfish, *J. Fish Biol.,* 31 (Suppl. A), 35, 1987.
12. Kallman, K. D., Genetics of tissue transplantation in teleostei, *Transplant. Proc.,* 2, 263, 1970.
13. Matsuzaki, T. and Shima, A., Number of major histocompatibility loci in inbred strains of the fish *Oryzias latipes, Immunogenetics,* 30, 226, 1989.
14. Stet, R. J. M. and Egberts, E., The histocompatibility system in teleostean fishes: from multiple histocompatibility loci to a major histocompatibility complex, *Fish Shellfish Immunol.,* 1,1, 1991.
15. Du Pasquier, L. and Chardonnes, X., Genetic aspects of the tolerance to allografts induced at metamorphosis in the toad *Xenopus laevis, Immunogenetics,* 1, 482, 1975.
16. Du Pasquier, L., Miggiano, V. C., Kobel, H. R., and Fischberg, M., The genetic control of histocompatibility reactions in natural and laboratory-made polyploid individuals of the clawed toad *Xenopus, Immunogenetics,* 5, 129, 1977.
17. Kaastrup, P., Stet, R. J. M., Tigchelaar, A. J., Egberts, E., and Van Muiswinkel, W. B., A major histocompatibility locus in fish: serological identification and segregation of transplantation antigens in the common carp (*Cyprinus carpio L.*), *Immunogenetics,* 30, 284, 1989.
18. Cooper, A.J., Ammocoete lymphoid cell populations *in vitro,* in *Proceedings of the 4th Annual Leukocyte Culture Conference,* McIntyre, O.R., Ed., Appleton-Century-Crofts, New York, 1971, 137.
19. Raison, R.L., Gilbertson, P., and Wotherspoon, J., Cellular requirements for mixed leukocyte reactivity in the cyclostome *Eptatretus stoutii, Immunol. Cell. Biol.,* 65, 183, 1987.
20. Ellis, A.E., Ontogeny of the immune response in *Salmo salar.* Histogenesis of the lymphoid organs and appearance of membrane immunoglobulin and mixed leucocyte reactivity, in *Developmental Immunobiology,* Solomon, J.B. and Horton, J.D., Eds., Elsevier/North-Amsterdam, Holland Biomedical Press, 1977, 225.

21. Smith, P.D. and Braun-Nesje, R., Cell mediated immunity in the salmon: lymphocyte and macrophage stimulation, lymphocyte/macrophage interactions, and the production of lymphokine-like factors by stimulated lymphocytes, *Dev. Comp. Immunol.*, 2, 233, 1982.
22. Etlinger, H.M., Hodgins, H.O., and Chiller, J.M., Evolution of the lymphoid system.II. Evidence for immunoglobulin determinants on all rainbow trout lymphocytes and demonstration of mixed leukocyte reaction, *Eur. J. Immunol.*, 7, 881, 1977.
23. Miller, N., Deuter, A., and Clem, L.W., Phylogeny of lymphocyte heterogeneity: the cellular requirements for the mixed leucocyte reaction with channel catfish, *Immunology*, 59, 123, 1986.
24. Caspi, R.R. and Avtalion, R.R., The mixed leukocyte reaction (MLR) in carp: bidirectional and unidirectional MLR responses, *Dev. Comp. Immunol.*, 8, 631, 1984.
25. Grondel, J.L. and Harmsen, E.G.M., Phylogeny of interleukins: growth factors produced by leucocytes of the cyprinid fish, *Cyprinus carpio L.*, *Immunology*, 52, 477, 1984.
26. Gloudemans, A.G.M., Cohen, N., Boersbroek, G.E., Grondel, J.L., Egberts, E., and Van Muiswinkel, W.B., The major histocompatibility complex (Mhc) in fish. Carp family studies using the mixed leucocyte reaction, in *Selection, Hybridization and Genetic Engineering in Aquaculture*, Tiews, K., Ed., Heeneman Verlag, Berlin, 1987, 265.
27. Kaastrup, P., Nielsen, B., Hørlyck, V., and Simonsen, M., Mixed lymphocyte reactions (MLR) in Rainbow trout (*Salmo gairdneri*) sibling, Dev. Comp. Immunol., 12, 801, 1988.
28. Shalev, A., Greenberg, A. H., Lögdberg, L., and Björck, L., β_2-Microglobulin-like molecules in low vertebrates and invertebrates, *J. Immunol.*, 127, 1186, 1981.
29. Shalev, A., Caspi, R. R., Lögdberg, L., Björck, L., Segal, S., and Avtalion, R. R., Specific activity of carp antisera against β_2m homologue in carp (*Cyprinus carpio*), *Dev. Comp. Immunol.*, 8, 639, 1984.
30. Lögdberg, L. and Björck, L., Binding of mammalian β_2-microglobulin by glycoproteins in fish serum, *Mol. Immunol.*, 20, 885, 1983.
31. Kaufman, J. F., Ferrone, S., Flajnik, M. F., Kilb, M., Völk, H., and Parisot, R., Mhc-like molecules in some nonmammalian vertebrates can be detected by some cross-reactive monoclonal antibodies, *J. Immunol.*, 144, 2273, 1990.
32. Kaufman, J. F., Skjoedt, K., Salomonsen, J., Simonsen, M., Du Pasquier, L., Parisot, R., and Riegert, P., Mhc-like molecules in some nonmammalian vertebrates can be detected by some cross-reactive xenoantisera, *J. Immunol.*, 144, 2258, 1990.
33. Hashimoto, K., Nakanishi, T., and Kurosawa, Y., Isolation of carp genes encoding major histocompatibility complex antigens, *Proc. Natl. Acad. Sci. U.S.A.*, 87, 6863, 1990.
34. Grimholt, U., Hordvik, I., Fosse, V. M., Olsaker, I., Endresen, C., and Lie, Ø., Molecular cloning of major histocompatibility complex class I cDNAs from Atlantic salmon (*Salmo salar*), *Immunogenetics*, 37, 469, 1993.
35. Takeuchi, H., Figueroa, F., O'hUigin, C., and Klein, J., Cloning and characterization of class I Mhc genes of the zebrafish, *Brachydanio rerio*, *Immunogenetics*, 42, 77, 1995.
36. Dixon, B., van Erp, S.H.M., Rodrigues, P.N.S., Egberts, E., and Stet, R.J.M., Fish major histocompatibility complex genes: an Expansion, *Dev. Comp. Immunol.*, 19, 109, 1995.
37. Klein, J. and O'hUigin, C., The conundrum of non-classical major histocompatibility complex genes, *Proc. Natl. Acad. Sci. U.S.A.*, 91, 6251, 1994.

38. Ono, H., Figueroea, F., O'hUigin, C., and Klein, J., Cloning of the β_2-microglobulin gene in the zebrafish, *Immunogenetics,* 38, 1, 1993.
39. Dixon, B., Stet, R. J., van Erp, H. M., and Pohajdak, B., Characterization of β_2-microglobulin transcripts from two teleot species, *Immunogenetics,* 38, 27, 1990.
40. Hordvik, I., Grimholt, U., Fosse, V. M., Lie, Ø., and Endresen, C., Cloning and sequence analysis of cDNAs encoding the Mhc class II β chain in Atlantic salmon (*Salmo salar*), Immunogenetics 37, 437, 1993.
41. Glamann, J., Complete coding sequence of rainbow trout Mhc II b chain, *Scand. J. Immunol.*, 41, 365, 1995.
42. Ono, H., O'hUigin, C., Vincek, V., Stet, R. J. M., Figueroea, F., and Klein, J., New β chain-encoding Mhc class II genes in the carp, *Immunogenetics,* 38, 146, 1993c.
43. Ono, H., Klein, D., Vincek, V., Figueroea, F., O'hUigin, C., Tichy, H., and Klein, J., Major histocompatibility complex class II genes of zebrafish, *Proc. Natl. Acad. Sci. U.S.A.*, 89, 11886, 1992.
44. Ono, H., O'hUigin, C., Vincek, V., and Klein, J., Exon-intron organization of fish major histocompatibility complex class II B genes, *Immunogenetics,* 38, 223, 1993d.
45. Walker, R.A. and McConnell, T.J., Variability in an Mhc class II b chain-encoding gene in striped bass (*Morone saxitilis*), *Dev. Comp. Immunol.*, 18, 325, 1994.
46. Juul-Madsen, H., Glamann, J., Madsen, H. O., and Simonsen, M., Mhc class II beta-chain expression in the rainbow trout, *Scand. J. Immunol.,* 35, 687, 1992.
47. Rodrigues, P.N.S., Hermsen, T.T., Rombout, J.H.W.M., Egberts, E., and Stet, R.J.M., Detection of Mhc class II transcripts of the common carp (*Cyprinus carpio L.*), *Dev. Comp. Immunol.*, 19, 483, 1995.
48. Sultmann, H., Mayer, W. E., Figueroea, F., O'hUigin, C., and Klein, J., Zebrafish Mhc class II α chain-encoding genes: polymorphism, expression and function, *Immunogenetics,* 38, 408, 1993.
49. Hardee, J.J., Godwin, U., Benedetto, R., and McConnell, T.J., Major histocompatibility complex class II A gene polymorphism in the striped bass, *Immunogenetics*, 41, 229, 1995.
50. Kasahara, M., Vazquez, M., Sato, K., and Churchill McKinney, E., Evolution of the major histocompatibility complex: isolation of class II cDNA clones from the cartilaginous fish, *Proc. Natl. Acad. Sci. U.S.A.,* 89, 6688, 1992.
51. Partula, S., Fellah, J.S., de Guerra, A., and Charlemagne, J., Identification of cDNA clones encoding the T-cell receptor β chain in the rainbow trout (*Oncorhynchus mykiss*), *C. R. Acad. Sci. Paris, Life Sciences*, 317, 765, 1994.
52. Rast, J.P., Haire, R.N., Litman, R.T., Pross, S., and Litman, G.W., Identification and characterization of T-cell antigen receptor-related genes in phylogenetically diverse species, *Immunogenetics*, 42, 204, 1995.
53. Rast, J. and Litman, G.W., T-cell receptor gene homologs are present in the most primitive jawed vertebrates, *Proc. Natl. Acad. Sci. U.S.A.*, 91, 9248, 1994.
54. Hildemann, W. H., Transplantation immunity in fishes: Agnatha, Chondrihyhyes and Osteichthyes, *Tranplant. Proc.,* 2, 253, 1970.
55. Hashimoto, K., Nakanishi, T., and Kurosawa, Y., Identification of a shark sequence resembling the major histocompatibility complex class I α3 domain, *Proc. Natl. Acad. Sci. U.S.A.,* 89, 2209, 1992.
56. Betz, U.A., Mayer, W.E., and Klein, J., Major histocompatibility complex class I genes of the coelacanth *Latimeria chalumnae*, *Proc. Natl. Acad. Sci. U.S.A.*, 91, 11065, 1994.

57. Jonsson, K., McDevitt, D., McGavin, M. H., Patti, J.M., and Hook, M., Staphylococcus aureus expresses a major histocompatibility complex class II analog, *J. Biol. Chem.*, 270, 37, 21457, 1995.
58. Kaufmann, J., Salomonsen, J., and Flajnik, M., Evolutionary conservation of Mhc class I and class II molecules — different yet the same, *Seminars in Immunology*, 6, 411, 1994.
59. Klein, D., Ono, H., O'hUigin, C., Vincek, V., Goldsmidt, T., and Klein, J., Extensive Mhc variability in cichlid fishes of Lake Malawi, *Nature,* 364, 330, 1993.
60. Grimholt, U., Olsaker, I., de Vries Lindstrøm, C., and Lie, Ø., A study of variability in the Mhc class II β1 and class I α2 domain exons of Atlantic salmon (*Salmo salar*), *Anim. Genet.,* 25, 147, 1994.
61. Stet, R. J., van Erp, S. H. M., Hermsen, T., Sultmann, H., and Egberts, E., Polymorphism and estimation of the number of Mhc-Cyca class I and class II genes in laboratory strains of the common carp (*Cyprinus carpio L.*), *Dev. Comp. Immunol.,* 17, 141, 1993.
62. Rebiere, M. C., Marche, P., and Kindt, T. J., A rabbit class I major histocompatibility complex gene with a T cell specific expression pattern, *J. Immunol.,* 139, 2066, 1987.
63. Satz, M. L., Wang, L. C., Singer, D. S., and Rudikoff, S., Structure and expression of two porcine genomic clones encoding class I Mhc antigens, *J. Immunol.,* 135, 2167, 1985
64. Guillemot, F., Billault, A., Pourquie, O., Behar, G., Chausse, A.-M., Zoorob, R., Kreibich, G., and Auffray, C., A molecular map of the chicken major histocompatibility complex: the class II β genes are closely linked to the class I genes and the nucleolar organizer, *EMBO J.,* 7, 2775, 1988.
65. Steinmetz, M., Winoto, A., Minard, K., and Hood, L., Clusters of genes encoding mouse transplantation antigens, *Cell,* 28, 489, 1982.
66. Srivastava, R., Duceman, B. W., Biro, P. A., Sood, A. K., and Weissman, S. M., Molecular organization of the class I genes of human major histocompatibility complex, *Immunol. Rev.,* 84, 93, 1985.
67. Bodmer, J. G. and Bodmer, W. F., Histocompatibility testing 1984, *Immunol. Today,* 5, 251, 1984.
68. Kaufman, J. F., Flajnik, M. F., and Du Pasquier, L., *Xenopus* Mhc class II molecules. II. Polymorphism as determined by two-dimensional gel electrophoresis, *J. Immunol.,* 134, 3258, 1985.
69. Lie, Ø., Slettan, A., Grimholt, U., Lundin, M., Syed, M., and Olsaker, I., Fish gene maps and their implications for aquaculture, *Animal Biotechnol.*, 5, 209, 1994.
70. Trowsdale, J., "Both man & bird & beast": comparative organization of Mhc genes, *Immunogenetics*, 41, 1, 1995.
71. Postlethwait, J. H., Johnson, S. L., Midson, C. N., Talbot, S. W., Gates, M., Ballinger, E. W., Africa, D., Andrews, R., Carl, T., Eisen, J. S., Horne, S., Kimmel, C. B., Hutchinson, M., Johnson, M., and Rodriguez, A., A genetic linkage map for the zebrafish, *Science*, 264, 699, 1994.
72. Williams, A. F. and Barclay, A. N., The immunoglobulin superfamily: domains for cell surface recognition, *Annu. Rev. Immunol.,* 6, 381, 1988.
73. Hunkapillar, T. and Hood, L., Diversity of the immunoglobulin gene superfamily, *Adv. Immunol.,* 44, 1, 1989.
74. Barclay, A. N., Johnson, P., McCaughan, G. W., and Williams, A. F., In *The T-Cell Receptors,* Mak, T. W., Ed., Plenum Press, New York, 1988, 53.

75. Flajnik, M. F., Canel, C., Kramer, J., and Kasahara, M., Evolution of the major histocompatibility complex: molecular cloning of major histocompatibility complex class I from the amphibian *Xenopus*. *Proc. Natl. Acad. Sci. U.S.A.*, 88, 537, 1991.
76. Lalanne, J. L., Delarbre, C., Gachelin, G., and Kourilsky, P., A cDNA clone containing the entire coding sequence of a mouse H-$2K^d$ histocompatibility antigen, *Nucleic Acid Res.*, 11, 1567, 1983.
77. Parham, P., Lawlor, D. A., Lomen, C. E., and Ennis, P. D., Diversity and diversification of HLA class I alleles, *J. Immunol.*, 142, 3937, 1989.
78. Xu, Y., Pitcovski, J., Peterson, L., Auffray, C., Bourlet, Y., Gerndt, B. M., Nordskog, A. W., Lamont, S. J., and Warner, C. M., Isolation and characterization of three class II Mhc genomic clones from the chicken, *J. Immunol.*, 142, 2122, 1989.
79. Sato, K., Flajnik, M. F., Du Pasquier, L., Katagiri, M., and Kasahara, M., Evolution of the Mhc: isolation of class II beta-chain cDNA clones from the amphibian *Xenopus laevis*, *J. Immunol.*, 150, 2831, 1993.
80. Malissen, M., Hunkapiller, T., and Hood, L., Nucleotide sequence of a light chain gene of the mouse I-A subregion: $A\beta^d$, *Science*, 221, 750, 1983.
81. Larhammer, D., Schenning, L., Gustavson, K., Wiman, K., Claesson, L., Rask, L., and Peterson, P. A., Complete amino acid sequence of an HLA-DR antigen-like β chain as predicted from the nucleotide sequence: similarities with immunoglobulins and HLA-A, -B, and -C antigens, *Proc. Natl. Acad. Sci. U.S.A.*, 79, 3687, 1982.
82. Jonsson, A. K., Andersson, L., and Rask, L., Complete sequences of DQW4 specificity, *Immunogenetics*, 30(3), 232, 1989.
83. Lawrance, S. K., Das, H. K., Pan, J., and Weissman, S. M., The genomic organization and nucleotide sequence of the *HLA-SB* (*DP*) alpha gene, *Nucleic Acids Res.*, 13, 7515, 1985.
84. Schamboeck, A., Korman, A. J., Kamb, A., and Strominger, J. L., Organization of the transcriptional unit of a human class II histocompatibility antigen: *HLA-DR* heavy chain, *Nucleic Acids Res.*, 11, 8663, 1983.

CHAPTER 3

The Chicken Major Histocompatibility Complex

James F. Kaufman and Susan J. Lamont

CONTENTS

I. THE INITIAL IDENTIFICATION OF Mhc MOLECULES AND CHARACTERIZATION OF THE Mhc

The *B* complex or major histocompatibility complex (Mhc) of the chicken was initially identified by Briles et al.[1] as an extensively polymorphic erythrocyte antigen or blood group system. The gene locus symbol was assigned as *B* to indicate that it was the second chicken blood group system discovered. Gilmour[2] independently discovered a blood group locus that was later demonstrated to be identical to the *B* system of Briles. Extensive multiple alleles of the *B* system, even in closed and

0-8493-8032-4/96/$0.00+$.50

inbred flocks, suggested a selective advantage of *B* heterozygosity for fitness and survival.[2,3]

The biological mechanism by which the *B* system influenced fitness remained a mystery until the discovery by Schierman and Nordskog[4] that the *B* blood group system is linked to the major histocompatibility complex. In skin grafts between partially inbred birds having like and unlike blood types, grafts between birds of the same *B* type were accepted while those of different *B* blood types were rejected. This confirmed the linkage of *B* with the major skin graft locus. Additional evidence that the *B* system represented the chicken Mhc was provided by the identification of *B* as the locus controlling graft-versus-host (GVH) splenomegaly,[5] GVH-chorio-allantoic membrane (CAM) pocks,[6] and the mixed lymphocyte reaction (MLR).[7]

The reactions and cross-reactions of alloantisera on blood cells of different strains of chickens proved to be quite complex, and, after much confusion and work, it became clear that there are three major groups of these alloantigens — B-F, B-L, and B-G antigens — each with an enormous serological complexity.[8-9]

As in mammals, the study of the chicken Mhc has in large part depended on the availability of partially inbred, congenic, and recombinant strains. In particular, the inbred Mhc congenic strains CB and CC along with the recombinants R1 and R2 (also called CB.R1 and CC.R1, now called B12r1 and B4r1) played an important role. A pair of recombinants involving B15 and B21 (so-called R4 and R5, now called B15r1 and B21r1) also have been used. Certain strains that share some alloantigens but not others (so-called "natural recombinants" such as GB-1 and CB which share B-F12 but not B-G) have been viewed as the product of unobserved recombination events. Finally, the use of these strains in defined crosses has been important, most recently in the large backcross families that are being used to construct linkage maps in the chicken.[10]

The first identification of Mhc molecules in a nonmammalian vertebrate was made in a remarkable series of papers by Pink and collaborators.[11-14] This work depended on Mhc congenic strains (including CB and CC) and their recombinants as well as a so-called "natural recombinant" (GB-1). They used alloantisera to identify the three classes of Mhc molecules: B-F (class I), B-L (class II), and B-G antigens. They determined the molecular weights of these antigens after reduction by sodium dodecyl sulfate (SDS) gel electrophoresis: a large band of 40 to 43 kD and a small band of 11 to 12 kD for B-F, two bands of around 30 kD for B-L, and two bands of 31 and 42 kD for B-G antigens. In contrast to the other two alloantigens, the B-G molecules without reduction had apparent molecular weights of 135 and 160 kD. They determined the blood cell expression pattern by immunofluorescence: both erythrocytes and leukocytes for B-F, immunoglobulin-bearing lymphocytes and monocytes for B-L, and erythrocytes only for B-G. They determined the location of the genes for these alloantigens with respect to recombination sites on the chicken Mhc chromosome: *B-F* and *B-L* in one region and *B-G* in another. Finally, they showed that the rapid allograft rejection, GVH reaction, and MLR were determined by the region encoding B-F and B-L antigens. Based on these data, they proposed a three-locus model for the chicken B complex, as well as suggesting that the chicken Mhc might have only one gene locus coding for a class I-like product, and thus be simpler in structure than mammalian Mhc.

II. THE STRUCTURE OF THE CHICKEN Mhc AND RELATED REGIONS

Since the work of Pink and collaborators, much has been learned about the structure of B-F, B-L, and B-G molecules and the genes that encode them. It will be easiest to approach this data after discussing the current picture of the chicken Mhc and certain other related regions (Figure 1); aspects of this work have been recently reviewed.[15-20]

The first important step was to establish the chromosomal location. Bloom and Bacon[21] analyzed aneuploid chickens that had an extra microchromosome (apparently number 16 in order of size) bearing the nucleoli organizer region (NOR) as visualized using a silver staining procedure. They found that these chickens also had an extra Mhc haplotype as detected by serology. They estimated this microchromosome to be 8 Mbp, of which some 6 Mbp was NOR encoding roughly 400 copies of tandemly repeated rRNA clusters.[22] Recently,[23] the β_2m gene was shown by fluorescence *in situ* hybridization (FISH) to be on a larger microchromosome (number 9 or 10 in size).

The next important steps were mainly the work of the laboratory of Charles Auffray. First, they isolated a class II β pseudogene by cross-hybridization with a human class II β cDNA.[24] They used this chicken class II β gene to isolate cosmids from a library made from DNA of the CB chicken strain (*B12* haplotype). These cosmids could be overlapped into four clusters, two of which were later connected by an overlapping lambda genomic DNA clone. They then identified potential genes by Southern blots of cosmid DNA hybridized with radioactive probes made from the total cDNA of B and T cell tumors as well as liver. The fragments bearing expressed genes were used to probe cDNA libraries and the cDNA clones isolated were identified by sequence. They identified two class II β gene fragments and four class I α gene fragments along with other genes on cluster I, two class II β gene fragments, and two class I α gene fragments along with the 17.8 gene on cluster II/IV, and a class II β gene fragment along with the 13.8 gene and some rRNA genes on cluster III.[25]

The genes on these cosmid clusters are not organized as in a typical mammalian Mhc. For example, some genes have been identified based on sequence, including the 12.3 gene that encodes for a G protein-like molecule,[26] the 8.5 gene that encodes a B-G molecule,[19] the 21.7 gene that encodes a TAP molecule (Kaufman and Bumstead, unpublished), and the 17.8 gene that encodes a member of the C-type animal lectin superfamily.[27] None of these identified genes correspond to the genes that are characteristic of mammalian class III regions (for instance, the tumor necrosis factor genes, the inducible gene for HSP-70, and complement components factor B, C2 and C4).[28,29] This is consistent with the evidence that electrophoretic polymorphisms of glyoxalase 1 (GLO)[30] and chicken factor B[31] do not segregate with the *B* complex; however, these polymorphisms could be due to post-translational modifications. With only 20 kB between the class II β genes and the class I α genes on cosmid cluster I, Guillemot and collaborators[25] found no evidence for a large class III region that separates the class I and class II genes into two distinct regions, and they proposed that the chicken Mhc is smaller and simpler than the Mhc of mammals. However, a DNA polymorphism detected by a newly discovered gene,

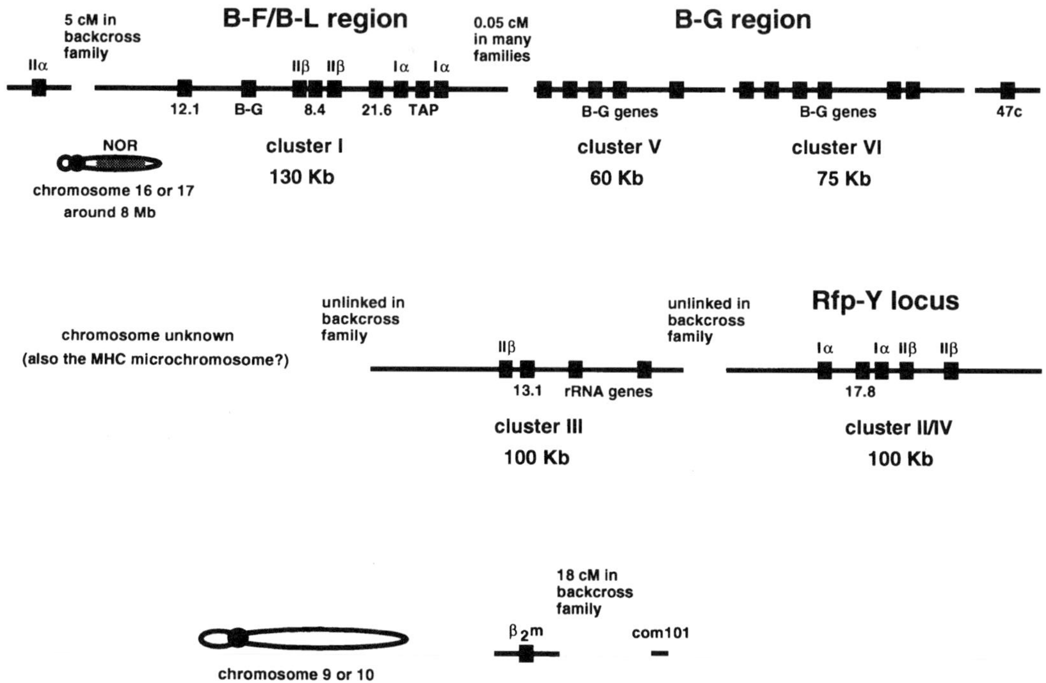

Figure 1 A current view of the chicken Mhc and genomic regions containing related genes. Gene assignments on cosmid clusters I, III, and II/IV are from Guillemot et al.[25] with modifications from Kaufman (unpublished). Gene assignments on cosmid clusters V and VI are modified from Kaufman et al.[81] Genetic distances between the *B-F/B-L* and *B-G* regions are from Crone and Simonsen[40] and from Simonsen et al.,[39] between the class II α gene and *B-F/B-L* region from Kaufman et al.,[60] between the β_2m gene and other genes on the linkage map from Riegert et al.,[23] and between cosmid clusters III, IV, and the rest of the linkage map from Miller et al.[35,36] and Kaufman and Bumstead (unpublished). Chromosomal location of the *B* complex is from Bloom and Bacon[21] and of the β_2m gene is from Riegert et al.[23]

G9a,[32] of the class III region of the human Mhc has been recently mapped to the chicken Mhc.[33] A *Pst*1 restriction fragment length polymorphism (RFLP) cosegregated with the EaB in the Compton reference population providing the first evidence that at least some part of the class III region may be associated with the chicken Mhc. In addition, the chicken C4 gene has been located between the class II β and class I α genes in cluster I (S. Beck and J. Kaufman, unpublished). Thus a remnant of the class III region may exist on the Mhc microchromosome.

It was originally thought that all of these cosmid clusters were located in the chicken Mhc, but in fact only cluster I corresponds to the *B* complex as defined by serology. The laboratory of Marcia Miller used Southern blots of DNA from certain F2 families from Briles to show that some class I α and class II β bands do not cosegregate with the *B* complex as defined by serology. Instead, these bands correspond to another locus which they named the *Rfp-Y* locus, with five alleles by RFLP analysis.[34] Using a large backcross family from Nat Bumstead (Compton) that had been used to establish a linkage map of the chicken genome, they also showed that the *Rfp-Y* locus failed to segregate with any other marker in the linkage map. Finally, they showed that this locus corresponded to cosmid cluster II/IV.[35,36] Probes from cosmid cluster III also fail to segregate with cosmid cluster I or any other linkage groups in the Compton backcross family (Kaufman and Bumstead, unpublished). It is now clear that cluster II/IV and cluster III are located on the Mhc microchromosome but separated from cosmid cluster I by a long distance or a highly recombinogenic region (such as the repetitive NOR).[175]

None of the genes identified in the three cosmid clusters corresponds to a chicken class II α gene, and the Auffray laboratory expended quite some energy in looking for it.[17] Eventually, the major (and perhaps only) class II α cDNA was identified by PCR using degenerate oligonucleotides based on the protein sequence, and indeed it fails to hybridize to any of the cosmid clusters. Using this cDNA as a probe, Salomonsen and Kaufman found that there is apparently a single class II α gene and that this gene cosegregates with neither the *B* complex nor the *Rfp-Y* locus in the Briles F2 family, but is located about 5 cm away from the *B* complex in the Compton reference family. Thus far, they have been unable to place this gene with respect to the recombinants, since there are no RFLPs in the CB and CC Mhc congenic chickens and too many RFLP variants in the B15 and B21 recombinant chicken strains[37] (Salomonsen and Kaufman, unpublished).

The β_2m gene was not expected in these cosmid clusters, and, when cDNA clones were identified by screening with oligonucleotides based on the protein sequence,[38] they failed to hybridize to any of the cosmids. Using these cDNAs as probes, a single β_2m gene that is located outside of the Mhc based on RFLP in backcross families was found; it is some 18 cm from a single marker in the Compton backcross family. By *in situ* hybridization, the β_2m gene is located on non-Mhc microchromosome.[23]

In contrast, B-G erythrocyte alloantigens have long been known to be very closely linked to the B-F/B-L region, within approximately 0.05 cm based on 30000 matings,[39,40] although certain families from Briles have higher levels of recombination sometimes giving complex recombinants.[41] Several groups have isolated cDNA clones corresponding to B-G molecules. Based on Southern blots using these cDNA clones as probes, the *B-G* genes form an extensive multigene family. There were a few *B-G* RFLPs that were not polymorphic with the restriction enzymes used, and most of the RFLPs followed the *B-G* region in the CB/CC Mhc congenics and the B15/B21 recombinants.[42,43] This is consistent with the fact that all of the erythrocyte antigens immunoprecipitated with rabbit antisera to B-G molecules and analyzed by SDS gel electrophoresis are determined by the B-G region.[44] However, one *B-G* RFLP was found on the *B-F*/*B-L* side and this corresponds to the 8.5 gene in cosmid cluster I which encodes a B-G molecule found on B cells.[19,43,45] Kaufman and colleagues used B-G cDNA clones to isolate cosmid clones from the CB library made by Guillemot. They identified two authentic cosmid clusters — cluster V with five *B-G* genes and cluster VI with six *B-G* genes; neither cosmid cluster has class I or class II genes. In addition, there is at least one cDNA from the CB haplotype (the 47c from a cecal tonsil library) that has not been placed on these cosmids.[19,46]

It is not yet clear whether the map of the Mhc from CB chickens provides a good model for other chicken strains or other species of birds. RFLP using class I α, class II β, and *B-G* probes has suggested gene expansion and contraction.[47-55] In particular, it has been suggested that there are more class I α genes in the B21 haplotype.[47,52] Certain chicken strains have extra TAP bands (Kaufman and Bumstead, unpublished).

The use of RFLP analysis to biochemically characterize the existing variation in the Mhc is a valuable addition to serological analysis. The RFLP analysis can examine subregions and genotypes for which the production of serological reagents has been elusive, analyze Mhc chromosomal recombinants, and categorize newly identified genotypes. As has already been done for serological analysis,[9] standard nomenclature should be defined for RFLP analysis of Mhc haplotypes.

III. CLASS I AND CLASS II GENES AND MOLECULES

Several chicken class I α cDNAs (F10 from CB and F3 from B19) and one class I α gene (B-FIV encoding F10) have been reported along with a few N-terminal sequences of protein isolated from erythrocytes.[25,38,56] A number of chicken β_2m cDNAs and genes have been sequenced, along with the mature β_2m protein isolated from erythrocytes.[23,38,57]

Based on these sequences, the chicken B-F molecule is clearly a classical class I molecule. The α chain gene is organized like mammalian classical class I genes with an exon encoding the signal sequence, three exons encoding the extracellular domains (α1, α2 and α3), an exon encoding the transmembrane (TM) region, and three exons contributing to the cytoplasmic tail, the last also encoding the 3′ untranslated region (3′UT). The β_2m gene is also organized like the mammalian β_2m genes, with one exon for the signal sequence, one for the majority of the protein, one with the last four amino acids of the protein as well as part of the 3′UT, and the last finishing the 3′UT. The α1 and α2 peptide-binding domains (PDBs) are polymorphic between chickens but have diverged to different extents compared to mammals (α1 roughly 40% identical amino acids compared to mammalian classical class I molecules, α2 roughly 55%). All but one of the residues that are thought to be involved in coordinating the N- and C-terminal ends of the antigenic peptide are identical to those in mammals, and nearly all of the polymorphic residues are located in positions that should interact with the peptide. The Ig-like domains, β_2m and the α3 domain, are invariant or nearly so between different chicken strains; they also have diverged to different extents (α3 roughly 33%; β_2m roughly 45%). The surface of α3 is almost completely diverged, so that only a few residues of the CD8 interaction site defined in mammals are conserved. The cytoplasmic tail bears a few residues that are conserved, including some involved in kinase sites — a tyrosine in exon 6 and two serines in exon 7.

There is evidence to suggest that there is only one class I gene product (or several that are extremely similar) expressed at high levels on chicken blood and spleen cells. Of the six CB cosmid fragments that hybridized with the class I α cDNA F10, only one (B-FIV of cosmid cluster I) hybridized with an oligonucleotide from the 3′ untranslated region (3′UT), suggesting that only this gene was expressed.[25] The four class I gene fragments in cosmid cluster I actually represent two genes; both of these are found as cDNAs (Kaufman unpublished). Using polymerase chain reaction (PCR), the transcripts of two such genes have been amplified from cDNA of many chicken strains. However, only one of these genes is expressed at high levels regardless of cell type (Kaufman unpublished). Certainly, there is another class I molecule that is found on B cells;[38] this CB-3 antigen represents around 10% of the class I on B cells based on flow cytometry (Kaufman, unpublished). It is also likely that transcripts encoded by the Rfp-Y locus will be expressed, although at low levels. There are multiple class I size and charge variants that immunoprecipitate with monoclonal antibodies to class I α chain (F21-21) and β_2m (F21-2), but these evidently differ in the C-terminal cytoplasmic regions, based on the collapse of the complicated patterns to a few spots after trypsin treatment.[59] Rabbit antisera directed to two different cytoplasmic exon peptides suggest that the top band in SDS gels represents a molecule with both exon 6 and exon 7 domains, the middle band with only exon 6 domain, and the bottom band with neither.[38] Many *B-F* cDNAs have been isolated that are identical except for the presence or absence of exon 7 (Kaufman, unpublished). Another point is that antigenic peptide eluted from all molecules isolated with the monoclonal antibody to chicken β_2m conform to a single motif, as though there is only one major class I molecule (or two extremely similar molecules) expressed.[60]

Class II β genes and pseudogenes from several chicken strains, along with many cDNAs derived from PCR amplification, have been reported.[24,61-65] Very recently, several class II α cDNAs have been isolated[60] (Salomonsen and Kaufman, unpublished). These sequences show that the B-L molecule is a classical class II molecule; the β chains are not especially close to any particular mammalian isotype, whereas the α chain is most similar to DR/E.

The β genes are organized like mammalian class II β genes, with an exon encoding the signal sequence, a polymorphic β1 PBR domain, a nonpolymorphic β2 Ig-like domain, a transmembrane domain, and two cytoplasmic exons, the last including the 3′UT. (Figure 2) There appear to be three isotypes of class II β genes, based on sequences of the five genes from the CB cosmid clusters (one from a lambda clone of unknown haplotype), and of fragments of cDNA derived from many chicken Mhc haplotypes (most of the genes and cDNAs were amplified by PCR). The two class II β genes (B-LBI and II) in cluster I are closely related, and the cDNAs show that they are highly polymorphic and highly expressed. The two genes located on cosmid cluster II/IV (B-LBIII and V, recently renamed Rfp-LIII and V) and the gene on cosmid cluster III (B-LBIV) are virtually identical to each other but quite different from the cluster I genes; they are nearly monomorphic in the strains examined and are poorly expressed in the tissues examined. Among the cDNA clones were other class II β sequences that formed a third group; these were named B-LBVI. As yet no B-LBVI gene has been reported.

The α chain cDNAs have an α1 PBR domain, an α2 Ig-like domain, a transmembrane domain, and a short cytoplasmic tail. All class II α chain coding sequences appear the same and there appears to be a single gene by southern blots. Two-dimensional gel electrophoresis of immunoprecipitates using a monoclonal antibody to chicken class II molecules show one monomorphic and two polymorphic chains in certain chicken strains;[66] this is consistent with the class II α chain in association with the two abundantly expressed β chains encoded on cosmid cluster I.

These class II β, class I α, and β_2m genes are compact, with some unexpected sequence elements in the promoters. The introns are all much smaller than the mammalian counterparts, with many as small as 100 bp (see Figure 2). Some portions of the genes (and the class II α cDNA) are G+C rich compared to the mammalian counterparts, particularly some exons and introns as well as the promoters. For instance, β_2m promoter, exon I, and intron I are all greater than 80% G+C nucleotides. Exon II of the β_2m gene and the exon 4 of the *B-F* gene encoding the α3 domain are both over 97% G+C in the third codon position, with the particular amino acids used biased toward those with G+C-rich codons. However, exons only 100 bp away might have no such bias, such as exon 3 of the B-F gene encoding the α2 domain with only 67% G+C in the third codon position. The promoters for class II β, class I α, and β_2m genes have G+C-rich regions near the start-point of transcription (instead of TATA boxes found in mammalian counterparts), and have X and Y boxes (expected for class II β but not class I α or β_2m genes); the promoters of class I α and β_2m also contain interferon response elements (IRE).[23,56,64] It is currently not clear whether there are functional meanings for these differences between chicken and mammalian genetic regulatory elements.

Research on functional activity of chicken Mhc regulatory elements is in the initial stages. The conserved IRE in the chicken Mhc class I promoter region has been shown to induce gene expression.[67] Lamont and colleagues[68] have characterized a functional chicken major histocompatibility complex (Mhc) class II gene promoter by using the chloramphenicol acetyltransferase (CAT) reporter system. A 0.7-kb DNA fragment from the 5′ flanking region of a class II gene was cloned into vectors upstream of the CAT structural gene and transiently transfected into the MQ-NCSU[69] chicken macrophage cell line. Positive promoter activity of the chicken DNA was evident when an SV40 enhancer was present in the reporter construct, but was weak if the SV40 enhancer was absent. The results indicate that the 0.7-kb chicken DNA fragment contains a functional promoter, but may require additional enhancer(s) outside of this immediate 5′ upstream region for efficient initiation of transcription. Deletion analysis of this 0.7-kb 5′ DNA fragment revealed a short fragment in the

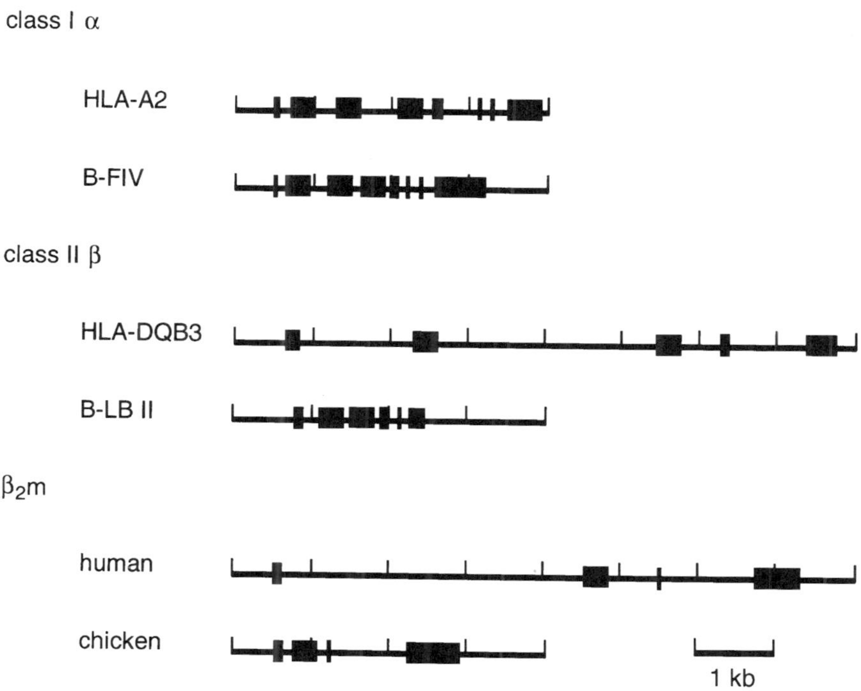

Figure 2 Comparison of the structural organization of human and chicken genes (boxes indicate exons). The maps are derived from the following sources: HLA-A2, Koller and Orr;[172] B-F12 IV, Kroemer et al.;[56] HLA-DQB3, Boss and Strominger;[173] B-LB II, Zoorob et al.;[64] human β_2m, Güssow et al.;[174] chicken β_2m, Riegert et al.[23]

3′ end that was crucial for the promoter function, and negative regulatory elements located further upstream. The conserved Mhc class II X and Y boxes surprisingly did not have a significant influence on promoter activity. Sequence analysis of the 0.7-kb class II gene upstream region suggests involvement of interferon, ETS-related proteins, and other factors in regulating this promoter. Glucocorticoids were shown to decrease interferon-induced gene expression in the same transient transfection assay system of chicken class II promoter construct and CAT reporter gene (Chen et al., unpublished).

IV. B-G GENES AND MOLECULES

B-G proteins, cDNAs, and genes have been examined in the last few years, and the work has been reviewed in great detail.[19] The B-G molecules were initially of interest because of their remarkable serological polymorphism; in fact, it is likely that alloantibodies to B-G antigens on erythrocytes were responsible for the discovery of the chicken Mhc. It is now clear that there are multiple B-G molecules present on erythrocytes, and that there are very similar molecules present on thrombocytes and lymphocytes as well as related molecules on certain stromal cells in cecal tonsil, bursa, thymus, and the epithelial cells of small intestine.[44,45,70-72] The B-G molecules on blood cells are generally disulfide-linked dimers;[14,44,72,73] it is not clear whether they are homo- or heterodimers (although we have found that the 8.5 gene transfected into L cells encodes chains that form homodimers). These B-G chains lack N-linked and apparently O-linked glycans and vary significantly in size (35 to 55 kD) depending on the length of the cytoplasmic tails. The B-G molecules on stromal cells in bursa, thymus, and cecal tonsil and on epithelial cells in the small intestine can be larger than the blood cell B-G molecules, and some of them may bear N-linked glycans.[44-45,71,73]

The sequences of cDNA and genomic clones[20,43,44,46,74] encoding blood cell B-G molecules led to a much clearer understanding of B-G protein structure. A signal sequence is presumably cleaved off just before a glutamine in all but one of the known B-G sequences; this glutamine can cyclize to a pyroglutamic acid, which may explain why all erythrocyte B-G preparations have blocked N-termini. A single extracellular immunoglobulin-like domain has the length and many sequence features of a variable (V)-like domain, but lacks N-linked glycosylation sites. There are three cysteines, two of which form an intradomain disulfide bridge like the one found in immunoglobulin V domains and one that participates in the interchain disulfide bond.[44] Based on the position of the cysteine responsible for the interdomain disulfide bridge, on the residues present in certain key positions, and on the positions that are conserved between B-G molecules, it seems unlikely that the V-like domains in B-G molecules are associated via the five-strand face, the way that V regions in antibodies are associated.[46] The B-G extracellular region is followed by a hydrophobic transmembrane region and then by a cytoplasmic region that is long but variable in length; some of this variability is due to alternative splicing. The cytoplasmic tail consists of seven amino acid repeats with all the sequence features of α-helical coiled-coils; presumably the tails from the two chains wrap around each other. In the 8.5 gene, the signal sequence, the extracellular V-like domain, and the transmembrane region are each encoded by a separate exon, while the cytoplasmic region is broken up into many exons of 18, 21, and 24 bp, with some longer exons at the very end. The 5′UT is not so clear, since many splicing variants have been observed (Figure 3).

The function (or functions) of B-G molecules remains the outstanding question. Based on the structure, they would appear to be typical adhesion molecules, with the extracellular region involved in interactions with other cells or with extracellular matrix and the cytoplasmic tail interacting with the cytoskeleton. This accords well with their sequence relationship with myelin-oligodendrocyte glycoprotein.[75] On the other hand, it is not obvious why such molecules would be so highly polymorphic nor would it explain certain immunological phenomena — the so-called "adjuvant effect", the rapid and strong antibody response, and the reported presence of natural antibodies.[19,76-80]

The sequence variability in the extracellular V-like domain is known for at least 12 *B-G* genes from the CB chicken that were cloned from cosmids and from cDNA libraries; in addition there are three complete B21 cDNAs and one partial cDNA from an unknown haplotype.[20,46,74] In general, it is not clear which of these might represent alleles (or in fact which of the putative CB genes besides the 8.5 gene are

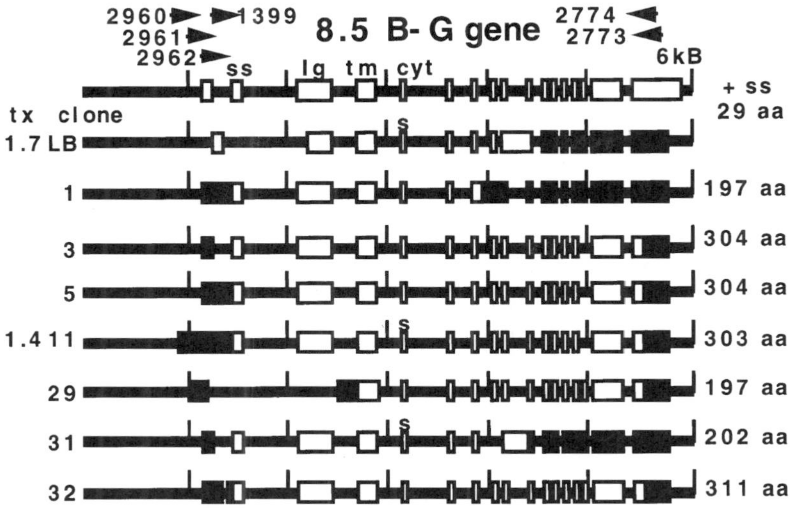

Figure 3 The structural organization of the 8.5 *B-G* gene in cosmid cluster I and transcribed alternative splice variants. Top line represents the gene (open boxes represent putative exons; ss, signal sequence; Ig, extracellular immunoglobulin-like domain; tm, transmembrane region; cyt, cytoplasmic region); the genomic sequence was determined from a 10-kB Nru I-Hha I subclone derived from the cosmid cβ12. The other lines represent the transcription products (boxes indicate portions of the genomic sequence found in a particular cDNA clone; open boxes indicate potentially translated sequence after the first AUG codon; filled boxes indicate untranslated regions; s indicates a splice version of the exon that is shorter by three nucleotides; aa, number of amino acids that might be translated after removal of the signal sequence). The linearized 10-kB Nru I-Hha I subclone and a plasmid carrying the neo gene were used to cotransfect mouse Ltk⁻ cells; stable transfectants were isolated based on cell surface expression of an antigen that reacted with the monoclonal antibody g6.[45] Single-stranded cDNA was prepared from poly-A+ RNA that was derived from total RNA of two independent L cell subclones (1.4 and 1.7). Amplification by polymerase chain reaction using primers based on the genomic sequence (primers 1399, 2960, 2961, 2962, 2773, and 2774 indicated with arrows) yielded products of discrete size that were cloned into Bluescript vector and then sequenced (clones LB, 1, 3, and 5 from transfectant 1.7; clones 11, 29, 31, and 32 from transfectant 1.4). Data from Kaufman, Döhring, Riegert and Guillemot, unpublished.

expressed), so it is not yet possible to discuss the residues involved in allelic polymorphism. However, the available sequences show that at least half of the positions can vary and these positions are located mostly in bends and on the five-strand face. Each variable position has only a few different residues, so the V-like sequences appear as a patchwork of epitopes with low Kabat and Wu variability as well as low replacement to silent mutation ratios. This may indicate that the polymorphism is not selected, but that conclusion really depends on comparison of true alleles or at least clearly orthologous genes (that is, genes with the same functions).

The most distantly related *B-G* sequence is a complete cDNA (clone 47c) that was isolated from cecal tonsil of CB chickens; it is absolutely identical over the region that can be compared to the partial cDNA from embryonic erythrocytes of an unknown haplotype. Therefore, seems that not all B-G molecules are polymorphic;

in fact it was reported years ago that among the B-G molecules immunoprecipitated by a monoclonal antibody was one invariant array, detected by two-dimensional gels, that was present in most Mhc haplotypes examined.[70]

If the polymorphism is not selected at the protein level, then it is presumably maintained either by a high mutation rate or by genetic hitchhiking.[81] There is no evidence for a high mutation rate among *B-G* genes, but genetic hitchhiking is not impossible, given the high level of linkage disequilibrium between the *B-F*/*B-L* region and the *B-G* region. If the polymorphism is selected, then the likeliest driving force is immunological. Several possibilities have been suggested, including presentation to T lymphocytes (similar to Mhc molecules), binding of antigen (similar to antibodies), and selection of the B-cell repertoire in the bursa. The first alternative seems unlikely now, based on the lack of structural similarity to Mhc molecules, the failure to stimulate lymphocytes *in vitro* or graft rejection *in vivo*, and the absence of bound peptides. The second alternative seems less likely based on the evidence that the V-like dimers do not associate in the manner of antibody Fv fragments. We are still in the process of analyzing the third possibility.

V. THE CHICKEN B COMPLEX AS A "MINIMAL ESSENTIAL Mhc"

To summarize, the current data indicate that the few class I α and class II β genes located in cosmid cluster I are the only ones that encode polymorphic molecules that are expressed at a high level, and this cluster is in fact the *B-F*/*B-L* locus, representing the Mhc of the chicken. Certain other genes, such as the TAP gene(s) involved in transport of antigenic peptides destined for association with class I molecules across the membrane of the endoplasmic reticulum, are also located in the Mhc. In addition, both a *B-G* gene expressed on B cells and the whole *B-G* region are closely linked to cosmid cluster I. Although this seems quite different from what is known in mammals, recently it was shown that two mammalian genes with some homology to *B-G* genes, myelin-oligodendrocyte glycoprotein and butyrophilin, are located on the edge of the Mhc in mouse and man.[75] This raises the possibility that the association between the Mhc and B-G-like molecules is in fact ancient.

In contrast, certain other genes that are located within the Mhc of mammals and amphibians are located outside of the chicken Mhc or missing in the chicken genome altogether. Most genes characteristic of the mammalian class III region have not been identified in the chicken cosmids isolated thus far, and a factor B polymorphism is determined by a gene located outside of the Mhc. However, one gene (G9a) known to map in the human Mhc class III region has been linkage-mapped to the chicken Mhc, suggesting that at least some part of the class III region remains associated with the chicken Mhc.[33] Class II α genes have virtually always been found right next to their partner class II β genes, but the chicken class II α gene is located outside the Mhc. Similarly, the class II β genes located in cosmid cluster II/IV (the *Rfp-Y* locus) and cosmid cluster III are located on the same microchromosome as the Mhc, but separated by a fair physical distance, a highly repetitive region that promotes recombination (like the NOR), or by smaller recombinational hotspots. These class II β genes are not highly expressed in the tissues examined and not highly polymorphic in the strains examined. The same is likely to be true for the class I α genes in cosmid clusters II/IV and III since there is no evidence for more than one locus that determines rapid tissue allograft rejection in the chicken. This is not to say that the class II β and class I α genes outside of the Mhc are without function; in mammals some nonpolymorphic, nonclassical class I α genes are involved in specialized antigen presentation.[82] An interesting hypothesis suggested by Miller and collaborators[35] is that the *Rfp-Y* region is involved in natural killer recognition, based on the involvement of class I and animal C-type lectin (like 17.8 gene) molecules in such phenomena.

Thus, the chicken Mhc is quite compact, in the order of 30 to 100 Kb long compared to up to 4 Mbp in mammals, and the number of members in Mhc-like multigene families is quite small (depending on how one is counting, some two class II β, one class II α, and one or two class I α genes in chickens compared to two to seven classical class II β, two to four classical class II α, and two or three classical class I α genes in mice and humans).[28,29] The chicken Mhc might be viewed as a "minimum essential Mhc", which leads to three interesting points.

The first point is based on the small size and lack of recombination within the chicken Mhc compared to mammals. Those genes within the *B* complex can be regarded as under some selection to travel together in evolution, for instance, in the sense of a supergene as envisioned by Bodmer (a group of polymorphic genes in which combinations of a particular allele at each locus are favored as haplotypes). The very low recombination in this region certainly could give rise to stable haplotypes. For example, in mammals the TAP genes are located in the class II region far away from their class I genes;[28,29] in contrast, chicken TAPs are located right next to the polymorphic class I genes in such a way that one could envision a special TAP allele servicing each class I allele. It will be very interesting to determine whether other genes, such as the LMP genes encoding proteosome components, nonclassical class II genes that are essential for class II antigen presentation, and the inducible HSP70 genes are located in the chicken Mhc.

A second point is based on the relatively few members of the Mhc multigene families in the chicken (one class I α and two class II β genes that are highly expressed and polymorphic in the *B* locus) compared to mammals. Despite the key roles played by Mhc molecules in the immune response, and the clear possibility of selection for polymorphism of Mhc molecules in response to mutation by pathogens, there are in fact few good examples of resistance or susceptibility to infectious pathogens determined by the Mhc,[83] and most of them are in the chicken (for example, Marek's disease and Rous Sarcoma virus[84,85]). It seems possible that a minimal Mhc, stripped of the confounding influences of many multigene family members and unrelated disease resistance genes, might allow one to see the effects of selection more clearly. A single dominant motif is present for most strains in the antigenic peptides bound to class I molecules isolated with a monoclonal antibody to β_2m. In some instances, the motifs suggest that the class I α genes directly contribute to the resistance determined by particular chicken Mhc haplotypes.[60]

The third point is that the unusual structure of the chicken Mhc may be part of the explanation for the appearance of microchromosomes and the evolutionary origins of birds.[23,81,86] The simplest explanation for the compact nature of the chicken Mhc is that it is located on a microchromosome. This is hardly an unusual occurrence for a chicken gene, since for most birds there are only 5 to 8 macrochromosomes that are roughly the size of mammalian chromosomes and around 30 microchromosomes ranging in size from 1 Mb on up. Most birds have a very similar karyotype (even to the banding patterns in the macrochromosomes) and those that are different are considered to be the result of subsequent chromosomal evolution. In contrast, amphibians and dipnoid fish have relatively large chromosomes like mammals. Moreover, the Mhc of amphibians appears to be more similar to mammals than birds — multiple class II α and β genes travel together with a class I α gene as well as the typical class III region genes C4 and HSP-70.[87-89]

We propose that, during evolution, some unknown mechanism removed DNA from ancestral macrochromosomes to produce microchromosomes, the result of which was frozen in some bottlenecked lineage leading to birds. If the removal of DNA was random, then the elimination of essential genes must have been fatal for most individuals. The descendants of the few survivors who founded the avian lineage would be especially homogeneous, as appears to be the case.

On microchromosomes, the wholesale removal of DNA would lead to short introns, short intergenic distances, and an absence of "nonessential genes" including unnecessary members of multigene families. In addition, the elimination of DNA sequences that control the G+C/A+T content of large stretches of DNA (isochores) might lead to microchromosomes with very skewed G+C content, G+C-rich promoter regions, biased codon usage, and changes in amino acid composition. This appears to be true for many Mhc genes as well as the β_2m gene, all of which are located on microchromosomes. On macrochromosomes, this postulated DNA removal would have occurred at a slower rate or not at all, leading to most genes being similar to mammalian or amphibian genes, as occurs for genes such as the ovalbumin gene.

VI. THE CHICKEN Mhc AND IMMUNORESPONSIVENESS

A critical function of the chicken major histocompatibility complex (Mhc) is the regulation of cellular communication in the immune response. The cell surface proteins of the Mhc, which serve to distinguish "self" from "nonself" are unique to each genetically unique individual. This allows immune reaction to occur against foreign antigens of an almost infinite spectrum while simultaneously preserving the organism from self-destruction. The control of the Mhc over communication events can occur at of two levels: molecular interactions at the cell surface and cellular cooperation in an immune response.[90] The Mhc cell surface glycoproteins interact with both the antigen and with complementary structures of other immune cells, generating an immune response that is specific for the inducing antigen. For some of the different cell types of the immune system to communicate effectively, they must share at least one Mhc haplotype, a phenomenon known as Mhc restriction.

The Mhc restriction may function at the level of interactions between mature cells in the immune response or of immune cell differentiation.[91,92]

The cyclophosphamide (CP) model has been effectively used by Vainio and the Toivanens to study T-B cell interaction in the chicken to determine the genetic requirements. Administration of CP in the neonatal period causes a permanent depletion of the B-cell compartment, but T-cell function is normal 3 to 4 weeks after CP treatment. Transferred allogeneic bursal cells which express only Mhc products of the donor genotype can repopulate the CP-treated host. Therefore, antigen-presenting cells and T cells share Mhc antigens with each other, but not with the grafted allogeneic bursal-derived cells. Antibody production to a T-cell-dependent antigen (*Brucella abortus*) is restored by allogeneic or syngeneic bursal cells, but only syngeneic cells restore antibody response to a thymus-dependent antigen (sheep red blood cells).[93,94] The use of Mhc-recombinant chicken lines demonstrated that class II (B-L) antigens alone were the necessary restriction elements in T-B cell cooperation for antibody production.[95]

The T-B cell cooperation needed for generation of germinal centers in the spleen requires identity at the Mhc.[96] The use of Mhc-recombinant chicks identified the restriction element to be the class II region of the Mhc.[95] The interaction of antigen-presenting cells and T cells was evaluated by the *in vitro* proliferation of T cells in response to specific antigen in the presence of antigen-presenting cells. It was demonstrated that the B-L (class II) gene products are also the restriction elements for antigen-presenting-cell-T cell interaction.[97]

The Mhc restricts T cell cytotoxic reactions with some virus-infected or transformed cells, in addition to restricting the interaction of different types of immune cells.[98] When cytotoxic T cells are combined with reticuloendotheliosis virus (REV) infected cells, significant cytotoxicity occurs only with syngeneic effector and target cell combinations.[99] An Ia-positive T cell with antigen-restricted cytotoxic properties was identified as the effector cell responsible for the Mhc-restricted cell-mediated cytotoxicity.[100] The Mhc- and virus-restricted cytotoxic cells may be an important immune surveillance mechanism to stop proliferation of virally induced tumors.[101]

The Mhc has profound effects on genetic control of immunoresponsiveness via its role as a restriction element or via specific immune response genes. Antibody production against a variety of antigens is linked to the chicken Mhc. Immune responses to synthetic polypeptides such as (T, G)-A — L,[102] GAT,[103] and GT[104] are linked to the Mhc in the chicken. Antibody titers to other soluble antigens (e.g., bovine serum albumin, BSA) and to cellular antigens (e.g., *Salmonella pullorum* bacterin, whole sheep erythrocytes) are also associated with the chicken Mhc.[105,106] Total serum IgG levels are also under genetic control, partially associated with the *B* complex.[107]

Genetic control of immunoresponsiveness can also be evaluated by assays of cell-mediated immunity. The delayed wattle reaction (DWR) is commonly used as a measure of cell-mediated immunity in the chicken.[108] The reaction inducer can either be a polyclonal activator of T cells, such as phytohemagglutinin,[109] or it may be more specific, such as the antigen of *Staphylococcus aureus*,[110] an important pathogen of poultry. Both polyclonal and specific inducers show a cell-mediated immunity associated with the B complex.

Although the complement system does not fit the classical definition of exhibiting the memory and specificity of the immune system, complement proteins are extremely important in the immune reaction as a means to destroy invading pathogens. Complement components that are activated by antibodies or bacteria can lyse bacteria by destroying their cell walls. The levels of total serum hemolytic complement have been associated with the Mhc in chickens,[111] although specific components were not found to segregate with the Mhc.[31]

The use of *B*-congenic lines of birds is valuable in the genetic analysis of Mhc control of immune response. The B-congenic lines differ from each other only in the microchromosome, or fraction of it, which bears the Mhc. Immune response differences between these lines are therefore attributable to the Mhc-bearing chromosome. Partially developed (two to five backcross generations) $15I_5$-*B* congenic lines showed significantly different hematocrits and hemolytic complement levels but no significant difference in immunoglobulin M or G levels or differential blood cell counts.[112] These lines also had differences in antibody responses to sheep red blood cells, *Brucella abortus* antigen, and infectious bursal disease virus vaccine.[113] Qureshi and colleagues[114,115] showed that chemotactic activity of chicken blood mononuclear leukocytes, and activity and recruitment to the peritoneal cavity of macrophages differed among the $15I_5$-*B*-congenic lines. Lines of chicken congenic for the Mhc were also used to demonstrate Mhc associations with CD4 and CD8 lymphocyte percentages and ratios.[116]

As an alternate approach to examining immune characteristics associated with the Mhc in *B*-congenic lines, the Mhc can also be studied in lines selected for various traits of immunoresponsiveness. Long-term selection experiments for antibody response to sheep erythrocytes have been conducted independently by Siegel and colleagues in Virginia and van der Zijpp and colleagues in the Netherlands. Upon analysis of the Mhc haplotypes in these lines, correlated changes in Mhc allelic frequencies were detected with the antibody selection.[117,118] The Mhc genotype explains part of the total variation in antibody levels, but background genome also has a substantial effect.[119,120] In meat-type lines divergently selected for early antibody response to *Escherichia coli* vaccination, differences in frequency of Mhc class IV RFLP bands were present.[121] In Leghorn lines divergently selected for an index of immunocompetence traits (antibodies, inflammatory response, and reticuloendothelial activity) differences occurred in Mhc genotype frequencies.[122] Diverse facets of the immune response, therefore, are under Mhc genetic control.

VII. THE Mhc AND ECONOMIC TRAITS IN CHICKENS

Studies of both the basic biology and the practical application of the chicken Mhc have been motivated by the association of the chicken Mhc with resistance to disease. Several reviews[123-128] have surveyed this subject. The most significant association of the Mhc with resistance to a specific disease is that of Marek's disease, a herpes virus-induced lymphoma. This Mhc influence has been mapped, using recombinants, to the *B-F*/*B-L* region.[129,130] The Mhc is associated with incidence of tumor formation, mortality, and transient paralysis induced by Marek's disease virus.

One specific haplotype, B^{21}, seems to convey Marek's resistance to many different genetic backgrounds. Variation in response to Marek's disease virus by sublines that are identical at the *B* locus, however, illustrates the influence of non-Mhc genes on Marek's response.[123] Genetic complementation of Mhc alleles with other genes such as *Ir-GAT* may affect the susceptibility to Marek's disease.[131] The Mhc also influences the response to other virally induced diseases, including Rous sarcoma virus tumors,[101] avian leukosis,[132] and infectious bursal disease. Other categories of disease, besides virally mediated, also have been demonstrated to be associated with the Mhc.[127] These include fowl cholera,[133] obese strain spontaneous autoimmune thyroiditis,[134] and coccidiosis.[135] Similar to immune response traits, Mhc congenic lines have played an important role in defining Mhc associations for diseases as diverse as Marek's disease,[136-137] tumors induced by Rous sarcoma virus[138] and v-*src*,[139] and *S. aureus*.[140] The range and variety of diseases influenced by the chicken Mhc is, therefore, extensive.

An important observation that emerges from studies on the Mhc and disease, however, is that there is not a single haplotype that performs optimally in all genetic backgrounds in response to all disease challenges. This is not surprising given the many different immune mechanisms involved in resistance to disease and the many levels at which the Mhc exerts genetic control. The repeated associations of resistance with the Mhc, but not with one Mhc haplotype, may suggest an important role for non-Mhc genes, or currently unidentified Mhc genes, located within the complex. Genetic associations with the Mhc suggest that there are genes for disease response within the Mhc, but the lack of consistent association with specific haplotypes (with the exception of B^{21} and Marek's disease) suggests that genes other than the currently identified Mhc genes may have a major influence. Specific resistance gene identification and utilization of genetic engineering may allow the construction of resistant haplotypes that have not been produced by conventional breeding.

The B complex has long been associated with genetic variation in production traits such as general mortality.[2,3] Traits of economic importance as diverse as fertilization rate, embryonic mortality, hatchability, juvenile and adult mortality, body weight, and egg production are influenced by Mhc genotype.[123,141,142] Several independent studies have shown that long-term selection for egg production traits and disease resistance have significantly altered Mhc allelic frequencies.[143-145] Although different combinations of B haplotypes were present in these selection studies, some general patterns emerge from examination of haplotypes that exhibit consistent changes under several selection regimes. From these long-term selection studies, it is suggested that B^2 and B^{21} may show overall beneficial associations with economic traits in chickens. A variety of other studies have demonstrated Mhc associations with particular production traits, but not necessarily consistency of specific Mhc haplotypes or associated traits.[146-150]

The mechanism by which the Mhc influences production traits is unknown. The effects may be indirect, via the Mhc influence on immunocompetency and disease resistance, or the effects may be more directly based on the role of the Mhc in mediating cellular development and communication, or on existence of chicken homologs of the mammalian t-complex.[151]

Aneuploid (trisomic or tetrasomic) birds for the Mhc-bearing chromosome have been used to map the Mhc to a specific microchromosome and examine effects of Mhc gene dosage on several biological systems. The chicken Mhc (*B* complex) and the ribosomal ribonucleic acid (rRNA) genes (nucleolar organizer region, NOR) were shown to be linked on a microchromosome of approximately size 16.[21,152] Excess rRNA genes in aneuploid birds appear to be inactivated, but the Mhc genes on the same chromosome are expressed in a dose-dependent fashion. Aneuploids have a greater concentration of erythrocyte cell-surface Mhc glycoproteins than do disomic chickens.[153] A stepwise increase in the amount of B-L antigen on trisomic and tetrasomic cells was shown by flow cytometry of bursal cells.[154] The dosage of Mhc genes alters primary immune organ development. Aneuploid chickens have smaller bursae and thymuses, compared to disomic chickens, and reduction in lymphocyte counts in these primary organs.[155] It is hypothesized that some of the Mhc dosage effects are a result of delay or extension of the lymphocyte differentiation period. Macrophages also exhibit changes associated with Mhc dosage. Level of Mhc class II antigen expression on macrophages is greater in aneuploid than disomic birds; however, this was not accompanied by a concomitant increase in antigen-presentation capacity by aneuploid macrophages.[156] Aneuploid macrophages did exhibit a cell-contact-dependent increase in killing tumor target cells, compared to diploid macrophages.[157] That the extra Mhc copies are not inactivated has important implications for the feasibility of producing chickens with increased copy numbers of the Mhc genes to enhance health or production characteristics.[124,158] Utilization of natural variation or genetic engineering techniques could be used to produce such birds.

VIII. THE TURKEY Mhc

Another economically important avian species in which significant progress has been made in characterizing the Mhc is the turkey. Graft-versus-host (GVH) reactivity, measured by the chorioallantoic membrane test, segregates in turkeys.[159] The turkey (GVH) splenomegaly response is controlled by two independently segregating loci.[160] An early study was unable to demonstrate that turkey erythrocytes have blood group markers of the Mhc.[161] A recent report, however, indicates that histocompatibility reactions such as mixed lymphocyte response, GVH reaction, and skin graft survival are all linked to the turkey Mhc.[162] The turkey Mhc genotypes were determined by hybridization of turkey DNA with chicken Mhc class II clones.[63,163] Thus, there is sufficient similarity between the Mhc genes of chicken and turkeys to utilize some gene clones across species for genotyping the Mhc.

The similarity in Mhc genes and antigens among avian species also has been used to extend definition of specific Mhc haplotypes to other species, for example, the ring-necked pheasant.[164] An erythrocyte alloantigen system in ring-necked pheasants was identified, by parallel cross-reactivity with specific Mhc subregional chicken anti-chicken reagents, as homologous with the Mhc.

IX. APPLICATIONS OF THE Mhc IN THE POULTRY INDUSTRY

Scientific knowledge about the avian immune system has progressed to a sufficient level to begin integration of immune management into poultry production systems.[165] Industrial use of Mhc information in chickens is in two main areas. One use is pedigree confirmation. Because of the separation of hens from progeny in commercial poultry breeding and the enormous number of eggs and chicks that may be simultaneously handled at a hatchery, even a very small percentage of pedigree errors could result in large numbers of incorrectly pedigreed chicks. Line-specific genetic markers can be used to assure purity of lines. If markers are identified in lines in which they should not exist, some contamination of that line can be assumed, originating from the line unique for the detected markers. Blood group antigens, including the B blood group within the Mhc, are excellent candidates to be used as genetic markers because they are quickly and economically identifiable from a small blood sample.

Another contemporary use of Mhc information in the poultry breeding industry is to selectively alter Mhc allelic frequencies for improvement of correlated traits, primarily disease resistance and vaccine response. Associations of certain Mhc alleles have been identified with Marek's disease vaccination efficacy in commercial genetic backgrounds.[166] Beneficial Mhc alleles can be increased in frequency or fixed within a parent line. The extreme polymorphism of the Mhc (over 50 allelic forms in Leghorns, more in meat-type birds) allows many options for choice of specific *B* alleles in parent lines and allelic combinations in commercial birds.

New industrial applications of Mhc manipulation should incorporate new biotechnologies.[167] Antigen-binding specificity is critical to effective poultry vaccine design, and knowledge of specific Mhc types in the flock could allow specific choices to optimize the vaccine-host Mhc combination or allow design of targeted recombinant vaccines.[98,166,168,169] With the use of specific gene probes, DNA can be directly examined for allelic forms of the Mhc genes. Direct analysis of class I and II genes will avoid the possibility of misclassifying allelic forms caused by recombinations between serologically determined class IV genes and the class I and II genes. Economic technologies for typing large sample numbers will be needed. Initial work has begun to address this need.[170] Although definition of variation in structural genes is advancing well, understanding of the genetic regulatory elements of the Mhc is minimal. The factors controlling gene expression must be understood to selectively enhance expression.[167] As allelic diversity of each individual class of Mhc genes is defined, its relationship with disease resistance must be determined. After identifying Mhc-disease associations, but before applying such knowledge in commercial practice, the effects of that Mhc variation on important economic traits must be defined. Additionally, the Mhc-linked, non-class I, non-class II genes need to be further characterized to determine the role they play in host defense.

Increasing gene copy number is an alternate possibility for enhancement of performance related to the Mhc. Bloom and colleagues have shown that the gene products of the Mhc are expressed in a dose-dependent fashion.[154,158] This suggests that adding additional Mhc gene copies may be a feasible approach to expanding the immune response repertoire. This increased Mhc copy number could be accomplished by genetic engineering or by utilizing naturally occurring variations. Because many traits for disease resistance associated with the Mhc are dominant traits,[127] a single individual with extra gene copies might then be resistant to a greater variety of diseases.

Improving immunoresponsiveness and disease resistance via genetic selection for the Mhc is a desirable approach.[164,171] The net progress per generation may be small, but it is heritable and therefore cumulative over time. Genetic selection can be made by marker-assisted selection rather than by challenge with disease agents. Improved immunoresponsiveness conserves natural resources by decreasing the production efficiency losses that are due to disease, and animal welfare is enhanced by possessing greater resistance to disease pathogens.

ACKNOWLEDGMENTS

We thank Rainer Mussmann for critical reading of the manuscript and Hans Spalinger for photography. The Basel Institute for Immunology was founded and is supported by Hoffman-La Roche, Basel, Switzerland. The poultry immunogenetics research program at Iowa State University is currently supported by state and Hatch funds, industrial funding, and grants from the USDA National Research Initiative Competitive Grants Program (95-37205-2191) and BARD (94-34339-1208). The excellent secretarial assistance of Ann Shuey is gratefully acknowledged. This is journal paper number J-16318 of the Iowa Agriculture and Home Economics Experiment Station. Project No. 2237, 2351, 2576, 3073, and 3290.

REFERENCES

1. Briles, W. E., McGibbon, W. H., and Irwin, M. R., On multiple alleles affecting cellular antigens in the chicken, *Genetics,* 35, 633, 1950.
2. Gilmour, D. G., Segregation of genes determining red cell antigens at high levels of inbreeding in chickens, *Genetics,* 44, 14, 1959.
3. Briles, W. E. and McGibbon, W. H., Heterozygosity of inbred lines of chickens at two loci effecting cellular antigens, *Genetics,* 33, 605, 1948.
4. Schierman, L. W. and Nordskog, A. W., Relationship of blood type to histocompatibility in chickens, *Science,* 134, 1008, 1961.
5. Jaffe, W. P. and McDermid, E. M., Blood groups and splenomegaly in chick embryos, *Science,* 137, 984, 1962.
6. Schierman, L. W. and Nordskog, A. W., Influence of the B blood group histocompatibility locus in chickens on a graft versus host reaction, *Nature,* 197, 511, 1963.
7. Miggiano, V. C., Birgen, J., and Pink, J. R. L., The mixed leukocyte reaction in chickens. Evidence for control by the major histocompatibility complex, *Eur. J. Immunol.,* 4, 397, 1974.
8. Pazderka, F., Longenecker, B. M., Law, G. R. J., and Ruth, R. F., The major histocompatibility complex of the chicken, *Immunogenetics,* 2, 101, 1975.

9. Briles, W. E., Bumstead, N., Ewert, D. L., Gilmour, D. G., Gogusev, J., Hala, K., Koch, C., Longenecker, B. M., Nordskog, A. W., Pink, J. R. L., Schierman, L. W., Simonsen, M., Toivanen, A., Toivanen, P., Vainio, O., and Wick, G., Nomenclature for chicken major histocompatibility (B) complex, *Immunogenetics*, 15, 441, 1982.
10. Bumstead, N. and Palyga, J., A preliminary linkage map of the chicken genome, *Genomics*, 13, 690, 1992.
11. Ziegler, A. and Pink, J. R. L., Characterization of major histocompatibility (B) antigens of the chicken, *Transplantation,* 20, 523, 1975.
12. Ziegler, A. and Pink, J. R. L., Chemical properties of two antigens controlled by the major histocompatibility complex of the chicken, *J. Biol. Chem.*, 251, 5391, 1976.
13. Hala, K., Vilhelmova, M., and Hartmanova, J., Probable crossing over in the B blood group system of the chicken, *Immunogenetics,* 3, 97, 1976.
14. Pink, J. R. L., Droege, W., Hala, K., Miggiano, V. C., and Ziegler, A., A three-locus model for the chicken major histocompatibility complex, *Immunogenetics,* 5, 203, 1977.
15. Guillemot, F. and Auffray, C., The molecular biology of the chicken major histocompatibility complex, *Crit. Rev. Poultry Biol.,* 2, 255, 1989.
16. Guillemot, F., Kaufman, J., Skjødt, K., and Auffray, C., The major histocompatibility complex of the chicken, *Trends Genet.*, 57, 300, 1989.
17. Kroemer, G., Bernot, A., Behar, G., Chause, A.-M., Gastinel, L.-N., Guillemot, F., Park, I., Thoraval, P., Zoorob, R., and Auffray, C., Molecular genetics of the chicken Mhc: current status and evolutionary aspects, *Immunol. Rev.*, 113, 118, 1990.
18. Kaufman, J., Skjødt, K., and Salomonsen, J., The Mhc molecules of nonmammalian vertebrates, *Immunol. Rev.,* 113, 83, 1990.
19. Kaufman, J., Skjødt, K., and Salomonsen, J., The B-G multigene family of the chicken major histocompatibility complex, *Crit. Rev. Immunol.*, 11, 113, 1991.
20. Miller, M. M., The major histocompatibility complex of the chicken, in *Phylogenesis of Immune Functions,* Warr, G. and Cohen, N., Eds., CRC Press, Boca Raton, FL, 1991, 151.
21. Bloom, S. and Bacon, L., Linkage of the major histocompatibility (B) complex and the nucleolar organizer region in the chicken, *J. Heredity,* 76, 146, 1985.
22. Delaney, M. E., Muscarella, D. E., and Bloom, S. E., Formation of nucleolar polymorphisms in trisomic chickens and subsequent microevolution of rRNA gene clusters in diploids, *J. Heredity,* 82(3), 213, 1991.
23. Riegert, P., Andersen, R., Bumstead, N., Döhring, C., Dominguez-Steglich, M., Engberg, J., Salomonsen, J., Schmidt, M., Schwager, J., Skjødt, K., and Kaufman, J., The chicken β_2-microglobulin gene is located on a nonMhc microchromosome: a small, G+C rich gene with X and Y boxes in the promoter, *Proc. Natl. Acad. Sci. U.S.A.,* 93, 1243, 1996.
24. Bourlet, Y., Behar, G., Guillemot, F., Frechin, N., Billault, A., Chausse, A.-M., Zoorob, R., and Auffray, C., Isolation of chicken major histocompatibility complex class II (B-L) β chain sequences, *EMBO J.*, 7, 1031, 1988.
25. Guillemot, F., Billault, A., Pourquie, O., Behar, G., Chausse, A.-M., Zoorob, R., Kreibich, G., and Auffray, C., A molecular map of the chicken major histocompatibility complex: the class II β genes are closely-linked to the class I genes and the nucleolar organizer, *EMBO J.,* 7, 2775, 1988.
26. Guillemot, F., Billault, A., and Auffray, C., Physical linkage of a guanine nucleotide-binding protein related gene to the chicken major histocompatibility complex, *Proc. Natl. Acad. Sci. U.S.A.,* 86, 4594, 1989.

27. Bernot, A., Zoorob, R., and Auffray, C., Linkage of a new member of the lectin supergene family to chicken Mhc genes, *Immunogenetics,* 39, 71, 1994.
28. Campbell, R.D. and Trowsdale, J., The human Mhc, *Immunol. Today,* 14, 349, 1993.
29. Trowsdale, J., Genomic structure and function in the Mhc, *Trends Genet.,* 9, 117, 1993.
30. Rubinstein, P., de Haas, L., Pevzner, I. Y., and Nordskog, A. W., Glyoxalase I (GLO) in the chicken: genetic variation and lack of linkage to the Mhc, *Immunogenetics,* 13, 493, 1981.
31. Koch, C., A genetic polymorphism of the complement component factor B in chickens is not linked to the major histocompatibility complex (Mhc), *Immunogenetics,* 23, 364, 1986.
32. Milner, C. M. and Campbell, R. D., The G9a gene in the human major histocompatibility complex encodes a novel protein containing ankyrin-like repeats, *Biochem. J.,* 290, 811, 1993.
33. Spike, C. A. and Lamont, S. J., Genetic analysis of 3 loci homologous to human G9a: evidence for linkage of a class III gene with the chicken Mhc, *Anim. Genet.,* 26, 185, 1995.
34. Briles, W. E., Goto, R. M., Auffray, C., and Miller, M. M., A polymorphic system related to but genetically independent of the chicken major histocompatibility complex, *Immunogenetics,* 37, 408, 1993.
35. Miller, M. M., Goto, R., Bernot, A., Zoorob, R., Auffray, C., Bumstead, N., and Briles, W. E., Two Mhc class I and two Mhc class II genes map to the chicken *Rfp-Y* system outside the B complex, *Proc. Natl. Acad. Sci. U.S.A.,* 91, 4397, 1994.
36. Miller, M. M., Goto, R., Zoorob, R., Auffray, C., and Briles, W. E., Regions of homology shared by *Rfp-Y* and major histocompatibility *B* complex genes, *Immunogenetics,* 39, 71, 1994.
37. Kaufman, J., Bumstead, N., Miller, M., Riegert, P., and Salomonsen, J., The chicken class II α gene is located outside of the B complex, in *Advances in Avian Immunology Research,* Davison, T. F., Burnstead, N., and Kaiser, P., Eds., Carfax, Abingdon, U.K., 1995, 119.
38. Kaufman, J., Andersen, R., Avila, D., Engberg, J., Lambris, J., Salomonsen, J., Welinder, K., and Skjødt, K., Different features of the Mhc class I heterodimer have evolved at different rates: chicken B-F and β_2-microglobulin sequences reveal invariant surface residues, *J. Immunol.,* 148, 1532, 1992.
39. Simonsen, M., Crone, M., Koch, C., and Hala, K., The Mhc haplotypes of the chicken, *Immunogenetics,* 16, 513, 1982.
40. Crone, M. and Simonsen, M., Avian major histocompatibility complex, in *Avian Immunology: Basis and Practice,* Vol. 2., Toivanen, A. and Toivanen, P., Eds., CRC Press, Boca Raton, FL, 1987.
41. Miller, M. M., Goto, R., and Briles, W. E., Biochemical confirmation of recombination within the B-G subregion of the chicken major histocompatibility complex, *Immunogenetics,* 27, 127, 1988.
42. Goto, R., Miyada, C. G., Young, S., Wallace, R. B., Abplanalp, H., Bloom, S. E., Briles, W. E., and Miller, M. M., Isolation of a cDNA clone from the B-G subregion of the chicken histocompatibility (*B*) complex, *Immunogenetics,* 27, 102, 1988.
43. Kaufman, J., Salomonsen, J., and Skjødt, K., B-G cDNA clones have multiple small repeats and hybridize to both chicken Mhc regions, *Immunogenetics,* 30, 440, 1989.
44. Kaufman, J., Salomonsen, J., Skjødt, K., and Thorpe, D., Size polymorphism of the chicken Mhc-encoded B-G molecules is due to length variation in the cytoplasmic heptad repeat region, *Proc. Natl. Acad. Sci. U.S.A.,* 87, 8277, 1990.

45. Salomonsen, J., Dunon, D., Skjødt, K., Thorpe, D., Vainio, O., and Kaufman, J., Chicken major histocompatibility complex-encoded B-G antigens are found on many cells that are important for the immune system, *Proc. Natl. Acad. Sci. U.S.A.*, 88, 1359, 1991.
46. Döhring, C., Riegert, P., Salomonsen, J., Skjødt, K., and Kaufman, J., The extracellular Ig V-like regions of the polymorphic B-G antigens of the chicken Mhc lack structural features expected for antibody variable regions, in *Avian Immunology in Progress*, Courdet, F., Ed., INRA, Paris, 1993, 145.
47. Miller, M. M., Abplanalp, H., and Goto, R., Genotyping chickens for the *B-G* subregion of the major histocompatibility complex using restriction fragment length polymorphisms, *Immunogenetics*, 28, 374, 1988.
48. Pitcovski, J., Lamont, S. J., Nordskog, A. W., and Warner, C. M., Analysis of *B-G* and immune response genes in the Iowa State University S1 chicken line by hybridization of sperm deoxyribonucleic acid with a major histocompatibility complex class II probe, *Poultry Sci.*, 68, 94, 1989.
49. Chen, Y. and Lamont, S. J., Major histocompatibility complex class I restriction fragment length polymorphism analysis in highly inbred chicken lines and lines selected for major histocompatibility complex and immunoglobulin production, *Poultry Sci.*, 71, 999, 1992.
50. Lamont, S. J., Warner, C. M., and Nordskog, A. W., Molecular analysis of the chicken major histocompatibility complex genes and gene products, *Poultry Sci.*, 66, 819, 1987.
51. Lamont, S. J., Gerndt, B. M., Warner, C. M., and Bacon, L. D., Analysis of restriction fragment length polymorphism of the major histocompatibility complex of $15I_5$-B-congenic chicken lines, *Poultry Sci.*, 69, 1195, 1990.
52. Chausse, A.-M., Coudert, F., Dambrine, G., Guillemot, F., Miller, M. M., and Auffray, C., Molecular genotyping of four chicken *B*-complex haplotypes with B-Lβ, B-F and B-G probes, *Immunogenetics*, 29, 127, 1989.
53. Warner, C., Gerndt, B., Xu, Y., Bourlet, Y., Auffray, C., Lamont, S., and Nordskog, A., Restriction fragment length polymorphism analysis of major histocompatibility complex class II genes from inbred chicken lines, *Anim. Genet.*, 20, 225, 1989.
54. Landesman, E., Uni, Z., and Heller, E. D., Designation by restriction fragment length polymorphism of major histocompatibility complex class IV haplotypes in meat-type chickens, *Anim. Genet.*, 24, 349, 1993.
55. Uni, Z., Hillel, J., Waiman, R., Cahaner, A., and Heller, E. D., Restriction fragment length polymorphism analysis of major histocompatibility complex class IV (B-G) genotypes in meat-type chickens, *Anim. Genet.*, 23, 379, 1992.
56. Kroemer, G., Zoorob, R., and Auffray, C., Structure and expression of a chicken Mhc class I gene, *Immunogenetics*, 31, 405, 1990.
57. Welinder, K., Jespersen, H., Walther-Rasmussen, J., and Skjødt, K., Amino acid sequences and structures of chicken and turkey $beta_2$-microglobulin, *Mol. Immunol.*, 28, 177, 1991.
58. Pickle, J., Chen, C.-L., and Cooper, M., An avian B lymphocyte protein associated with β_2-microglobulin, *Immunogenetics*, 32, 1, 1990.
59. Møller, L., Kaufman, J., Verland, S., Salomonsen, J., Avila, D., Lambris, J., and Skjødt, K., Variations in the cytoplasmic region evidently account for the chicken Mhc class I (B-F) heterogeneity, *Immunogenetics*, 34, 110, 1991.
60. Kaufman, J., Völk, H., and Wallny, H.-J., A "minimal essential Mhc" and an "unrecognized Mhc": two extremes in selection for polymorphism, *Immunol. Rev.*, 143, 63, 1995.

61. Pharr, G. T., Bacon, L. D., and Dodgson, J. B., Analysis of *B-Lβ*-chain gene expression in two chicken cDNA libraries, *Immunogenetics,* 37, 381, 1993.
62. Sung, A. M., Nordskog, A. W., Lamont, S. J., and Warner, C. M., Isolation and characterization of cDNA clones for chicken major histocompatibility complex class II molecules, *Anim. Genet.,* 24, 227, 1993.
63. Xu, Y., Pitcovski, J., Peterson, L., Auffray, C., Bourlet, Y., Gerndt, B. M., Nordskog, A. W., Lamont, S. J., and Warner, C. M., Isolation and characterization of three class II Mhc genomic clones from the chicken, *J. Immunol.,* 142, 2122, 1989.
64. Zoorob, R., Behar, G., Kroemer, G., and Auffray, C., Organization of a functional chicken class II B gene, *Immunogenetics,* 31, 179, 1990.
65. Zoorob, R., Bernot, G., Renoir, D. M., Choukri, F., and Auffray, C., Chicken major histocompatibility complex class II B genes: analysis of interallelic and inter-locus sequence variance, *Eur. J. Immunol.,* 23, 1139, 1993.
66. Guillemot, F., Turmel, P., Charron, D., Le Douarin, N. M., and Auffray, C., Structure, polymorphism and biosynthesis of chicken Mhc class II (*B-L*) antigens and associated molecules, *J. Immunol.,* 137, 1251, 1986.
67. Zoller, B., Ozato, K., Kroemer, G., Auffray, C., and Jungwirth, C., Interferon induction of chicken Mhc class I gene expression: phylogenetic conservation of the interferon-responsive element, *Virology,* 191, 141, 1992.
68. Chen, Y., Carpenter, S., and Lamont, S. J., Transfection of a putative chicken Mhc class II gene promoter region into chicken macrophages, in *Advances in Avian Immunological Research,* Davison, J. F., Ed., Carfax, Abingdon, U.K., 1995.
69. Qureshi, M. A., Miller, L., Lillehoj, H. S., and Ficken, M. D., Establishment and characterization of a chicken mononuclear cell line, *Vet. Immunol. Immunopathol.,* 26, 237, 1990.
70. Miller, M. M., Goto, R., and Abplanalp, H., Analysis of the B-G antigens of the chicken Mhc by two-dimensional gel electrophoresis, *Immunogenetics,* 20, 373, 1984.
71. Miller, M. M., Goto, R., Young, S., Liu, J., and Hardy, J., Antigens similar to major histocompatibility complex B-G are expressed in the intestinal epithelium in the chicken, *Immunogenetics,* 32, 45, 1990.
72. Kline, K., Briles, W. E., Bacon, L., and Sanders, B. G., Characterization of two distinct disulfide-linked B-G molecules in the chicken, *J. Hered.,* 79, 249, 1988.
73. Salomonsen, J., Skjødt, K., Crone, M., and Simonsen, M., The chicken erythrocyte-specific antigen. Characterization and purification of the B-G antigen by monoclonal antibodies, *Immunogenetics,* 25, 373, 1987.
74. Miller, M. M., Goto, R., Young, S., Chirivella, J., Hawke, D., and Miyada, C. G., Immunoglobulin variable-region-like domains of diverse sequence within the major histocompatibility complex of the chicken, *Proc. Natl. Acad. Sci. U.S.A.,* 88, 4377, 1991.
75. Pham-Dinh, D., Mattei, M., Nussbaum, J., Roussel, G., Pontarroti, P., Roekel, P., Mather, I., Artzt, K., Lindahl, K. F., and Dautigny, A., Myelin-oligodendrocyte glycoprotein is a member of a subset of the immunoglobulin superfamily encoded with the major histocompatibility complex, *Proc. Natl. Acad. Sci. U.S.A.,* 90, 7990, 1993.
76. Schierman, L. W. and McBride, R. A., Adjuvant activity of erythrocyte isoantigens, *Science,* 156, 658, 1967.
77. Hala, K., Plachy, J., and Schulmannova, J., Role of the B-G-region antigen in the humoral immune response to the B-F-region antigen of chicken Mhc, *Immunogenetics,* 14, 393, 1981.

78. Salomonsen, J., Eriksson, H., Skjødt, K., Lundgren, L., Simonsen, M., and Kaufman, J., The "adjuvant effect" of the polymorphic B-G antigens of the chicken Mhc analyzed using purified molecules incorporated in liposomes, *Eur. J. Immunol.*, 21, 649, 1991.
79. Longenecker, B. M., Mosmann, T. R., and Shiozawa, C., A strong, preferential response of mice to polymorphic antigenic determinants of the chicken Mhc, analyzed with mouse hybridoma (monoclonal) antibodies, *Immunogenetics,* 9, 137, 1979.
80. Longenecker, B. M. and Mosmann, T. R., "Natural" antibodies to chicken Mhc antigens are present in mice, rats, humans, alligators and allogeneic chickens, *Immunogenetics,* 11, 293, 1980.
81. Kaufman, J., Salomonsen, J., Riegert, P., and Skjødt, K., Using chicken class I sequences to understand how xenoantibodies crossreact with Mhc-like molecules in nonmammalian vertebrates, *Am. Zool.,* 31, 570, 1991.
82. Shawar, S., Vyas, J., Rodgers, J., and Rich, R., Antigen presentation by major histocompatibility complex class I-B molecules, *Annu. Rev. Immunol.,* 12, 839, 1994.
83. Klein, J., Origin of major histocompatibility complex polymorphism: the trans-species hypothesis, *Hum. Immunol.,* 19, 155, 1987.
84. Schat, K., Immunity in Marek's disease and other tumors, in *Avian Immunology: Basis and Practice,* Vol. 2, Toivanen, A. and Toivanen, P., Eds., CRC Press, Boca Raton, FL, 1987, 101.
85. Plachy, J., Pink, J. R. L., and Hala, K., Biology of the chicken Mhc (*B* complex), *Crit. Rev. Immunol.,* 12, 47, 1992.
86. Kaufman, J. and Salomonsen, J., What in the dickens is with those chickens? *Res. Immunol.,* 144, 495, 1993.
87. Du Pasquier, L., Schwager, J., and Flajnik, M., The immune system of *Xenopus, Annu. Rev. Immunol.,* 7, 251, 1989.
88. Nakamura, T., Sekizawa, A., Fujii, T., and Katakiri, C., Cosegregation of the polymorphic C4 with Mhc in the frog *Xenopus laevis, Immunogenetics,* 23, 1181, 1986.
89. Salter-Cid, L., Kasahara, M., and Flajnik, M. F., *Hsp70* genes are linked to the *Xenopus* major histocompatibility complex, *Immunogenetics,* 39, 1, 1994.
90. Dietert, R., Introduction: the major histocompatibility complex as a communication gene complex, *Poultry Sci.,* 66, 774, 1987.
91. Vainio, O. and Toivanen, A., Cellular cooperation in immunity, in *Avian Immunology: Basis and Practice,* Vol. 2, Toivanen, A. and Toivanen, P., Eds., CRC Press, Boca Raton, FL, 1987.
92. Vainio, O., Toivanen, P., and Toivanen, A., Major histocompatibility complex and cell cooperation, *Poultry Sci.*, 66, 795, 1987.
93. Vainio, O., Peck, R., Koch, C., and Toivanen, A., Origin of peripheral blood macrophages in bursa cell-reconstituted chickens: further evidence for Mhc-restricted interactions between T and B lymphocytes, *Scand. J. Immunol.*, 17, 193, 1983.
94. Toivanen, P., Toivanen, A., and Vainio, O., Complete restoration of bursa-dependent immune system after transplantation of semi-allogeneic stem cells into immunodeficient chicks, *J. Exp. Med.,* 139, 1344, 1974.
95. Vainio, O., Koch, C., and Toivanen, A., B-L antigens (class II) of the chicken major histocompatibility complex control T-B cell interaction, *Immunogenetics,* 19, 131, 1984.
96. Toivanen, A. and Toivanen, P., Histocompatibility requirements for cellular cooperation in the chicken: generation of germinal centers, *J. Immunol.,* 118, 431, 1977.

97. Vainio, O., Veromaa, T., Eerola, E., Toivanen, P., and Ratcliffe, M., Antigen presenting cell-T cell interaction in the chicken is Mhc class II antigen restricted, *J. Immunol.*, 140, 2864, 1988.
98. Schat, K. A., Cell-mediated immune effector functions in chickens, *Poultry Sci.*, 73, 1077, 1994.
99. MacCubbin, D. L. and Schierman, L. W., Mhc-restricted cytotoxic response of chicken T cells: expression, augmentation, and clonal characterization, *J. Immunol.*, 136, 12, 1986.
100. Weinstock, D. and Schat, K. A., Virus specific syngeneic killing of reticuloendotheliosis virus transformed cell line target cells by spleen cells, in *Progress in Clinical and Biological Research, Avian Immunology*, Vol. 238, Weber, W. T. and Ewert, D. L., Eds., Alan R. Liss, New York, 1987.
101. Schierman, L. W. and Collins, W. M., Influence of the major histocompatibility complex on tumor regression and immunity in chickens, *Poultry Sci.*, 66, 812, 1987.
102. Gunther, E., Balcarova, J., Hala, K., Rude, E., and Hraba, T., Evidence for an association between immune responseness of chicken to (T, G)-A — L and the major histocompatibility system, *Eur. J. Immunol.*, 4, 548, 1974.
103. Benedict, A. A., Pollard, L. W., Morrow, P. R., Abplanalp, H. A., Mauer, P. H., and Briles, W. E., Genetic control of immune responses in chickens. I. Responses to a terpolymer of poly ($glu^{60}ala^{30}tyr^{10}$) associated with the major histocompatibility complex, *Immunogenetics*, 2, 313, 1975.
104. Pevzner, I. Y., Trowbridge, C. L., and Nordskog, A. W., B-complex genetic control of immune response to HSA, (T, G)-A — L, GT and other substances in chickens, *J. Immunogenetics* 6, 453, 1979.
105. Pevzner, I. Y., Nordskog, A. W., and Kaeberle, M. L., Immune response and the B blood group locus in chickens, *Genetics*, 80, 753, 1975.
106. Loudovaris, T., Brandon, M. R., and Fahey, K. J., The major histocompatibility complex and genetic control of antibody response to sheep red blood cells in chickens, *Avian Pathol.*, 19, 89, 1990.
107. Rees, M. J. and Nordskog, A. W., Genetic control of serum immunoglobulin G levels in the chicken, *J. Immunogenet.*, 8, 425, 1981.
108. Klesius, P., Johnson, W., and Kramer, T., Delayed wattle reaction as a measure of cell-mediated immunity in the chicken, *Poultry Sci.*, 56, 249, 1977.
109. Taylor, R. L., Jr., Cotter, P. F., Wing, T. L., and Briles, W. E., Major histocompatibility (B) complex and sex effects on the phytohemagglutinin wattle response, *Anim. Genet.*, 18, 343, 1987.
110. Cotter, P. F., Taylor, R. L., Wing, T. L., and Briles, W. E., Major histocompatibility (B) complex-associated differences in the delayed wattle reaction to Staphylococcal antigen, *Poultry Sci.*, 66, 203, 1987.
111. Chanh, T. C., Benedict, A. A., and Abplanalp, H., Association of serum hemolytic complement levels with the major histocompatibility complex in chickens, *J. Exp. Med.*, 144, 555, 1976.
112. Shen, P. F., Smith, E. J., and Bacon, L. D., The ontogeny of blood cells, complement, and immunoglobulins in 3- to 12-week-old $15I_5$-B congenic White Leghorn chickens, *Poultry Sci.*, 63, 1083, 1984.
113. Bacon, L. D., Ismail, N., and Motta, J. V., Allograft and antibody responses of $15I_5$-*B*-congenic chickens, in *Avian Immunology*, Weber, W. T. and Ewert, D. L., Eds., Alan R. Liss, New York, 1987.

114. Qureshi, M. A., Dietert, R. R., and Bacon, L. D., Genetic variation in the recruitment and activation of chicken peritoneal macrophages, *Proc. Soc. Exp. Biol. Med.,* 181, 560, 1986.
115. Qureshi, M. A., Dietert, R. R., and Bacon, L. D., Chemotactic activity of chicken blood mononuclear leukocytes from $15I_5$-B-congenic lines to bacterially-derived chemoattractants, *Vet. Immunol. Immunopathol.,* 19, 351, 1988.
116. Hala, K., Vainio, O., Plachy, J., and Bock, G., Chicken major histocompatibility complex congenic lines differ in the percentages of lymphocytes bearing CD4 and CD8 antigens, *Anim. Genet.,* 22, 279, 1991
117. Dunnington, E. A., Briles, R. E., Briles, W. E., Gross, W. B., and Siegel, P. B., Allelic frequencies in eight alloantigen systems of chickens selected for high and low antibody response to sheep red blood cells, *Poultry Sci.,* 63, 1470, 1984.
118. Pinard, M. H., van Arendonk, J. A. M., Nieuwland, M. G. B., and van der Zijpp, A. J., Divergent selection for humoral immune responsiveness in chickens: distribution and effects of major histocompatibility complex types, *Genet. Sel. Evol.,* 25, 1991, 1993.
119. Pinard, M. H. and van der Zijpp, A. J., Effects of major histocompatibility complex on antibody response in F1 and F2 crosses of chicken lines, *Genet. Sel. Evol.,* 24, 283, 1993.
120. Dunnington, E. A., Martin, A., Briles, R. W., Briles, W. E., Gross, W. B., and Siegel, P. B., Antibody responses to sheep erythrocytes for White Leghorn chickens differing in haplotypes of the major histocompatibility complex (B), *Anim. Genet.,* 20, 213, 1989.
121. Uni, Z., Gutman, M., Leitner, G., Landesman, E., Heller, D., and Cahaner, A., Major histocompatibility complex class IV restriction fragment length polymorphism markers in replicated meat-type chicken lines divergently selected for high or low early immune response, *Poultry Sci.,* 72, 1823, 1993.
122. Kean, R. P., Briles, W. E., Cahaner, A., Freeman, A. E., and Lamont, S. J., Differences in major histocompatibility complex frequencies after multitrait, divergent selection for immunocompetence, *Poultry Sci.,* 73, 7, 1994.
123. Bacon, L. D., Influence of the Mhc on disease resistance and productivity, *Poultry Sci.,* 66, 802, 1987.
124. Lamont, S. J., The chicken major histocompatibility complex in disease resistance and poultry breeding, *J. Dairy Sci.,* 72, 1328, 1989.
125. Lamont, S. J., Immunogenetics and the major histocompatibility complex, *Veter. Immunol. Immunopathol.,* 30, 121, 1991.
126. Lamont, S. J., The major histocompatibility complex, in *Manipulation of the Avian Genome*, Etches, R. J. and Verrinder, A. M., Eds., CRC Press, Boca Raton, FL, 1993, 185.
127. Gavora, J. S., Disease genetics, in *Poultry Breeding and Genetics*, Crawford, R. D., Ed., Elsevier, Amsterdam, 1990, chap. 33.
128. Stevens, L., *Genetics and Evolution of the Domestic Fowl*, Cambridge University Press, Cambridge, 1991, chap. 11.
129. Briles, W. E., Briles, R. W., Taffs, R. E., and Stone, H. A., Resistance to a malignant lymphoma in chickens is mapped to subregion of major histocompatibility (*B*) complex, *Science,* 219, 977, 1983.
130. Hepkema, B. G., Blankert, J. J., Albers, G. A. A., Tilanus, M. G. J., Egberts, E., van der Zijpp, A. J., and Hensen, E. J., Mapping of susceptibility to Marek's disease within the major histocompatibility (B) complex by refined typing of White Leghorn chickens, *Anim. Genet.,* 24, 283, 1993.

131. Steadham, E. M., Lamont, S. J., Kujdych, I., and Norskog, A. W., Association of Marek's disease with Ea-B and immune response genes in subline and F2 populations of the Iowa State S1 Leghorn line, *Poultry Sci.,* 66, 571, 1987.
132. Yoo, B. H. and Sheldon, B. L., Association of the major histocompatibility complex with avian leukosis virus infection in chickens, *Br. Poultry Sci.,* 33, 613, 1992.
133. Lamont, S. J., Bolin, C., and Cheville, N., Genetic resistance to fowl cholera is linked to the major histocompatibility complex, *Immunogenetics,* 25, 284, 1987a.
134. Rose, N. R., Avian models of autoimmune disease: lessons from the birds, *Poultry Sci.,* 73, 984, 1994.
135. Lillehoj, H. S., Ruff, M. D., Bacon, L. D., Lamont, S. J., and Jeffers, T. K., Genetic control of immunity to *Eimeria tenella*: interaction of Mhc genes and genes within the genetic background influence levels of disease susceptibility, *Vet. Immunol. Immunopathol.* 20, 135, 1989.
136. Schat, K. A., Taylor, R. L., Jr., and Briles, W. E., Resistance to Marek's disease in chickens with recombinant haplotypes of the major histocompatibility (B) complex, *Poultry Sci.,* 73, 502, 1994.
137. Bacon, L. D. and Witter, R. L., Influence of turkey herpesvirus vaccination on the B-haplotype effect on Marek's disease resistance in 15.B-congenic chickens, *Avian Dis.,* 36, 378, 1992.
138. White, E. C., Briles, W. E., Briles, R. W., and Taylor, R. L., Jr., Response of six major histocompatibility (B) complex recombinant haplotypes to rous sarcomas, *Poultry Sci.,* 73, 836, 1994.
139. Taylor, R. L., Jr., Ewert, D. L., England, J. M., and Halpern, M. S., Major histocompatibility (B) complex control of the growth pattern of v-src DNA-induced primary tumors, *Virology,* 191, 477, 1992.
140. Cotter, P. F., Taylor, R. L., Jr., and Abplanalp, H., Differential resistance to *Staphylococcus aureus* challenge in major histocompatibility (B) complex congenic lines, *Poultry Sci.,* 7(1), 1873, 1992.
141. Dietert, R. R., Taylor, R. L., Jr., and Dietert, M. F., Biological function of the chicken major histocompatibility complex, *Crit. Rev. Poultry Biol.,* 3, 111, 1991.
142. Sander, J. F., The major histocompatibility complex and its role in poultry production, *World's Poultry Science J.,* 49, 132, 1993.
143. Simonsen, M., Kolstad, N., Edfors-Lilja, I., Liledahl, L. E., and Sorensen, P., Major histocompatibility genes in egg-laying hens, *Am. J. Reprod. Immunol.,* 2, 148, 1982.
144. Gavora, J. S., Simonsen, M., Spencer, J. L., Fairful, R. W., and Gowe, R. S., Changes in the frequency of major histocompatibility haplotypes in chickens under selection for both high egg production and resistance to Marek's disease, *Z. Tierzucht Zuchtungsbiol.,* 103, 218, 1986.
145. Lamont, S. J., Hou, Y.-H., Young, B. M., and Nordskog, A. W., Research note: differences in major histocompatibility complex gene frequencies associated with feed efficiency and laying performance, *Poultry Sci.,* 66, 1064, 1987.
146. Dunnington, E. A., Larsen, A. S., O'Sullivan, N. P., and Siegel, P. B., Growth and egg production traits in chickens as influenced by major histocompatibility types and background genomes, *J. Anim. Breed. Genet.,* 109, 188, 1992.
147. Kim, C. D., Lamont, S. J., and Rothschild, M. F., Genetic associations of body weight and immune response with the major histocompatibility complex in White Leghorn chicks, *Poultry Sci.,* 66, 1258, 1987.
148. Kim, C. D., Lamont, S. J., and Rothschild, M. F., Associations of major histocompatibility complex haplotypes with body weight and egg production traits in S1 White Leghorn chickens, *Poultry Sci.,* 68, 464, 1989.

149. Abplanalp, H., Sato, K., Napolitano, D., and Reid, J., Reproductive performance of inbred congenic Leghorns carrying different haplotypes for the major histocompatibility complex (B-blood groups), *Poultry Sci.,* 71, 9, 1992.
150. Sato, K., Ablanalp, H., Napolitano, D., and Reid, J., Effects of heterozygosity of major histocompatibility complex haplotypes on reproduction and viability of Leghorn hens sharing a common inbred population, *Poultry Sci.,* 71, 18, 1992.
151. Klein, J., *Natural History of the Major Histocompatibility Complex,* John Wiley & Sons, New York, 1986, chap. 9.
152. Dominguez-Steglich, M., Auffray, C., and Schmid, M., Linkage of the chicken Mhc to the nucleolus organizer region visualized using nonisotopic *in situ* hybridization, *J. Heredity,* 82(6), 503, 1991.
153. Bloom, S. E., Briles, W. E., Briles, R. W., Delaney, M. E., and Dietert, R. R., Chromosomal localization of the major histocompatibility (B) complex (Mhc) and its expression in chickens aneuploid for the major histocompatibility complex/ribosomal deoxyribonucleic acid microchromosome, *Poultry Sci.,* 66, 782, 1987.
154. Delaney, M. E., Dietert, R. R., and Bloom, S. E., Mhc-chromosome dosage effects: evidence for increased expression of Ia glycoprotein and alteration of B cell subpopulations in neonatal aneuploid chickens, *Immunogenetics,* 27, 24, 1988.
155. Hemendinger, R. A., Putnam, J. R., and Bloom, S. E., Mhc dosage effects on primary immune organ development in the chicken, *Dev. Comp. Immunol.,* 16, 175, 1992.
156. Lin, H.-K., Bloom, S. E., and Dietert, R. R., Partitioning of cell surface class II expression and antigen presenting capacity of macrophages in an Mhc-gene dosage model, *Anim. Biotech.,* 2(2), 123, 1991.
157. Lin, H.-K., Bloom, S. E., and Dieter, R. R., Macrophage anti-tumor functions in a chicken Mhc chromosome dosage model, *Anim. Biotech.,* 4(2), 121, 1993.
158. Bloom, S. E., Delaney, M. E., Muscarella, D. M., and Dietert, R. R., Gene expression in chickens aneuploid for the Mhc-bearing chromosome, in *The Molecular Biology of the Major Histocompatibility Complex of Domestic Animal Species,* Warner, C. M, Rothschild, M. F., and Lamont, S. J., Eds., Iowa State University Press, Ames, 1988.
159. Shoffner, R. N., A histocompatibility locus in the turkey, *Poultry Sci.,* 43, 1364, 1964.
160. Cherms, F. L., The histocompatibility system in turkeys, *Proc. Natl. Breeders Roundtable,* 26, 24, 1977.
161. Palmer, D. K. and Nordskog, A. W., The major histocompatibility complex in the turkey: do erythrocytes bear serologically detectable H antigens?, *Poultry Sci.,* 59, 2158, 1980.
162. Emara, M. G., Nestor, K. E., and Bacon, L. D., The turkey major histocompatibility complex: characterization by mixed lymphocyte, graft-vs-host splenomegaly, and skin graft reactions, *Poultry Sci.* 72, 60, 1993.
163. Emara, M. G., Nestor, K. E., Foster, D. N., and Lamont, S. J., The turkey histocompatibility complex: identification of class II genotypes by restriction fragment length polymorphism analysis of deoxyribonucleic acid, *Poultry Sci.,* 71, 2083, 1992.
164. Jarvi, S. I. and Briles, W. E., Identification of the major histocompatibility complex in the ring-necked pheasant, *Phasicenus colchicus, Anim. Genet.,* 23, 211, 1992.
165. Dietert, R. R. and Lamont, S. J., Avian immunology: from fundamental immune mechanisms to the integrative management of poultry, *Poultry Sci.* 73, 975, 1994.
166. Bacon, L. D. and Witter, R. L., B haplotype influence on the relative efficacy of Marek's disease vaccines in commercial chickens, *Poultry Sci.,* 73, 481, 1994.
167. Lamont, S. J., Poultry immunogenetics, which way do we go, *Poultry Sci.,* 73, 1044, 1994.

168. Pharr, G. T., Hunt, H. D., Bacon, L. D., and Dodgson, J. B., Identification of class II major histocompatibility complex polymorphisms predicted to be important in peptide antigen presentation, *Poulty Sci.,* 72, 1312, 1993.
169. Witter, R. L. and Hunt, H. D., Poultry vaccines of the future, *Poultry Sci.,* 73, 1087, 1994.
170. Heath, E. M., Hopf, N. D., and Shuman, R. M., Developing DNA-based Mhc typing reagents for commercial white Leghorn and broiler strains, *Poultry Sci.,* 73(suppl 1), 5, 1994.
171. Lamont, S. J. and Dietert, R. R., New directions in poultry immunogenetics, in *Poultry Breeding and Genetics*, Crawford, R. D., Ed., Elsevier, Amsterdam, 1990, chap. 22.
172. Koller, B. H. and Orr, H. T., Cloning and complete sequence of an HLA-A2 gene: analysis of two HLA-A alleles at the nucleotide level, *J. Immunol.*, 134, 2727, 1985.
173. Boss, J. M. and Strominger, J. L., Cloning and sequence analysis of the human major histocompatibility complex gene *DC-3β, Proc. Natl. Acad. Sci. U.S.A.,* 81, 5199, 1984.
174. Güssow, D., Rein, R., Ginjaar, I., Hochstenbach, F., Seeman, G., Kottman, A., and Ploegh, H., Sequence of the human β_2-microglobulin gene and definition of the transcriptional unit, *J. Immunol.,* 139, 332, 1987.
175. Miller, M. M., Goto, R. M., Taylor, R. L., Zoorob, R., Auffray, C., Briles, R. W., Briles, W. E., and Bloom, S. E., Assignment of Rfp-Y to the chicken Mhc/NOR microchromosome and evidence for high frequency recombination associated with the NOR, *Proc. Natl. Acad. Sci. U.S.A.,* in press, 1996.

CHAPTER 4

Genetic Organization, Polymorphism, and Function of the Bovine Major Histocompatibility Complex

Harris A. Lewin

CONTENTS

0-8493-8032-4/96/$0.00+$.50

I. INTRODUCTION

The history of research on the *Bos taurus* and *Bos indicus* ("bovine") major histocompatibility complex (Mhc) cannot be understood without a general appreciation for the origins of immunogenetics and the discoveries that led to the elucidation of the mouse *H-2* complex and the human leukocyte antigen (*HLA*) system.[1] Like *HLA*, research on the bovine Mhc has deep roots in classical serology and transplantation genetics. The discovery of the Mhc of cattle is attributed to Amorena and Stone,[2] and Spooner and co-workers.[3] The genetic region they and others defined using serological reagents produced by skin transplants and alloimunization was named the bovine lymphocyte antigen (*BoLA*) system.[4] Workers in the field unanimously agreed to retain the name *BoLA* rather than *Bota*[5] because of the long and rich history of research conducted on the Mhc of cattle, and to avoid unnecessary confusion. Among all species studied, the *BoLA* system ranks next in terms of depth of knowledge compared with *HLA* and *H-2*.

The serological approach in cattle led to the biochemical characterization of determinant-bearing molecules having the same properties as products of the human and mouse Mhc.[6,7] Histogenetic methods, such as the mixed lymphocyte reaction (MLR)[8] and allospecific cytotoxic T lymphocytes (CTL),[9] were used to identify determinants that were eventually shown to be the property of Mhc class I and class II antigens.[10] The modern history of *BoLA* research began with the application of Southern blotting and restriction fragment length (RFLP) analysis. Using *HLA* genes as probes, nucleic acid hybridizations were used successfully to identify orthologous genes, reveal the extensive polymorphism of the class II genes, and delineate *BoLA* haplotypes.[11,12] The discovery that *HLA* probes would cross-hybridize to *BoLA* genes led to the molecular cloning and sequencing of genes encoding *BoLA* class I and class II molecules.[13,14] Determination of genomic sequences permitted the application of PCR technology to the identification of sequence variation in coding exons of *BoLA* genes.[15,16] All these methods have been applied to research topics that range from comparative evolution[17] to disease association.[18] The scope of this review is to retrace the history summarized above, emphasizing gene mapping, sequence-based typing methods, haplotype diversity, function, and disease association.

II. GENETIC ORGANIZATION OF THE BOVINE Mhc (*BoLA*)

A. The Genetic Map

The bovine Mhc and at least 30 structural genes have been mapped to bovine autosome 23 (BTA23), making this chromosome the most well characterized of the species.[19] In addition to the structural genes, 11 microsatellite markers have been mapped to BTA23.[20,21] The majority of structural genes were assembled into linkage maps using family studies and sperm typing (Figure 1).

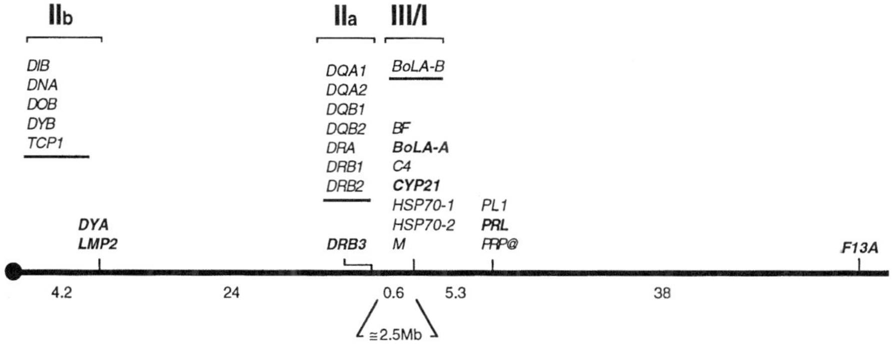

Figure 1 The genetic map of structural genes located on bovine chromosome 23. Map distances are expressed in centimorgans and were estimated using the Illinois Reference/Resource families, a large collection of paternal half-sib families used for the construction of a bovine linkage map.[157] Loci shown in bold were typed in these families. Map positions of other loci were assigned on the basis of studies cited in the text.

Two class II subregions have been defined on the basis of genetic mapping.[22,23] These two subregions are separated by a relatively large distance, spanning at least 15 centimorgans (cM) from *DYA* in the class IIb region to *DRB3* in the class IIa region (Figure 1). Thus, while it may be convenient to refer to *BoLA* genes as a "system", they are organized in clusters; some loci are tightly linked and others are relatively distant companions. The large genetic distance between the two class II regions distinguishes the bovine Mhc from those of the mouse and human. Recently, the *LMP2* proteosome subunit gene was mapped to the class IIb region,[24] providing evidence of a major chromosomal rearrangement, insertion, or recombination hot spot compared with *HLA* and *H-2*.

The orientation of *BoLA* genes relative to the centromere was determined by multilocus gene-centromere mapping using PCR analysis of secondary oocytes and first polar bodies; the class IIb *DYA* locus was shown to be located proximal to the centromere.[25] A recent study by Park et al.[26] identified large variation in recombination rate in the class IIa-IIb interval using sperm typing, with one bull having nearly double the recombination rate between *DYA* and the gene for prolactin (*PRL*), which is 2.5 to 4 cm telomeric to *DRB3*. The studies by Jarrell et al.[25] and Park et al.[26] suggest the presence of a polymorphic recombination hot spot located in the class IIa-IIb interval that might involve four-strand double crossovers.

The *BoLA* class I genes have not been shown to be separable by recombination.[27] The *BoLA-A* locus, the major expressed class I gene, is located telomeric to the class IIa cluster.[28] In addition to the class I and class II genes, the class III genes *CYP21*,[20,21,28,29] *BF*,[30] *HSP70*,[20] and *C4*,[22] also have been mapped to the bovine Mhc, and lie telomeric to the class IIa region. However, it has not been resolved as to whether the class III genes are centromeric or telomeric to the class I genes. Thus, while the order of genes in the major *BoLA* cluster is still somewhat speculative, the available data suggest that the general organization is probably identical to that of man, centromere-class II-class III-class I.

B. The Physical Map

Long range physical mapping of the bovine Mhc was accomplished by Bensaid and co-workers.[31] These studies employed pulsed field gel electrophoretic techniques, using DNA from *BoLA-A* typed homozygous and heterozygous Boran (*B. indicus*) and N'dama (*B. taurus*) cattle and γ-irradiated or chemically mutagenized lymphoblastoid cell lines. The large number of class I genes and the lack of locus-specific probes has made it difficult to determine conclusively the organization of the class I genes. However, using cDNA probes specific for two different expressed *BoLA* class I genes, it has been determined by field inversion gel electrophoresis (FIGE) that these genes are within 200 kb of each other.[32] The overall size of the class I region is no less than 770 kb and probably no more than 1650 kb as determined by FIGE, similar to the organization of the *HLA* class I genes.[31]

The availability of homologous, locus-specific probes for the class II genes enabled a more detailed physical map to be constructed using FIGE.[31] The *DQ* and *DR* genes are probably not more than 400 kb apart in most haplotypes. *DRA* and *DRB* genes are separated by no more than 340 kb, and at least one of the *DRB* genes is on a 200-kb fragment that also contains the *DQA* and *DQB* genes. The *DQA* and *DQB* genes are no more than 170 kb apart, consistent with the finding of a phage lambda clone with 11 kb between *DQA* and *DQB* genes.[33] In most haplotypes, the *DQA* and *DQB* genes have been duplicated,[34] and the assumption is that the duplicated cluster is tightly linked, although this has not been proven formally.

Bensaid and co-workers[31] also have tried to resolve the position of the *C4* gene relative to the class I and class II loci using FIGE. The *C4* gene was found on a 1000 kb *Not*I fragment that included the *DR* and *DQ* genes in most haplotypes. However, the class I genes could not be physically linked to the *DR*/*DQ*/*C4* cluster. Combining the physical and genetic maps for this region indicates a ratio of about

1% recombination per megabase of DNA (Figure 1). These are the first available data for correlation of the genetic and physical maps in this species and are consistent with the expected ratio based upon estimated genome size and chiasma counts.[21]

III. CLASS I GENES AND GENE PRODUCTS

A. Expression

On the basis of the number of fragments obtained on Southern blots using a human class I cDNA probe, the number of class I genes in the cattle genome is between 10 and 20.[27] Serological studies employing alloimmune reagents supported the existence of at least one major expressed class I locus, designated *BoLA-A* (see below).

The first convincing evidence for the existence of a second expressed polymorphic class I locus was reported by Ennis and co-workers,[13] who isolated two distinct class I cDNA clones from the *BoLA*-homozygous lymphoblastoid cell line BL3.[35] These clones were distinguished by a small 6-bp deletion in the 3′ untranslated region. On the basis of similar locus-specific deletions in other species, the authors concluded that the two cDNAs represented nonallelic class I genes. Recent cDNA cloning and sequencing[32,36,37] support these initial findings.

One-dimensional isoelectric focusing (1D-IEF), using monoclonal antibodies (mAb) to phylogenetically conserved monomorphic class I determinants, revealed the presence of more than one class I gene product expressed by lymphocytes from homozygous individuals.[38,39] Subsequently, mouse L cells transfected with genomic DNA from a homozygous cow produced by sire-daughter mating were shown to express two distinct class I products, as detected with allospecific CTL and mAb to determinants on the *KN104* (African Boran class I) and *A10* molecules.[40] Bensaid and co-workers[32] followed up this study by cloning unique class I cDNA from the homozygous individual and then correlating clone-specific sequences with the expression of either *A10* or *KN104* in transfected L cells. Although these experiments appeared to provide formal proof for the expression of more than one class I locus, Joosten and co-workers[41] were not able to distinguish with 1D-IEF the *KN104* and *A10* alleles when expressed from the same haplotype.

Recent studies have employed cDNA cloning and sequencing to more directly address the question of the number and polymorphism of expressed *BoLA* class I genes.[36,37,42] Garber and co-workers[37] used gene tree analysis to group nine available full-length class I cDNA sequences into four clusters. The two main clusters, A and B, differ by a 6-bp deletion in exon 5, thus encoding a short transmembrane region. A single sequence defined both the clusters C (KN104) and D (BSX). It is not clear from the sequence analysis whether KN104 is a distant member of the A, B, or D allelic series or is unique to the Boran breed from which it was isolated. The ability of Garber and co-workers[37] to clone up to six distinct cDNAs from a single heterozygous bull seems to indicate that there are three class I genes transcribed in bovine lymphocytes; however, the functionality of these genes has yet to be clearly established. At least one clone appears to encode a molecule that is not expressed at the cell surface.[37] Peptide mapping of charge variants of class I molecules separated by IEF also supports the expression of three class I loci.[43] However, until the cDNA sequences are correlated with the available serological and biochemical data, the number of functional class I genes in cattle will remain an open question. Given the available information, one cannot exclude the possibility that the number of expressed class I molecules differs between haplotypes.

B. Polymorphism

1. Serology

Polymorphism of class I antigens has been studied using the classical immunogenetic approaches: transplantation, serology, allospecific T-cells, and one-dimensional isoelectric focusing (1D-IEF). The serological approach for elucidating *BoLA* polymorphism was highly successful, resulting in the initial characterization of the system.[2,3] Sera produced by skin transplants, injection of leukocytes, and stimulation by pregnancy revealed the presence of a highly polymorphic Mhc for the species.[44,45] International workshops have played a vital role in standardization of serological reagents, typing methods, and nomenclature of the *BoLA* system. The five international *BoLA* workshops during the period 1978 to 1992[4,46-49] elevated the status of the *BoLA* system to unparalleled leadership among the domestic animal species (Table 1). More than 50 *BoLA* antigens have been recognized by the international community, nearly all of which behave as alleles of a single locus. The Fifth International BoLA Workshop (BoLA5) held at Interlaken Switzerland in 1992,[49] produced acceptance of 27 full-status *BoLA-A* specificities (e.g., *BoLA-A2*), 25 provisional workshop (w) *BoLA* specificities, and 1 specificity that may be encoded by a closely linked locus. All full-status *BoLA-A* antigens are defined by sera from at least two laboratories and have been recognized on two or more cells submitted to one of the international cell exchanges. Eleven specificities are "supertypic," i.e., they are shared by more than one allelic product. The *BoLA* specificities exhibit distinct differences in frequency in the different cattle breeds.[47,50]

A lingering question is whether the sera commonly employed for *BoLA* typing are specific for the allelic products of a single locus or contain antibodies to the products of other class I loci as well. It has been suggested that typing sera are haplotype specific[43,51] due to a high frequency of gametic phase disequilibrium between class I alleles at the population level. Within a cluster of sera that defines a *BoLA-A* specificity an individual serum may have extra reactions that correspond to a second polymorphic class I locus.[43] However, because this has never been shown for an entire cluster of *BoLA* typing sera, it is generally assumed that the primary specificities defined by these sera are alleles of a single locus. The large diversity of breeds represented in the international workshops, which one would expect to contain the class I alleles in different haplotypic combinations, have not been able to dissect the vast majority of sera used in the comparison tests. Therefore, although some sera that define a *BoLA-A* specificity may themselves be polyspecific, the *BoLA-A* specificities behave as alleles of a single locus. If antibodies to more than one locus product were present in the majority of the sera used to define a cluster,

Table 1 The International BoLA Workshops

Workshop	Location	No. labs	Format	Methods	No. new Ag/alleles	No. alleles[a]	Ref.
First, 1978	Edinburgh, United Kingdom	9	Sera exchange	Serology	11	11	Spooner et al.[4]
Second, 1980	Wageningen, Netherlands	9	Cell exchange	Serology	6	17	Anonymous[46]
Third, 1986	Helsinki, Finland	20	Sera exchange	Serology	16	33	Bull et al.[47]
Fourth, 1990	East Lansing, Michigan	15	Cell exchange	Serology	17	50	Bernoco et al.[48]
				RFLP			
				DQA	6	25	
				DQB	6	30	
				DRB	5	30	
				IEF	2	13	
Fifth, 1992	Interlaken, Switzerland	20	Cell exchange	Serology			
				Class I	3	53	Davies et al.[49]
				Class II	5	5	Davies et al.[92]
				RFLP			
				DQA	5	31	
				DRB	12	46	
				DQB	1	31	
				DIB	0	3	
				DOB	0	2	
				DYA	0	2	
				IEF (DRBF)	0	12	
				PCR-RFLP			
				DRB3	1	35	
				DYA	0	2	
				Microsatellite			
				DRB3	6	20	

[a] Total number of alleles for DNA typing methods represents new alleles in addition to alleles described in the literature.

a large number of cells in the workshops would have more than two specificities. Clearly, this has not been the case,[47-49] although there is evidence for a few ancestral haplotypes (see below).

2. *Isoelectric Focusing*

One dimensional-isoelectric focusing (ID-IEF) has been successfully applied to problems of antigen and haploytype definition and for investigation of the number of expressed class I products (see above). Most studies have utilized mAb to monomorphic determinants on human class I antigens or β2 microglobulin. However, this approach has its limitations. At least three *BoLA-A* alleles are not recognized by the commonly used anti-human class I-specific mAb W6/32.[38,41,52] Furthermore, W6/32 is not locus-specific,[41] indicating the desirability of using species-specific reagents whenever possible. Using 1D-IEF, two specific sets of bands are precipitated from most homozygous cells, whereas three or more bands are obtained from most heterozygotes, supporting the presence of one or two immunodominant class I genes in most haplotypes.[38,39] ID-IEF has been most useful for characterization of alleles not recognized by serological reagents and for defining new allelic subtypes of serologically defined antigens.[49,52] Coprecipitation studies with mAb and conventional serological typing reagents should allow for the assignment of IEF bands as *BoLA-A* products, thereby permitting the first description of the polymorphism of *BoLA-B* and other class I loci.[41,43] The expression of *BoLA-B* or other class I loci may be analogous to *HLA-C*, which is less polymorphic and expressed at a much lower level.

3. *Sequence Polymorphism*

Surprisingly, the extensive serological characterization of class I polymorphism has not been complemented by DNA sequence data. As of the last quarter of 1994, 11 class I sequences have been reported.[36,37,40,42,53] Only one sequence, *A10*, has been correlated with an internationally accepted *BoLA-A* specificity.[40] The class I cDNA clone KN104 encodes a product that is recognized by the mAb ILA-4,[36] but the specificity so far exists only in African Boran cattle and has not been scrutinized by an international workshop. This cDNA carries a 3-bp deletion in the exon encoding the transmembrane region that distinguishes it from all other published *BoLA* class I sequences. The BL3-6 and BL3-7 clones[13] were produced from a *BoLA* homozygous cell line,[35,54] but it is unknown which clone corresponds to the *BoLA-w33* specificity possessed by these cells. Similarly, although the *BoLA-A* genotype (*A2/A30*) of the animal used for preparation of cDNA was known by Garber and co-workers,[37] locus specificity could not be assigned definitively to the six clones sequenced from this animal.

Consistent with the expressed class I genes of other species, polymorphisms of *BoLA* class I molecules are clustered at sites putatively involved in the binding of peptide fragments.[55] In particular, binding-associated residues 9, 24, 62, 65 to 67, 69, 70, 73, 76, 77, 80, and 82 in the α1 domain show extensive sequence variability. In the α2 domain, binding site-associated residues 95, 97 114, 116, 143, 147, 152,

155, 156, and 163 are also highly polymorphic.[37] Other polymorphic residues and deletions appear to distinguish the products of the different class I loci. As functional studies progress, it will be important to elucidate the structure-function relationships of the different *BoLA* class I alleles. A concerted cDNA sequencing effort using serologically and biochemically defined cells is needed to clarify relationships among the different sequence clusters that have been identified.

4. *Other Typing Methods*

Other class I typing techniques, such as cell-mediated lympholysis,[56] immunoblotting,[57] and Southern blotting,[27] have been used for demonstration of class I polymorphism. There are problems in standardizing these methods both for man and cattle. The number of RFLPs on a Southern blot hybridized with a full-length class I cDNA probe makes it difficult to assign an RFLP to a single locus.[27] The complex patterns are useful for identification of homozygotes at the population level or for performing linkage studies because they behave as haplotypes. Recently, a highly polymorphic microsatellite associated with a class I gene was identified.[58] This marker should be very useful for population studies and haplotype definition. However, at the present time, serology remains the most universally accepted and easily performed method for class I typing.

C. Regulation of Expression

Very little is known about how *BoLA* class I genes are regulated. Harms and Splitter[59] used deletion mutants to probe the regulation of class I gene expression. Using γ-irradiation and immunoselection, a variant of the BL3 cell line was isolated that expressed three to fivefold less class I mRNA and cell surface class I. Class I expression was restored by treatment with γ-interferon and lipopolysaccharide. The defect in the cell line was presumed to be at the level of transcription initiation, as mRNA stability was not altered and no obvious change in RFLP patterns was observed. In another study, a powerful class I enhancer, BoLAenh-9, that overlaps with an interferon response sequence was isolated from the BL3 cell line.[60]

Alternatively spliced transcripts lacking exon 7 have been identified by PCR-cloning and sequencing from *Theileria annulata* transformed cell lines.[36] Indirect evidence suggests that these alternatively processed transcripts, which encode a 16 amino acid-shorter transmembrane domain, are expressed in the transformed cell lines. Results from this study suggested that class I expression may also be post-transcriptionally controlled on the basis of the observation that equal amounts of mRNA for each expressed locus were present within activated cells.[36]

D. Function

The *BoLA* class I gene products appear to have all the functional attributes of classical class I molecules. Bovine cytotoxic T-cells (CTL) are restricted by class I molecules, i.e., CTL require co-recognition of class I and peptide for effector function. The restriction of antigen-specific CTL by *BoLA* class I was first shown by

Teale and co-workers[9] who used *BoLA*-defined *T. parva* transformed lymphoblasts as targets. Cytotoxic T lymphocyte (CTL) function could be blocked by mAb to *BoLA* class I antigens.[10,61] The responding cells were subsequently shown to belong to the CD8+ population.[62-64] The genomic DNA transfection experiments performed by Toye and co-workers[40] demonstrated that cloned CTL were restricted by *w10* (*A10*) or *KN104*, but not both. Thus, the class I products encoded by *BoLA-A* and the putative *BoLA-B* loci could serve as restriction elements. T-cell responses to bovine herpes virus-1 were also shown to be restricted by *BoLA* class I (reviewed by Cook and Splitter).[65]

The peptide-binding capabilities of *BoLA* class I molecules are unknown. Given the extensive polymorphism of the *BoLA-A* locus, it is likely that a very large array of peptides can be bound to these molecules. With the importance attached to developing effective vaccines against economically important pathogens or selecting for resistance to infectious diseases, it is essential that studies advance to an understanding of peptide-Mhc interactions. Peptide elutions from class I molecules purified from genetically defined animals will offer a direct approach toward understanding of the requirements for effective peptide and genetically engineered vaccines.

Our understanding of the nonclassical class I genes and antigens of cattle is practically nonexistent. This should be a fertile ground for future investigations given the high frequency of γ/δ T-cells in peripheral blood of cattle[66] and other ruminants, and the postulated relationship between development and function of γ/δ T-cells and expression of nonclassical class I antigens in the gut.[67,68] Indeed, the ruminant should be an excellent model to explore fundamental issues in tolerance and mucosal immunity.

IV. CLASS II GENES AND GENE PRODUCTS

A. Expression

Expression of the classical class II heterodimers on cattle leukocytes was identified using cross-species-reactive mAb[69] The identification of mAb with apparent locus-specificity was first recognized on B-cells from a subset of cows infected with bovine leukemia virus (BLV).[70,71] These mAb were later confirmed to be specific for *DR* and *DQ* products by sequential immunoprecipitation and 1D-IEF.[72,73] While it is clear that distinct *DR* and *DQ* products are expressed on the cell surface, *DY*, *DI*, or *DO* products have not been identified. Ababou and coworkers[74] recently reported evidence for the expression of a third class II molecule that is cross-reactive with *HLA-DP*. The locus that encodes this product has not been established.

Cytoplasmic and cell-surface expression of class II antigens has been studied using immunofluorescence,[69] antibody-dependent complement-mediated cytotoxicity,[70] flow cytometry,[70,75] immunprecipitation,[7] and 1D-IEF.[76,77] Expression has been demonstrated on B-cells,[69] activated T- cells,[78,79] cell lines,[74] alveolar macrophages,[80] monocytes,[79,81] and mammary[82] and bronchial[83] epithelial cells. In contrast to the *HLA-DR* and *-DQ* products, which predicted the existence of discrete loci on the basis of serological and biochemical data, *BoLA-DR* and *-DQ genes* were detected first.[11,12] Expression of the class II genes has been studied at the transcriptional level using Northern blot analysis,[84,85] cDNA cloning,[54,86,87] and PCR-RFLP.[87] The class IIa region of cattle contains the known expressed class II genes: *DRA, DRB3, DQA, DQB1,* and *DQB2*. In at least three haplotypes with duplicated *DQ* genes, both *DQB* genes are expressed.[87] Expression of other genes in the class IIa region, *DRB1* (a pseudogene), *DRB2*, *DQA2*, and the class IIb region genes, *DYA*, *DYB*, *DIB*, and *DOB*, has not been demonstrated to date.[85]

B. Polymorphism

1. *Cellular Typing*

In cattle, one-way mixed lymphocyte culture tests performed on families of paternal half-sibs identified two lymphocyte-defined (LD) determinants.[88] At least one determinant was presumed to be encoded by the bovine equivalent of the Mhc. This was the first published evidence for an Mhc-like locus in cattle. The existence of Mhc-linked LD determinants was later shown by cosegregation of the MLR with *BoLA-A* antigens in full-sib families[8] and using genetically defined cells.[89] This locus was named *BoLA-D* after its human counterpart. Formal demonstration that MLR determinants are encoded by class II antigens was presented by Teale and co-workers.[90] They and others demonstrated that responder T-cells from MLRs and alloreactive cytotoxic T-cell clones with the BoT4+ (CD4) phenotype could be inhibited by mAb directed against class II antigens.[10,90] In a study designed to investigate the relationship between determinants recognized by alloreactive T-cells generated in MLRs, and 1D-IEF for *DR* antigens (see below), Glass and co-workers[91] showed that the two methods are correlated but that subtypes of EDF3 and EDF6 could be distinguished with cloned allospecific T-cell lines. Because the MLR is difficult to replicate and standardize, this method never became a widely used typing tool. Some utility can be envisioned for functional studies of class II antigens when DNA sequence information on the expressed allele(s) is not available.

2. *Serology*

The status of *BoLA* class II serology and its relationship to other typing methods has been considered in detail elsewhere,[48,51,92,93] thus it will be discussed only briefly. Injections of peripheral blood leukocytes and mononuclear cells and subcutaneous skin transplants have been the preferred routes to produce anti-class II sera;[94-97] antibodies to class II alloantigens do not frequently occur in parous sera as they do in humans (H. A. Lewin, unpublished data). These alloimmune sera tend to be highly complex and require extensive absorption with platelets to render them operational. Without the use of animals that were defined for class II on the basis of consanguineous mating, MLR, or DNA typing,[89,95] it might never have been possible to sort out the complexity of the class II alloantisera.

Despite the difficulties in producing serological typing reagents by alloimunization, several laboratories have been successful using this approach.[94-97] These reagents have been characterized in independent comparision tests[98,99] and in the fourth (BoLA4) and fifth international *BoLA* workshops.[48,92] Five class II specificities (Dw designation) were accepted by the international community.[92] Evidence for eight others was sufficient to warrant cluster designations (Dc series). Although it has been suggested that the Dw specificities are associated predominantly with *DQ* products, this has not been proven formally. A recent study by Nilsson and co-workers[100] showed that *Dw3*-specific antisera contained antibodies to determinants on both *DQ* and *DR* antigens. The advent of locus-specific biochemical and PCR-based DNA typing methods (see below) has diminished the emphasis on development of serological reagents for class II typing.

3. Isoelectric Focusing

Isoelectric focusing of immunoprecipitated class II antigens clearly demonstrated for the first time class II polymorphism at the protein level.[76,77] Although frought with problems of standardization between labs, the method was an important early tool for correlating expressed polymorphism with RFLPs detected at the DNA level.[101] Recently, discrimination of the products of the *DQ* and *DR* genes has been achieved using locus-specific mAb.[73] Correlations between alleles detected by 1D-IEF typing and those identified using other typing methods are generally good. In all, 12 IEF variants (DRBF designations) have been accepted by the international community.[92] The international workshops have demonstrated conclusively that 1D-IEF for *DRB3* alleles underestimates expressed polymorphism as judged against PCR-RFLP typing.[92] This is not surprising given the essential limitation of the method, i.e., only charge variants can be detected leaving undetected transitions that cause neutral and conserved amino acid substitions.

4. Sequence Polymorphism

a. Restriction Fragment Length Polymorphism

The first genuine description of *BoLA* class II polymorphism was made using Southern blotting and human cDNA clones as probes.[11,12,102] The restriction fragment patterns obtained permitted an estimation of the number of class II genes in each subregion and the degree of polymorphism within and around each locus. The *DRA* gene is present in one copy per haploid genome and posesses limited polymorphism, much like its murine and human orthologs.[12,103] The *DRB* genes exhibit highly complex RFLP patterns, indicating extreme underlying genetic variability.[12] Using exon-specific *DRB* probes for RFLP studies, the *DRB* genes were shown to be duplicated; molecular cloning revealed the presence of three *DRB* genes.[14,15,33,84] Thus, *DRB* RFLP types detected using probes containing exon 3 are haplotype-specific and the complex patterns can be very difficult to determine without family data.[48,92] A total of 46 *DRB* RFLP haplotyes have been identified.[92]

The *DQB* genes were also shown to be highly polymorphic by RFLP studies,[11,102,104] with 26 allelic RFLP patterns identified to date.[92] Use of exon-specific probes showed that the *DQB* genes are duplicated in most haplotypes,[34] contributing to the complexity of the RFLP patterns. The duplicated *DQB* loci are both highly polymorphic,[105] which may be related to the fact that both loci are expressed.[87] Although easier to interpret than the *DRB* patterns, discriminating the different patterns in heterozygous individuals often requires family data. The *DQA* genes are also duplicated in haplotypes containing the *DQB* duplication, but, unlike human and mouse orthologs, *BoLA-DQA* genes are highly polymorphic. Examination of widely divergent cattle breeds has resulted in the identification of 31 *DQA* RFLP types.[92]

The number of class IIb region genes detected by RFLP indicates much less variability at these loci, *DYA, DOB,* and *DIB* having 2, 2, and 3 alleles, respectively.[22,85,106] None of these polymorphisms has been shown to reside in exons.

b. DNA Sequencing

Cloning and sequencing of bovine class II genes has paved the way for detailed molecular analyses, evolutionary studies, and improved typing methods. Molecular techniques arrived relatively late to the *BoLA* scene; thus, the first class II sequences were not published until 1988.[14] Since then, an avalanche of class II sequences have been reported but have not yet been formally compiled. The BoLA Nomenclature Committee of the International Society of Animal Genetics is presently collecting and analyzing these sequences. A brief historical description and summary is given below.

Genomic sequences encoding *BoLA* class II α-chains were first reported by van der Poel and co-workers.[103] Exons with high percentages of sequence similarity to *HLA-DRA* and *HLA-DQA* permitted the assignment of bovine orthologs. One clone (R2) had no clear ortholog but ~75% similarity to *BoLA-DQA*.[22] This gene was designated *DYA*, with the assumption of correspondence to the independently segregating α-chain gene detected with a human *DQA* probe. A polymorphism in *DYA* exon 2 was identified by direct sequencing of PCR products.[106] Using this polymorphism it was then shown by sperm typing that *DYA* is located in the class IIb region[23] (Figure 1). Due to the low polymorphism of *DRA*, as indicated by RFLP typing,[12] sequences of *DRA* have not been intensively studied. Conversely, the unusually high level of polymorphism of *DQA* genes detected by RFLP analysis prompted additional studies. A nearly full-length cDNA clone encoding *DQA1*02* was isolated from the homozygous BL3 cell line and sequenced.[86] A definitive assignment to *DQA1* could be made because the *DQA* gene is not duplicated in the DQ^2 haplotype. A total of 11 *DQA* sequences have been identified to date.[107]

In studies aimed at isolating genes encoding class II β-chains, sequencing of bovine genomic clones demonstrated the existence of three genes with a clear relationship to *HLA-DRB*.[14,33,108] The first sequenced *DRB*-like gene (*DRB1*)[14,33] was found to have stop codons in exons encoding the β_1 and transmembrane domains. Thus, *DRB1* is a pseudogene. The second *DRB*-like gene sequenced (*DRB2*)[108] clearly differed in its genomic organization from *DRB1* and did not contain a premature stop codon in the exons sequenced. However, expression of this gene could not be demonstrated by Northern blot analysis.[84] A third *DRB*-like gene (clone A1),[33] distinct from *DRB1* and *DRB2* in its genomic organization and with about 85% homology to *HLA-DRB1*, was also identified. Isolation of a cDNA clone with strong sequence similarity to the A1 clone demonstrated that this gene is transcriptionally active.[84] Subsequent PCR cloning and sequencing of approximately 38 exon 2 alleles of this gene, named *DRB3*, revealed extensive polymorphism,[15,109,110] suggesting that *BoLA-DRB3* is the functional equivalent of *HLA-DRB1*.[15] On the basis of PCR-RFLP studies and haplotype analysis (see below), there are at least 50 *DRB3* alleles. The large number of *DRB3* alleles identified by PCR-RFLP that have not been sequenced indicates that there is considerably more sequencing work to be done to gain a full understanding of *DRB3* polymorphism and function.

Two distinct genomic *DQB*-like sequences have been identified.[33,111] Isolation of a *DQB* cDNA clone,[54] Northern blot analysis,[85] and RT-PCR[87] have demonstrated expression of the *DQB* gene(s). Detailed analysis of cloned PCR fragments from genomic DNA representing haplotypes with duplicated and nonduplicated *DQ* genes indicate four *DQB* "subtypes", *DQB1* to *DQB4*.[105] *DQB1* and *DQB2* are closely related and are found together in several haplotypes.[105] In all, 16 *DQB1* and 5 *DQB2* alleles have been identified.[33,54,105] *DQB3* and *DQB4* appear to be distantly related to *DQB1*. However, assignment of these sequences to distinct loci is tenuous because only one sequence each for *DQB3* and *DQB4* was described.[105]

On the basis of nucleotide and amino acid similarity, Stone and Muggli-Cockett[112] described a novel β-chain gene, *DIB*, that maps to the class IIb region.[85] The sequence is most closely related to *DQB* (75%) but was not found to be expressed in stimulated or unstimulated peripheral blood lymphocytes using Northern blot analysis.[85] Sequences for *DYB* and the putative bovine orthologs of *HLA-DOB* and *-DNA* have not been identified.

c. *PCR-RFLP*

The availability of DNA sequence information for class II genes enabled the development of PCR-based typing methods. In cattle, restriction site polymorphism is correlated with the high degree of sequence polymorphism in *DRB3* exon 2,[16] thus enabling the combined use of PCR and RFLP for detection of *DRB3* alleles. The strategy developed for PCR-RFLP of *DRB3* alleles was to select the minimum number of enzymes that discriminate the maximum number of alleles.[16,113] To date, 35 alleles have been identified in a 284-bp segment of *DRB3* exon 2 using *Rsa*I, *BstY*I, and *Hae*III[16] (H. A. Lewin, unpublished results). Two restriction sites are within codons for peptide binding residues. In BoLA5, PCR-RFLP was shown to be the most powerful *DRB3* typing method and extremely useful for haplotype discrimination.[92] Although the method works best when family data are available, population studies are feasible because of the minimal overlap in fragment patterns. Not unexpectedly, not all *DRB3* alleles can be discerned by PCR-RFLP; however, the method can easily be extended by adding more enzymes for alleles not split by *Rsa*I, *BstY*I, and *Hae*III. Coupling PCR-RFLP with serology or 1D-IEF can also be of some value in identifying the relatively small number of alleles not discriminated using the three enzymes. When used in conjunction with class I serology, *DRB3* PCR-RFLP is extremely valuable for identification of new alleles for sequencing,[109] for characterization of *BoLA* haplotypes[16,92] (Table 2), and for disease association studies.[109] The method is also applicable to other ungulate mammals, because the primers have been shown to amplify orthologous genes from ten different species.[114]

Table 2 ***BoLA-A-DRB3*** **Haplotypes**[a]

Breed	*BoLA-A*	*DRB3*
Angus	2	7
	3	7
	5	22
	22	4
	24	12
	43	25
	w50	7
	Blank	18
	Blank	23
Ayrshire	2	7
	6	8
	7	2
	7	18
	10	**3**
	10	18
	11	**24**
	18	3
	18	**28**
	19	12
	23	28
	26	**21**
	26	24
	26	28
	34	14
British Friesian	1	16
	11	11
	11	24
	15	3
	16	7
	18	3
	18	22
	w27	3
	w27	18
	31	21
	31	28
	32	15
	32	22
	36	8
	40	11
	99	3
Brown Swiss	**2**	**7**
	9	7
	9	**28**
	10	12
	14	10
	14	**11**
	20	10
	22	**22**
	22	28
	31	**10**
	34	**9**
Danish Black	9	18
	11	24
	12	16
	15	22
	18	22
	19	24
	19	31
	23	28
	26	16
	w27	3
	31	7
	32	22
	36	8
	44	7
	w53	7
Danish Red	16	27
	18	11
	18	15
	37	10
	w39	22
	40	11
	52	15
South Devon	**2**	**7**
	2	8
	2	20
	9	**3**
	10	20
	11	24
	14	**10**
	14	19
	18	16
	18	32
	23	8
	23	12
	24	12
	29	9
	30	9
	30	20
	32	9
	44	11
	Blank	8
	Blank	9
	Blank	10
	Blank	20
	Blank	23
Gelbvieh	2	1
	2	**7**
	7	**7**
	8	12
	9	3
	9	**28**

Table 2 (continued) *BoLA-A-DRB3* Haplotypes[a]

Breed	*BoLA-A*	*DRB3*
Gelbvieh (cont.)	10	8
	10	18
	11	**7**
	11	24
	12	1
	12	5
	12	7
	14	22
	16	7
	16	28
	19	18
	20	**7**
	20	8
	22	5
	23	**8**
	24	28
	31	1
	49	28
	UR80	28
	Blank	5
	Blank	7
	Blank	23
	Blank	32
Guernsey	6	3
	10	**3**
	18	**3**
	18	10
Holstein	5	28
	7	16
	7	22
	10	**3**
	10	7
	10	18
	10	22
	10	23
	10	25
	10	26
	11	**24**
	12	**16**
	12	28
	13	8
	13	11
	13	**23**
	14	**11**
	14	27
	15	3
	15	**22**
	16	**7**
	18	22
	19	8
	19	9
	19	11
	19	22
	19	23
	19	**24**
	20	**8**
	26	28
	31	12
	32	22
	34	11
	35	28
	41	12
	44	7
	44	11
	44	27
	49	12
	49	22
Jersey	3	10
	w4	18
	6	17
	6	22
	6	28
	9	4
	9	7
	9	21
	11	28
	12	**7**
	12	18
	12	20
	12	21
	21	**10**
	22	**15**
	22	23
	23	**9**
	24	28
	w28	20
	31	8
	49	17
Simmental	2	7
	2	**28**
	3	7
	3	18
	3	24
	5	6
	5	12
	5	18
	7	**7**
	7	16
	7	18
	9	3
	10	13
	10	15
	11	7
	11	8

Table 2 (continued) *BoLA-A-DRB3* Haplotypes[a]

Breed	*BoLA-A*	*DRB3*	Breed	*BoLA-A*	*DRB3*
Simmental (cont.)	11	24		21	7
	11	29		22	18
	14	8		**24**	**12**
	14	15		Blank	5
	19	**18**		Blank	10
	20	1		Blank	21
	20	2		Blank	22
	20	16		Blank	23
	21	**2**			

[a] Haplotype data for British Friesian and Danish Red was compiled from the Fifth International BoLA Workshop. For all other breeds, haplotypes were compiled by the laboratory of the author. Haplotypes in bold are frequent within the breed indicated, with evidence for linkage or gametic disequilibrium.

The PCR-RFLP technique has also been shown to be useful for typing *DQB* alleles,[87] although the description of *DQB* polymorphism by PCR-RFLP is complicated by a duplication of *DQB* genes in most haplotypes and the lack of PCR primers to discriminate the *DQB* genes in genomic DNA. Combining reverse transcription and PCR (RT-PCR), Xu et al.[87] used the restriction enzymes *Eco*O109I and *Dra*III to demonstrate that both *DQB* genes are expressed in three of the haplotypes carrying a duplication. Unlike *DRB3*, the polymorphism in exon 2 of the *DQB* genes is not as easily exploited by RFLP analysis due to the paucity of polymorphic restriction sites. Other DNA typing methods, such as allele-specific oligonucleotides, may be required for a convenient and accurate typing system for *DQB* alleles.

The PCR-RFLP method has also been used for routine screening of a G $\rightarrow$ A transition in exon 2 of *DYA*.[106] The ^{219}G polymorphism has a perfect correlation with the 1.4 kb *Taq*I *DYA* RFLP at the population level,[92] indicating strong gametic disequilibrium between these polymorphisms.

5. Other Typing Methods

Russel[115] developed a *DRB3* typing method employing asymmetric PCR for amplification of exon 2 followed by analysis of single-strand comformational polymorphism in nondenaturing polyacrylamide gels. This method has significant potential for identifying new alleles within PCR-RFLP subtypes.

A highly polymorphic complex microsatellite located in intron 2 of *DRB3* has been shown to be a powerful approach for discriminating *BoLA* haplotypes.[116] The 20 recognized microsatellite alleles are closely associated with *DRB3* polymorphism.[92,116] In combination with microsatellites within or flanking other *BoLA* genes, the relative simplicity of microsatellite typing can be exploited for detailed haplotype determination. Fine discrimination of haplotypes can be useful for disease association studies, but such associations will ultimately depend on the characterization of functional genes within the haplotype.

C. Regulation of Expression

Expression of *BoLA* class II molecules is down-regulated by dexamethasone [83] and *Pasteurella haemolytica* leukotoxin,[81] whereas interferon-γ,[80] formalin killed *Streptococcus uberis* extract [82] and infection by *T. annulata*[117] and *T. parva*[118] up-regulate expression. TNF-α, interferon-α1, and cyclosporine elicited no consistent effects on class II expression.[83,100]

The effects of bovine leukemia virus (BLV) on class II expression has been analyzed in detail. Integration of BLV in cows with persistent lymphocytosis (PL) is associated with a loss in expression of *BoLA-DR*, whereas *DQ* expression is not affected.[71] Loss of *DR* has also been noted on lymphosarcoma cells.[71] Short-term culture of B-cells from *DR*-negative cows with PL was found to induce expression of *DR* antigen.[71] Recent studies have demonstrated that *DRA* and *DRB3* mRNA is present at abundant levels in the *DR*-negative PL B-cells (M.R. Teutsch and H.A. Lewin, manuscript in preparation). However, in tumor cells from one cow, no *DRA* mRNA was detected, providing a clear explanation for the defect in *DR* expression at the cell surface (M.R. Teutsch and H.A. Lewin, manuscript in preparation). The loss of *DR* expression may be related to progression from PL to lymphosarcoma in BLV-infected cattle.[119]

D. Function

Responses of bovine $CD4^+$ T-cells to peptides are dependent on the presentation of peptide ligands bound to class II molecules on the surface of antigen-presenting cells. The requirement of adherent class II^+ cells for peripheral blood T-cell proliferative responses to ovalbumin was first demonstrated by Glass and Spooner.[120] Responses were associated with class I haplotype,[121] implying a classical *Ir* gene effect associated with class II alleles. Convincing evidence for the role of *BoLA* class II in the presentation of ovalbumin was presented by Glass and co-workers[122] who used antigen-specific $CD4^+$ T-cell clones and autologous antigen-presenting cells to demonstrate that class II matching is required for *in vitro* responsiveness. Class II restricted T-cell responses to *T. annulata*[117] antigens also have been described.

Glass et al.[123] identified *BoLA* class II haplotypes associated with high responder and low responder status to synthetic foot and mouth disease virus (FMDV) peptides, as measured by antigen-specific T-cell responses and cloned $CD4^+$ cell lines. Results from that study illustrated two additional important points: (1) the IEF method used to define class II antigens was not particularly good at predicting responder phenotype and (2) although binding of FMDV peptides to class II molecules appeared promiscuous, not all individuals mounted strong responses to the peptides. These data indicate that *BoLA* typing is an important adjunct for evaluating the effectiveness of peptide vaccines, and that it is essential to determine the peptide binding motifs for the different *BoLA* class II antigens. Prediction of immune responses to synthetic and genetically engineered vaccines is, arguably, the most directly applicable information one can derive from *BoLA* typing. With the paucity of information on peptide binding currently available, there is a high probability that such studies will yield useful data.

V. HAPLOTYPE DIVERSITY

A. *DQ-DR* Haplotypes

The application of RFLP analysis to *DR* and *DQ* genes made it possible to elucidate *BoLA* haplotypes. In Swedish Red and White cattle, about 30 *DQ-DR* haplotypes were defined by segregation of allelic RFLP patterns in five half-sib families comprising 50 offspring.[102] In BoLA4[48] nine new *DR-DQ* haplotypes were identified among 46 animals representing five breeds. In BoLA5[92] 38 *DQ-DR* haplotypes were identified among 59 animals, of which at least 20 were new. Haplotype discrimination was enhanced by use of serology, 1D-IEF, PCR-RFLP, and microsatellite typing in addition to RFLP analysis. Thus, there are at least 59 *DQ-DR* haplotypes, although unique gene products may not be expressed by all haplotypes.[92] Within breeds, many alleles are restricted to a single *DR-DQ* haplotype. Among breeds, different haplotypic combinations are found, but certain haplotypes (e.g., *DR1A-DQ1A*) appear in several breeds. These ancestral haplotypes (see below) are the strongest evidence for true linkage disequilibrium in the *BoLA* system.

B. *DYA-DOB-DIB* Haplotypes

The *DYA*, *DOB*, and *DIB* genes are closely linked in the class IIb region. With the limited degree of polymorphism of these loci (see above) 12 haplotypes are possible; however, only 5 haplotypes were identified in 54 animals typed in BoLA5. Thus, there is evidence that genes in this region may be in linkage or gametic disequilibrium. Mapping of *LMP2* to the class IIb region,[24] and thus probably *LMP7* and *TAP* genes as well, might provide a functional basis for maintaining class IIb alleles in disequilibrium. As might be expected from the large genetic map distance between the class IIa and IIb genes, there is no evidence that alleles in the two clusters are in disequilibrium.

C. *BoLA-A-DRB3* Haplotypes

The advent of powerful DNA typing methods for class II genes permitted an examination of *BoLA-A*-class II haplotypes. An enormous number of haplotypes is theoretically possible given the extreme polymorphism of the *BoLA-A, DR,* and *DQ* loci. However, gametic or linkage disequilibrium appears to extend to the class I region, thus limiting the number of haplotypes observed. In BoLA5, 48 class I-class IIa haplotypes were defined using serology, 1D-IEF, RFLPs, PCR-RFLPs, and microsatellite typing. A listing of 170 *BoLA-A-DRB3* haplotypes identified in samples from most of the major dairy and beef breeds is given in Table 2. Looking at this table, it is abundantly clear that domesticated cattle have a huge reservoir of haplotypic diversity, although diversity in some breeds appears much more limited than in others. Differences could be due to genetic drift (small sample size for breeds with less apparent diversity). However, certain differences are probably more real than apparent, given the isolated geographic origin of the breeds and population sizes. For example, Holstein-Friesian probably has at least 40 *BoLA-A-DRB3* haplotypes, whereas Jersey and Ayrshire have about 21 and 15, respectively.

Several ancestral haplotypes can be deduced from *BoLA-A-DRB3* typing. For example, the haplotype *A2-DRB3*07* is present in Angus, Ayrshire, Brown Swiss, South Devon, Simmental, and Gelbvieh. Similarly, *A11-DRB3*24* is present in Ayrshire, British Friesian, Holstein-Friesian, Danish Black Pied, South Devon, Simmental, and Gelbvieh. An important point is that the disequilibrium between certain *BoLA-A* and *DRB3* alleles appears to be complete in some breeds (e.g., *A36(A20)-DRB3*08* in Holstein), whereas other *BoLA-A* alleles, such as *A10*, appear in combination with at least five *DRB3* alleles (only one subtype of *A10* has been identified). With these data it should be possible to elucidate the evolutionary relationships between cattle breeds and to relate the origins of these breeds to human migration patterns.[124] Furthermore, this information has obvious utility for disease association studies.

D. Extended *BoLA* Haplotypes

The combinations of alleles at linked genes on a chromosome can reveal important attributes of the evolutionary history of a species.[125] Significant deviations from expected frequencies of haplotypes may indicate founder effects due to close linkage (gametic phase disequilibrium) or effects due to selection (linkage disequilibrium). In discussing Mhc haplotype data, these two definitions are often confused or misconstrued. Provided the nature of cattle breeding, it may not be possible to distinguish the two effects. This is because mating is nonrandom and observed haplotype frequencies may vary widely from year to year due to the restricted number of bulls used for breeding. Although the evolutionary implications may be unclear, haplotype discrimination has become increasingly important because of the emphasis on dissecting Mhc effects on diseases. Analysis of haplotypes in disease association studies can be helpful in mapping disease resistance and susceptibility to localized regions within the Mhc.[109,126] Haplotype diversity is also a useful measure of genetic variability and can be used for genetic management of rare breeds and endangered species.[114]

VI. DISEASE ASSOCIATIONS

Research on the Mhc of domestic animal species is often justified in terms of using alleles or haplotypes for selection of disease-resistant animals. Under certain conditions, such as herds or geographic areas in which a specific disease is endemic, selection for resistance alleles could provide a favorable economic effect. In no case should breeding for Mhc homozygosity be contemplated on a herd or breed-wide basis. Nor should "haplocide"[127] be considered as a rational strategy for elimination of high-frequency susceptibility alleles. Haplotype diversity ensures that a breed will retain the maximum repertoire for peptide binding and antigen presentation. For diseases where resistance is the dominant trait, increasing the allele frequency up to 50% should be sufficient to realize any economic benefit, provided such an increase is not contraindicated by negative associations with production traits.

A review of the entire literature on trait associations with *BoLA* is beyond the scope of this article; others have reviewed this topic.[51,127,128] A synopsis of research on trait associations can be found in Table 3. The intent is to distill this literature and provide the reader with a synthesis of common themes. Examples of two diseases for which studies have been conducted by more than one laboratory are discussed below.

A. Mastitis

Studies of the relationship between *BoLA* and mastitis have been received with much interest because losses due to mastitis run into tens of millions of dollars annually. An essential limitation in studying diseases such as mastitis is that a complex etiology can obscure experimental results and comparisons between data sets.[127] Clinical mastitis can be caused by at least four organisms[129] and susceptibility has a strong environmental component. Therefore, it is impressive that there is some consistency between studies. In Norwegian Red cattle, *A11* and *A16* are associated with susceptibility and *A2* with resistance to clinical mastitis, whereas the *DQ1A* allele is weakly associated with susceptibility to clinical mastitis in Swedish Red and White cattle.[130,131] The association with *A11* in Norwegian Red cattle can be considered as confirmatory because *A11* is in linkage disequilibrium with *DQ1A* in the breed.[132] However, Weigel et al.[133] found *A11* to be associated with resistance to clinical mastitis in Holstein cattle; *A11* is also in strong disequilibrium with *DQ1A* in Holstein. These contradictory results point out the difficuly in interpretation when the infectious agent is not identified. Several groups have tried to detect effects on somatic cell count (SSC) in milk, a trait associated with subclinical mastitis.[134,135] Somatic cell count is an indirect measure of subclinical mastitis that is not absolutely correlated with the trait. Therefore, it is not surprising that effects on SSC are associated with different alleles in different breeds in different geographical locations. Resolution of the relationship between *BoLA* and mastitis will require more thoughtful experimental design than has been employed thus far.

B. Bovine Leukemia Virus

Bovine leukemia virus is a C-type retrovirus that causes lymphosarcoma and persistent lymphocytosis (PL) in about 1% and 30% of infected cattle, respectively. Lymphosarcoma and PL cause at least $86 M loss to the dairy industry each year.[136] Association between *A14* and resistance to PL and delay in seroconversion has been documented in several herds (reviewed by Lewin[127,137]). The *A14* allele was also associated with longevity in two Holstein herds with high prevalence of BLV infection.[136,138] Dominant resistance to PL has been mapped to the class IIa region,[126] and is closely associated with the presence of glutamic acid and arginine at positions 70 and 71 of the *DRB3* molecule.[109] Because resistance is dominant, and the *A14-DRB3*11* allele is present in moderate frequency (10 to 25% in most herds), it has been suggested that genetic resistance to PL can be selected for.[109] Increasing

Table 3 Trait Associations

Trait	Breed	*BoLA*	Effect	Ref.
Enzootic bovine leukosis				
Seroconversion	Holstein	A14	Late	Lewin et al.;[139] Palmer et al.[138]
	Holstein	A15	Rapid	Palmer et al.[138]
	Guernsey	A21	Late	Palmer et al.[138]
	Guernsey	DA6.2, A12	Rapid	Palmer et al.[138]
PL or B-cell numbers	Shorthorn	DA7	Resistance	Lewin and Bernoco[18]
	Shorthorn	DA12.3	Susceptibility	Lewin and Bernoco[18]
	I. Shorthorn	A6, EU28R	Susceptiblity	Stear et al.[140]
	I. Shorthorn	A8	Resistance	Stear et al.[140]
	Holstein	A12 and A15	Susceptibility	Lewin et al.[139]
	Holstein	A14 and A13	Resistance	Lewin et al.[139]; Zanotti et al.[158]
	Holstein	DRB2*2A	Resistance	van Eijk et al.[126]; Zanotti et al.[158]
	Holstein	DRB2*1C	Susceptibility	van Eijk et al.[126]; Zanotti et al.[158]
	Holstein	DRB3(ER motif)	Resistance	Xu et al.[109]; Zanotti et al.[158]
Mastitis				
Clinical mastitis	Norwegian Red	A2	Resistance	Solbu et al.;[141] Solbu and Lie;[142] Våge et al.;[131] Mejdell et al.[143]
	Norwegian Red	A16	Susceptibility	Larsen et al.[144]
	Norwegian Red	A11	Susceptibility	Solbu and Lie[142]
	Swedish Red and White	DQ1A	Susceptibility	Lundén et al.[130]
	Holstein	A11	Resistance	Weigel et al.[133]
SSC	Icelandic	A19	High	Oddgeirsson et al.[135]
	Simmental (S) or S × Red Holstein	A15	High	Arriëns et al.[134]
CMT	Holstein	A14	Low	Weigel et al.[133]
Parasitemia				
Nematodes	Belmont Red	A7, CA36	Resistance	Stear et al.[145]
	AfricanderXHereford	A9	Susceptibility	Stear et al.[146]
Ticks				
Boophilus microplus	Brahman XShorthorn	A6, CA31	Susceptibility	Stear et al.[147]
	Belmont Red	A19, A7	Resistance	Stear et al.[145]

Posterior spinal paresis	Holstein	A8	Susceptibility	Park et al.[148]
Ketosis	Norwegian Red	A2, A13	Resistance	Mejdell et al.[143]
Retained placenta	Dutch Friesian	Compatibility	Susceptibility	Joosten et al.[149]
Immune response				
Multiple traits	Holstein	Several alleles		Weigel et al.[150]
(T,G)-A — L	Norwegian Red	A8, A20	High response	Lie et al.[151]
Human serum albumin	Norwegian Red	A2	Low response	Lie et al.[151]
Human serum albumin	Norwegian Red	A6, A16	High response	Lie et al.[151]
Ovalbumin	Friesian	ED99, A10	Low responders	Glass et al.[121]
Ovalbumin	Friesian	A11, ED85	High responders	Glass et al.[121]
Ovalbumin	Friesian	EDF9	Nonresponder	Glass et al.[122]
T. parva	Boran, Friesian	A6, A10, A11	Predominant	Morrison et al.[62]
FMDV peptides	Friesian	EDF alleles	Variable	Glass et al.[123]
Production traits				
Growth	Nine breeds	Multiple alleles	Variable	Stear et al.[152]
Rib-eye area	Angus	A2 vs. A28	Larger	Beever et al.[153]
Milk yield	Holstein	A14	Increased	Weigel et al.[133]
Milk fat yield	Holstein	A11	Increased	Batra et al.[154]
Milk fat yield	Holstein	A11	Decreased	Weigel et al.[133]
Milk fat yield	Holstein	A15	Increased	Weigel et al.[133]
Milk fat percentage	Holstein	A10	Increased	Hines et al.[155]
Milk fat percentage	Holstein	A10, A11	Increased	Batra et al.[154]
Milk fat percentage	Holstein	A16	Decreased	Batra et al.[154]
Milk fat percentage	Holstein	A11, A31	Decreased	Weigel et al.[133]
Milk fat percentage	Holstein	A20	Increased	Weigel et al.[133]
Milk protein yield	Holstein	A19	Increased	Batra et al.[154]
Gestation length	Holstein	A14	Increased	Batra et al.[154]
Estrus	Holstein	w12.2	Decreased	Batra et al.[154]
Health costs	Holstein	A14	Lower	Weigel et al.[133]
Income/feed costs	Holstein	A14	Increased	Weigel et al.[133]
Paired testicular volume	Hereford	A19	Increased	Stear et al.[156]
Fertility	Norwegian Red	A2	Increased	Mejdell et al.[143]

the frequency of this haplotype to one copy per animal through the use of homozygous bulls could completely negate the economic impact of PL. The association of *A14* with significantly increased fat yield, milk yield, fat percentage, and income over feed costs and decreased scores on the California Mastitis Test[133] suggests that selection for *A14* in herds with a high prevalence of BLV infection would have multiple desirable effects.

VII. CONCLUDING REMARKS

The last 10 years has produced a wealth of descriptive knowledge of the bovine Mhc. While most of the fundamental discoveries about Mhc structure-function relationships have originated from our colleagues working on the human and mouse Mhc, practical utilization of this knowledge was first achieved for resistance to Marek's Disease by chicken breeders and may soon come to fruition with BLV in cattle. While it remains to be proven, it is fairly certain that peptide-binding motifs described for human and mouse class I and class II alleles will be shared with structurally similar bovine alleles. This may facilitate an efficient process for the development of broadly effective peptide subunit vaccines against bovine infectious diseases. Therefore, it is essential that sequencing of class I and class II alleles continues and that T-cell epitopes are identified for bovine pathogens. In addition to functional studies, comparative studies should be expanded.

Interspecies differences in the organization and expression of Mhc genes may help resolve long-standing questions about the evolution of immune function, e.g., the function of γδ T-cells, and why cattle are virtually free of autoimmunity and have very low incidence of cancer.

ACKNOWLEDGMENTS

This work was supported in parts by grants 92-37204-7695 and 94-37205-1165 from the United States Department of Agriculture and 1RO1CA591148 from the National Cancer Institute, National Institutes of Health. The author gratefully acknowledges the assistance of Colleen Olmstead, Elaine Sampson, and Chankyu Park in preparing the manuscript, and Jon Beever, Meco Bernoco, Elissa Gaynor, and Mike Mirsky for helpful suggestions.

REFERENCES

1. Klein, J., The story, in *Natural History of the Major Histocompatibility Complex*, John Wiley & Sons, New York, 1986, 1.
2. Amorena, B. and Stone, W. H., Serologically defined (SD) locus in cattle, *Science*, 201, 159, 1978.
3. Spooner, R. L., Leveziel, H., Grosclaude, F., Oliver, R. A., and Vaiman, M., Evidence for a possible major histocompatibility complex (BLA) in cattle, *J. Immunogenet.*, 5, 335, 1978.

4. Spooner, R. L., Oliver, R. A., Sales, D. I., McCoubrey, C. M., Millar, P., Morgan, A. G., Amorena, B., Bailey, E., Bernoco, D., Brandon, M., Bull, R. W., Caldwell, J., Cwik, S., van Dam, R. H., Dodd, J., Gahne, B., Grosclaude, F., Hall, J. G., Hines, H., Leveziel, H., Newman, M. J., Stear, M. J., Stone, W. H., and Vaiman, M., Analysis of alloantisera against bovine lymphocytes. Joint report of the 1st International Bovine Lymphocyte Antigen (BoLA) Workshop, *Anim. Blood Grps. Biochem. Genet.*, 15, 63, 1979.
5. Klein, J., Bontrop, R. E., Dawkins, R. L., Erlich, H. A., Gyllensten, U. B., Heise, E. R., Jones, P. P., Parham, P., Wakeland, E. K., and Watkins, D. I., Nomenclature for the major histocompatibility complexes of differerent species: a proposal, *Immunogenetics*, 31, 217, 1991.
6. Hoang-Xuan, M., Leveziel, H., Zilber, M.-T., Parodi, A. L., and Levy, D., Immunochemical characterization of major histocompatibility antigens in cattle, *Immunogenetics*, 15, 207, 1982.
7. Hoang-Xuan, M., Charron, D., Zilber, M.-T., and Levy, D., Biochemical characterization of class II bovine major histocompatibility complex antigens using cross-species reactive antibodies, *Immunogenetics*, 15, 621, 1982.
8. Usinger, W. R., Curie-Cohen, M., Benforado, K., Pringnitz, D., Rowe, R., Splitter, G. A., and Stone, W. H., The bovine major histocompatibility complex *(BoLA)*: close linkage of the genes controlling serologically defined antigens and mixed lymphocyte reactivity, *Immunogenetics*, 14, 423, 1981.
9. Teale, A. J., Morrison, W. I., Goddeeris, B. M., Groocock, C. M., Stagg, D. A., and Spooner, R. L., Bovine alloreactive cytotoxic cells generated *in vitro:* target specificity in relation to BoLA phenotype, *Immunology*, 55, 355, 1985.
10. Teale, A. J., Baldwin, C. L., Ellis, J. A., Newson, J., Goddeeris, B. M., and Morrison, W. I., Alloreactive bovine T lymphocyte clones: an analysis of function, phenotype, and specificity, *J. Immunol.*, 136, 4392, 1986.
11. Andersson, L., Böhme, J., Rask, L., and Peterson, P. A., Genomic hybridization of bovine class II major histocompatibility genes: I. Extensive polymorphism of DQα and DQβ genes, *Anim. Genet.*, 17, 95, 1986.
12. Andersson, L., Böhme, J., Peterson, P. A., and Rask, L., Genomic hybridization of bovine class II major histocompatibility genes: II. Polymorphism of DR genes and linkage disequilibrium in the DQ-DR region, *Anim. Genet.*, 17, 295, 1986.
13. Ennis, P. D., Jackson, A. P., and Parham, P., Molecular cloning of bovine class I Mhc cDNA, *J. Immunol.*, 141, 642, 1988.
14. Muggli-Cockett, N. E. and Stone, R. T., Identification of genetic variation in the bovine major histocompatibility complex DRβ-like genes using sequenced bovine genomic probes, *Anim. Genet.*, 19, 213, 1988.
15. Sigurdardóttir, S., Borsch, C., Gustafsson, K., and Andersson, L., Cloning and sequence analysis of 14 DRB alleles of the bovine major histocompatibility complex by using the polymerase chain reaction, *Anim. Genet.*, 22, 199, 1991.
16. van Eijk, M. J. T., Stewart-Haynes, J. A., and Lewin, H. A., Extensive polymorphism of the BoLA-DRB3 gene distinguished by PCR-RFLP, *Anim. Genet.*, 23, 483, 1992.
17. Andersson, L., Gustafsson, K., Jonsson, A.-K., and Rask, L., Concerted evolution in a segment of the first domain exon of polymorphic Mhc class II β loci, *Immunogenetics*, 33, 235, 1991.
18. Lewin, H. A. and Bernoco, D., Evidence for BoLA-linked resistance and susceptibility to subclinical progression of bovine leukaemia virus infection, *Anim. Genet.*, 17, 187, 1986.
19. Fries, R., Eggen, A., and Womack, J. E., The bovine genome map, *Mamm. Genome*, 4, 405, 1993.

20. Bishop, M. D., Kappes, S. M., Keele, J. W., Stone, R. T., Sunden, S. L. F., Hawkins, G. A., Toldo, S. S., Fries, R., Grosz, M. D., Yoo, J., and Beattie, C. W., A genetic linkage map for cattle, *Genetics*, 136, 619, 1994.
21. Barendse, W., Armitage, S. M., Kossarek, L. M., Shalom, A., Kirkpatrick, B. W., Ryan, A. M., Clayton, D., Li, L., Neibergs, H. L., Zhang, N., Grosse, W. M., Weiss, J., Creighton, P., McCarthy, F., Ron, M., Teale, A. J., Fries, R., McGraw, R. A., Moore, S. S., Georges, M., Soller, M., Womack, J. E., and Hetzel, D. J. S., A genetic linkage map of the bovine genome, *Nature Genet.*, 6, 227, 1994.
22. Andersson, L., Lundèn, A., Sigurdardottir, S., Davies, C. J., and Rask, L., Linkage relationships in the bovine Mhc region. High recombination frequency between class II subregions, *Immunogenetics*, 27, 273, 1988.
23. van Eijk, M. J. T., Russ, I., and Lewin, H. A., Order of bovine *DRB3, DYA,* and *PRL* determined by sperm typing, *Mamm. Genome*, 4, 113, 1993.
24. Shalhevet, D., Da, Y., Beever, J. E., van Eijk, M. J. T., Ma, R., Lewin, H. A., and Gaskins, H. R., Genetic mapping of the *LMP2* proteasome subunit gene to the *BoLA* class IIb region, *Immunogenetics*, 41, 44, 1995.
25. Jarrell, V. L., Lewin, H. A., Da, Y., and Wheeler, M. B., Gene-centromere mapping of bovine *DYA, DRB3* and *PRL* using secondary oocytes and first polar bodies: evidence for four-strand double crossovers between *DYA* and *DRB3*, *Genomics*, 27, 33, 1995.
26. Park, C., Russ, I., Da, Y., and Lewin, H. A., Genetic mapping of *F13A* to BTA23 by sperm typing: difference in recombination rate between bulls in the *DYA-PRL* interval, *Genomics,* 27, 113, 1995.
27. Lindberg, P.-G. and Andersson, L., Close association between DNA polymorphism of bovine major histocompatibility complex class I genes and serological BoLA-A specificities, *Anim. Genet.*, 19, 245, 1988.
28. van Eijk, M. J. T., Beever, J. E., Da, Y., Stewart, J. A., Nicholaides, G. E., Green, C. A., and Lewin, H. A., Genetic mapping *BoLA-A, CYP21, DRB3, DYA,* and *PRL* on BTA23, *Mamm. Genome*, 6, 151, 1995.
29. Gwakisa, P., Mikko, S., and Andersson, L., Close genetic linkage between *DRBP1* and *CYP21* in the Mhc of cattle, *Mamm. Genome*, 5, 731, 1994.
30. Teutsch, M. R., Beever, J. E., Stewart, J. A., Schook, L. B., and Lewin, H. A., Linkage of complement factor B gene to the bovine major histocompatibility complex, *Anim. Genet.*, 20, 427, 1989.
31. Bensaid, A., Young, J. R., Kaushal, A., and Teale, A. J., Pulsed-field gel electrophoresis and its application in the physical analysis of the bovine Mhc, in *Gene Mapping Techniques and Applications*, Schook, L. B., Lewin, H. A., and McLaren, D. G., Eds., Marcel Dekker, New York, 1991, 127.
32. Bensaid, A., Kaushal, A., Baldwin, C. L., Clevers, H., Young, J. R., Kemp, S. J., MacHugh, N. D., Toye, P. G., and Teale, A. J., Identification of expressed bovine class I Mhc genes at two loci and demonstration of physical linkage, *Immunogenetics*, 33, 247, 1991.
33. Groenen, M. A. M., van der Poel, J. J., Dijkhof, J. M. R., and Giphart, M. J., The nucleotide sequence of bovine Mhc class II *DQB* and *DRB* genes, *Immunogenetics*, 31, 37, 1990.
34. Andersson, L. and Rask, L., Characterization of the Mhc class II region in cattle. The number of *DQ* genes varies between haplotypes, *Immunogenetics*, 27, 110, 1988.
35. Romano, M. J., Stewart, J. A., and Lewin, H. A., Phenotypic characterization of bovine lymphoblastoid cell lines, *Vet. Immunol. Immunopathol.*, 23, 293, 1989.

36. Ellis, S. A., Braem, K. A., and Morrison, W. I., Transmembrane and cytoplasmic domain sequences demonstrate at least two expressed bovine Mhc class I loci, *Immunogenetics*, 37, 49, 1992.
37. Garber, T. L., Hughes, A. L., Watkins, D. I., and Templeton, J. W., Evidence for at least three transcribed *BoLA* class I loci, *Immunogenetics*, 39, 257, 1994.
38. Joosten, I., Oliver, R. A., Spooner, R. L., Williams, J. L., Hepkema, B. G., Sanders, M. F., and Hensen, E. J., Characterization of class I bovine lymphocyte antigens (BoLA) by one-dimensional isoelectrofocusing, *Anim. Genet.*, 19, 103, 1988.
39. Watkins, D. I., Shadduck, J. A., Stone, M. E., Lewin, H. A., and Letvin, N. L., Isoelectric focusing of bovine major histocompatibility complex class I molecules, *J. Immunogenet.*, 1683, 233, 1989.
40. Toye, P. G., MacHugh, N. D., Bensaid, A. M., Alberti, S., Teale, A. J., and Morrison, W. I., Transfection into mouse L cells of genes encoding two serologically and functionally distinct bovine class I Mhc molecules from a Mhc-homozygous animal: evidence for a second class I locus in cattle, *Immunology*, 70, 20, 1990.
41. Joosten, I., Teale, A. J., van der Poel, A., and Hensen, E. J., Biochemical evidence of the expression of two major histocompatibility complex class I genes on bovine peripheral blood mononuclear cells, *Anim. Genet.*, 23, 113, 1992.
42. Garber, T. L., Hughes, A. L., Letvin, N. L., Templeton, J. W., and Watkins, D. I., Sequence and evolution of cattle Mhc class I cDNAs: concerted evolution has not taken place in cattle, *Immunogenetics*, 38, 11, 1993.
43. Al-Murrani, S. W. K., Glass, E. J., Williams, J. L., and Oliver, R. A., 1D-IEF analysis of BoLA class I expression using allo-antisera reveals additional complexity, *Anim. Genet.*, 24, 427, 1993.
44. Newman, M. J. and Hines, H. C., Production of foetally stimulated lymphocytotoxic antibodies by primiparous cows, *Anim. Blood Grps. Biochem. Genet.*, 10, 87, 1979.
45. Hines, H. C. and Newman, M. J., Production of foetally stimulated lymphocytotoxic antibodies by multiparous cows, *Anim. Blood Grps. Biochem. Genet.*, 12, 201, 1981.
46. Anonymous, Proceedings of the Second International Bovine Lymphocyte Antigen (BoLA) Workshop, *Anim. Blood Grps. Biochem. Genet.*, 13, 33, 1982.
47. Bull, R. W., Lewin, H. A., Wu, M. C., Peterbaugh, K., Antczak, D., Bernoco, D., Cwik, S., Dam, L., Davies, C., Dawkins, R. L., Dufty, J. H., Gerlach, J., Hines, H. C., Lazary, S., Leibold, W., Leveziel, H., Lie, Ø., Lindberg, P. G., Meggiolaro, D., Meyer, E., Oliver, R., Ross, M., Simon, M., Spooner, R. L., Stear, M. J., Teale, A. J., and Templeton, J. W., Joint report of the Third International Bovine Lymphocyte Antigen (BoLA) Workshop, Helsinki, Finland, 27 July 1986, *Anim. Genet.*, 20, 109, 1989.
48. Bernoco, D., Lewin, H. A., Andersson, L., Arriens, M. A., Byrns, G., Cwik, S., Davies, C. J., Hines, H. C., Leibold, W., Lie, Ø., Meggiolaro, D., Oliver, R., Østergård, H., Spooner, R. L., Stewart, J. A., Teale, A. J., Templeton, J. W., and Zanotti, M., Joint report of the Fourth International Bovine Lymphocyte Antigen (BoLA) Workshop, East Lansing, Michigan, U.S.A., August 25, 1990, *Anim. Genet.*, 22, 477, 1991.
49. Davies, C. J., Joosten, I., Bernoco, D., Arriens, M. A., Bester, J., Ceriotti, G., Ellis, S., Hensen, E. J., Hines, H. C., Horin, P., Kristensen, B., Lewin, H. A., Meggiolaro, D., Morgan, A. L. G., Morita, M., Nilsson, P. R., Oliver, R. A., Orlova, A., Østergård, H., Park, C. A., Schuberth, H.-J., Simon, M., Spooner, R. L., and Stewart, J. A., Polymorphism of bovine Mhc class I genes. Joint report of the Fifth International Bovine Lymphocyte Antigen (BoLA) Workshop, Interlaken, Switzerland, 1 August 1992, *Eur. J. Immunogenet.*, 21, 239, 1994.

50. Stear, M. J., Pokorny, T. S., Muggli, N. E., and Stone, R. T., Breed differences in the distribution of BoLA-A locus antigens in American cattle, *Anim. Genet.*, 19, 171, 1988.
51. Andersson, L. and Davies, C. J., The major histocompatibility complex, in *Cell-Mediated Immunity in Ruminants*, Goddeeris, B. M. and Morrison, W. I., Eds., CRC Press, Boca Raton, FL, 1994, 37.
52. Oliver, R. A., Brown, P., Spooner, R. L., Joosten, I., and Williams, J. L., The analysis of antigen and DNA polymorphism within the bovine major histocompatibility complex: I. The class I antigens, *Anim. Genet.*, 20, 31, 1989.
53. Brown, P., Spooner, R. L., and Clark, A. J., Cloning and characterization of a *BoLA* class I cDNA clone, *Immunogenetics*, 29, 58, 1989.
54. Xu, A., Clarke, T. J., Teutsch, M. R., Schook, L. B., and Lewin, H. A., Sequencing and genetic analysis of a bovine DQB cDNA clone, *Anim. Genet.*, 22, 381, 1991.
55. Bjorkman, P. J., Saper, M. A., Samraoui, B., Bennett, W. S., Strominger, J. L., and Wiley, D. C., The foreign antigen-binding site and T cell recognition regions of class I histocompatibility antigens, *Nature*, 329, 512, 1987.
56. Spooner, R. L. and Morgan, A. L. G., Analysis of BoLA w6. Evidence for multiple subgroups, *Tissue Antigens*, 17, 178, 1981.
57. Viuff, B., Østergård, H., Aasted, B., and Kristensen, B., One-dimensional isoelectric focusing and immunoblotting of bovine major histocompatibility complex (BoLA) class I molecules and correlation with class I serology, *Anim. Genet.*, 22, 147, 1991.
58. Skow, L. C., Goy, J., and Honeycutt, D., Dinucleotide repeat polymorphism near a bovine Mhc class I sequence, *Anim. Genet.*, 25, 290, 1994.
59. Harms, J. S. and Splitter, G. A., Impairment of Mhc class I transcription in a mutant bovine B cell line, *Immunogenetics*, 35, 1, 1992.
60. Harms, J. S. and Splitter, G. A., The enhancer A/IRS region of a cattle Mhc class I promoter, *Immunogenetics*, 39, 372, 1994.
61. Li, W. and O'Reilly, K. L., Bovine major histocompatibility complex class I specific monoclonal antibodies characterized by flow cytometry, one- and two-dimensional electrophoresis, Western blot and inhibition of cytotoxic T lymphocyte function, *Vet. Immunol. Immunopathol.*, 33, 309, 1992.
62. Morrison, W. I., Goddeeris, B. M., Teale, A. J., Groocock, C. M., Kemp, S. J., and Stagg, D. A., Cytotoxic T-cells elicited in cattle challenged with *Theileria parva* (Muguga): evidence for restriction by class I Mhc determinants and parasite strain specificity, *Parasite Immunol.*, 9, 563, 1987.
63. Goddeeris, B. M., Morrison, W. I., and Teale, A. J., Generation of bovine cytotoxic cell lines, specific for cells infected with the protozoan parasite *Theileria parva* and restricted by products of the major histocompatibility complex, *Eur. J. Immunol.*, 16, 1243, 1986.
64. McKeever, D. J., Taracha, E. L. N., Innes, E. L., MacHugh, N. D., Awino, E., Goddeeris, B. M., and Morrison, W. I., Adoptive transfer of immunity to *Theileria parva* in the $CD8^+$ fraction of responding efferent lymph, *Proc. Natl. Acad. Sci. U.S.A.*, 91, 1959, 1994.
65. Cook, C. G. and Splitter, G. A., Comparison of bovine mononuclear cells with other species for cytolytic activity against virally-infected cells, *Vet. Immunol. Immunopathol.*, 20, 239, 1989.
66. Hein, W. R. and Mackay, C. R., Prominence of $\gamma\delta$ T cells in the ruminant immune system, *Immunol. Today*, 12, 30, 1991.
67. Pamer, E. G., Bevan, M. J., and Fischer Lindahl, K., Do nonclassical, class Ib Mhc molecules present bacterial antigens to T cells? *Trends Microbiol.*, 1, 35, 1993.

68. Gaskins, H. R., Immunological aspects of host/microbiota interactions at the intestinal epithelium, in *Gastrointestinal Microbial and Host Interactions*, Mackie, R. I., Isaacson, R. E., and White, B. A., Eds., Chapman and Hall, in press.
69. Lewin, H. A., Calvert, C. C., and Bernoco, D., Cross reactivity of a monoclonal antibody (H4) with bovine, equine, ovine and porcine peripheral blood B lymphocytes, *Am. J. Vet. Res.*, 46, 785, 1985.
70. Lewin, H. A., Davis, W. C., and Bernoco, D., Monoclonal antibodies that distinguish bovine T and B lymphocytes, *Vet. Immunol. Immunopathol.*, 9, 87, 1985.
71. Lewin, H. A., Nolan, T. J., and Schook, L. B., Altered expression of class II antigens on peripheral blood B lymphocytes from BLV-infected cows with persistent lymphocytosis, in *Antigen Presenting Cells: Diversity, Differentiation, and Regulation*, Schook, L. B. and Tew, J. G., Eds., Alan R. Liss, New York, 1987, 211.
72. Janzer-Pfeil, A. M. and Splitter, G. A., Identification of two independent Mhc class II antigens in a bovine lymphoblastoid cell line, *Vet. Immunol. Immunopathol.*, 22, 307, 1989.
73. Bissumbhar, B., Nilsson, P. R., Hensen, E. J., Davis, W. C., and Joosten, I., Biochemical characterization of bovine Mhc DQ allelic variants by one-dimensional isoelectric focusing, *Tissue Antigens*, 44, 100, 1994.
74. Ababou, A., Goyeneche, J., Davis, W. C., and Lévy, D., Evidence of the expression of three different BoLA-class II molecules on the bovine BL-3 cell line: determination of a non-DR non-DQ gene product, *J. Leukocyte Biol.*, 56, 182, 1994.
75. Davis, W. D., Marusic, S., Lewin, H. A., Splitter, G. A., Perryman, L. E., McGuire, T. C., and Gorham, J. R., The development and analysis of species specific and cross reactive monoclonal antibodies to leukocyte differentiation antigens and antigens of the major histocompatibility complex for use in the study of the immune system in cattle and other species, *Vet. Immunol. Immunopathol.*, 15, 337, 1987.
76. Watkins, D. I., Shadduck, J. A., Rudd, C. E., Stone, M. E., Lewin, H. A., and Letvin, N. L., Isoelectric focusing of bovine major histocompatibility complex class II molecules, *Eur. J. Immunol.*, 19, 567, 1989.
77. Joosten, I., Sanders, M. F., van der Poel, A., Williams, J. L., Hepkema, B. G., and Hensen, E. J., Biochemically defined polymorphism of bovine Mhc class II antigens, *Immunogenetics*, 29, 213, 1989.
78. Lalor, P. A., Morrison, W. I., Goddeeris, B. M., Jack, R. M., and Black, S. J., Monoclonal antibodies identify phenotypically and functionally distinct cell types in the bovine lymphoid system, *Vet. Immunol. Immunopathol.*, 13, 121, 1986.
79. Taylor, B. C., Choi, K. Y., Scibienski, R. J., Moore, P. F., and Stott, J. L., Differential expression of bovine Mhc class II antigens identified by monoclonal antibodies, *J. Leukocyte Biol.*, 53, 479, 1993.
80. Ohmann, H. B., Davis, W. C., and Babiuk, L. A., Surface antigen expression by bovine alveolar macrophages: functional correlation and influence of interferons *in vivo* and *in vitro*, *Immunobiology*, 171, 125, 1986.
81. Hughes, H. P. A., Campos, M., McDougall, L., Beskorwayne, T. K., Potter, A. A., and Babiuk, L. A., Regulation of major histocompatibility complex class II expression by *Pasteurella haemolytica* leukotoxin, *Infect. Immun.*, 62, 1609, 1994.
82. Fitzpatrick, J. L., Cripps, P. J., Hill, A. W., Bland, P. W., and Stokes, C. R., Mhc class II expression in the bovine mammary gland, *Vet. Immunol. Immunopathol.*, 32, 13, 1992.
83. Spurzem, J. R., Sacco, O., Rossi, G. A., Beckmann, J. D., and Rennard, S. I., Regulation of major histocompatibility complex class II gene expression on bovine bronchial epithelial cells, *J. Lab. Clin. Med.*, 120, 94, 1992.

84. Burke, M. G., Stone, R. T., and Muggli-Cockett, N. E., Nucleotide sequences and Northern analysis of a bovine major histocompatibility class II DRβ-like cDNA, *Anim. Genet.*, 22, 343, 1991.
85. Stone, R. T. and Muggli-Cockett, N. E., *BoLA-DIB*: species distribution, linkage with *DOB*, and Northern analysis, *Anim. Genet.*, 24, 41, 1993.
86. Xu, A., McKenna, K., and Lewin, H. A., Sequencing and genetic analysis of a bovine *DQA* cDNA clone, *Immunogenetics*, 37, 231, 1993.
87. Xu, A., Park, C., and Lewin, H. A., Both *DQB* genes are expressed in *BoLA* haplotypes carrying a duplicated *DQ* region, *Immunogenetics*, 39, 316, 1994.
88. Usinger, W. R., Curie-Cohen, M., and Stone, W. H., Lymphocyte-defined loci in cattle, *Science*, 196, 1017, 1977.
89. Davies, C. J. and Antczak, D. F., Mixed lymphocyte culture studies reveal complexity in the bovine Mhc not detected by class I serology, *Anim. Genet.*, 22, 31, 1991.
90. Teale, A. J. and Kemp, S. J., A study of BoLA class II antigens with BoT4$^+$ T lymphocyte clones, *Anim. Genet.*, 18, 17, 1987.
91. Glass, E. J., Oliver, R. A., Williams, J. L. W., and Millar, P., Alloreactive T-cell recognition of bovine major histocompatibility complex class II products defined by one-dimensional isoelectric focusing, *Anim. Genet.*, 23, 97, 1992.
92. Davies, C. J., Joosten, I., Andersson, L., Arriens, M. A., Bernoco, D., Bissumbhar, B., Byrns, G., van Eijk, M. J. T., Kristensen, B., Lewin, H. A., Mikko, S., Morgan, A. L. G., Muggli-Cockett, N. E., Nilsson, P. R., Oliver, R. A., Park, C. A., van der Poel, J. J., Polli, M., Spooner, R. L., and Stewart, J. A., Polymorphism of bovine Mhc class II genes. Joint report of the Fifth International Bovine Lymphocyte Antigen (BoLA) Workshop, Interlaken, Switzerland, 1 August 1992, *Eur. J. Immunogenet.*, 21, 259, 1994.
93. Davies, C. J., Andersson, L., Joosten, I., Mariani, P., Gasbarre, L. C., and Hensen, E. J., Characterization of bovine Mhc class II polymorphism using three typing methods: serology, RFLP and IEF, *Eur. J. Immunogenet.*, 19, 253, 1992.
94. Arriens, M. A., Hesford, F., Ruff, G., and Lazary, S., Production of alloantibodies against bovine B-lymphocyte antigens, *Anim. Genet.*, 22, 399, 1991.
95. Davies, C. J. and Antczak, D. F., Production and characterization of alloantisera specific for bovine class II major histocompatibility complex antigens, *Anim. Genet.*, 22, 417, 1991.
96. Williams, J. L., Oliver, R. A., Morgan, A. L. G., Glass, E. J., and Spooner, R. L., Production of alloantisera against class II bovine lymphocyte antigens (BoLA) by cross-immunization between class I matched cattle, *Anim. Genet.*, 22, 407, 1991.
97. Nilsson, P. R., Schreuder, G. M. T., Damaro, J., Joosten, I., Buis, R. C., Davies, C. J., and van der Poel, J. J., Serological definition of bovine Mhc class-II polymorphism in Holstein-Friesians, *Tissue Antigens*, 43, 229, 1994.
98. Lewin, H. A., Arnet, E. F., Lie, O., Bernoco, D., Davies, C. J., Nilsson, P. R., Ruff, G., and Lazary, S., First international comparison test of serological reagents for typing class II products of the bovine major histocompatibility complex. XXII International Conference on Animal Genetics, *Anim. Genet.*, 22 (Suppl. 1), 46, 1991 (Abstract).
99. Nilsson, P. H. R., van der Poel, J. J., van het Klooster, J. W., Davies, C. J., Joosten, I., Williams, J. L., and Arriens, M. A., Comparison of BoLA class II serotyping and 1D-IEF in a Dutch herd of Holstein-Friesians, *Anim. Genet.*, 22, 47, 1991 (Abstract).
100. Nilsson, P. R., van het Klooster, J. W., van der Poel, J. J., Davis, W. C., and Schreuder, G. M. T., Complexity of the bovine Mhc class-II specificity DW3 as defined by alloantisera, *Eur. J. Immunogenet.*, 21, 169, 1994.

101. Joosten, I., Hensen, E. J., Sanders, M. F., and Andersson, L., Bovine Mhc class II restriction fragment length polymorphism linked to expressed polymorphism, *Immunogenetics*, 31, 123, 1990.
102. Sigurdardóttir, S., Lundén, A., and Andersson, L., Restriction fragment length polymorphism of DQ and DR class II genes of the bovine major histocompatibility complex, *Anim. Genet.*, 19, 133, 1988.
103. van der Poel, J. J., Groenen, M. A. M., Dijkhoff, R. J. M., Ruyter, D., and Giphart, M. J., The nucleotide sequence of the bovine Mhc class II alpha genes: *DRA, DQA,* and *DYA*, *Immunogenetics*, 31, 29, 1990.
104. Teutsch, M. R., Stewart, J. A., Beever, J. E., Schook, L. B., and Lewin, H. A., Haplotypes of the *BoLA-A, Bf, CYP21* and *DQB* loci in Angus cattle, *Anim. Biotechnol.*, 1, 185, 1990.
105. Sigurdardóttir, S., Borsch, C., Gustafsson, K., and Andersson, L., Gene duplications and sequence polymorphism of bovine class II *DQB* genes, *Immunogenetics*, 35, 205, 1992.
106. van Eijk, M. J. T., Russ, I., Beever, J. E., and Lewin, H. A., Polymorphism in exon 2 of the bovine lymphocyte antigen (BoLA)-DYA gene, *Anim. Genet.*, 23, 476, 1992.
107. van der Poel, J. J., Davies, C. J., van Noort, J., Groenen, M. A. M., and Giphart, M. J., Determination of nucleotide variation at the BoLA-DQA locus and the use of sequence specific oligo's for BoLA-DQA typing, *Anim. Genet.*, 23, 39, 1992 (Abstract).
108. Muggli-Cockett, N. E. and Stone, R. T., Partial nucleotide sequence of a bovine major histocompatibility class II DRβ-like gene, *Anim. Genet.*, 20, 361, 1989.
109. Xu, A., van Eijk, M. J. T., Park, C., and Lewin, H. A., Polymorphism in *BoLA-DRB3* exon 2 correlates with resistance to persistent lymphocytosis caused by bovine leukemia virus, *J. Immunol.*, 151, 6977, 1993.
110. Russell, G. C., Marello, K. L., Gallagher, A., McKeever, D. J., and Spooner, R. L., Amplification and sequencing of expressed *DRB* second exons from *Bos indicus*, *Immunogenetics*, 39, 432, 1994.
111. Stone, R. T. and Muggli-Cockett, N. E., Nucleotide sequence of the second exon of a *BoLA-DQB* gene, *Anim. Genet.*, 23, 273, 1992.
112. Stone, R. T. and Muggli-Cockett, N. E., Partial nucleotide sequence of a novel bovine major histocompatibility complex class β-chain, gene *BoLA-DIB*, *Anim. Genet.*, 21, 353, 1990.
113. Lewin, H. A., Schmitt, K., Hubert, R., van Eijk, M. J. T., and Arnheim, N., Close linkage between bovine prolactin and BoLA-DRB3 genes: genetic mapping in cattle by single sperm typing, *Genomics*, 13, 44, 1992.
114. Lewin, H. A., Spevak, E. M., and Paige, K. N., Genetic management of herd health and productivity: molecular strategies, in *Proc. North American Public Bison Herds Symposium*, Walker R. E., Ed., 1993, 38.
115. Russell, G. C., Improved single-strand conformation polymorphism analysis by asymmetric polymerase chain reaction with end-labeled primers, *GATA*, 11, 24, 1994.
116. Ellegren, H., Davies, C. J., and Andersson, L., Strong association between polymorphisms in an intronic microsatellite and in the coding sequence of the *BoLA-DRB3* gene: implications for microsatellite stability and PCR-based *DRB3* typing, *Anim. Genet.*, 24, 269, 1993.
117. Glass, E. J. and Spooner, R. L., Parasite-accessory cell interactions in Theileriosis. Antigen presentation by *Theileria annulata*-infected macrophages and production of continuously growing antigen-presenting cell lines, *Eur. J. Immunol.*, 20, 2491, 1990.
118. DeMartini, J. C., MacHugh, N. D., Naessens, J., and Teale, A. J., Differential *in vitro* and *in vivo* expression of Mhc class II antigens in bovine lymphocytes infected by *Theileria parva*, *Vet. Immunol. Immunopathol.*, 35, 253, 1993.

119. Ostrand-Rosenberg, S., Thakur, A., and Clements, V., Rejection of mouse sarcoma cells after transfection of Mhc class II genes, *J. Immunol.*, 144, 4068, 1990.
120. Glass, E. J. and Spooner, R. L., Requirement for Mhc class II positive accessory cells in an antigen specific bovine T cell response, *Res. Vet. Sci.*, 46, 196, 1989.
121. Glass, E. J., Oliver, R. A., and Spooner, R. L., Variation in T cell responses to ovalbumin in cattle: evidence for Ir gene control, *Anim. Genet.*, 21, 15, 1990.
122. Glass, E. J., Oliver, R. A., and Spooner, R. L., Bovine T cells recognize antigen in association with Mhc class II haplotypes defined by one-dimensional isoelectric focusing, *Immunology*, 72, 380, 1991.
123. Glass, E. J., Oliver, R. A., Collen, T., Doel, T. R., Dimarchi, R., and Spooner, R. L., Mhc class II restricted recognition of FMDV peptides by bovine T cells, *Immunology*, 74, 594, 1991.
124. Machugh, D. E., Loftus, R. T., Bradley, D. G., Sharp, P. M., and Cunningham, P., Microsatellite DNA variation within and among European cattle breeds, *Proc. R. Soc. Lond. B*, 256, 25, 1994.
125. O'Brien, S. J., Genetic and phylogenetic analyses of endangered species, *Annu. Rev. Genet.*, 28, 467, 1994.
126. van Eijk, M. J. T., Stewart-Haynes, J. A., Beever, J. E., Fernando, R. L., and Lewin, H. A., Development of persistent lymphocytosis in cattle is closely associated with *DRB2*, *Immunogenetics*, 37, 64, 1992.
127. Lewin, H. A., Disease resistance and immune reponse genes in cattle: strategies for their detection and evidence of their existence, *J. Dairy Sci.*, 72, 1334, 1989.
128. Østergård, H., Kristensen, B., and Andersen, S., Investigations in farm animals of associations between the Mhc system and disease resistance, *Livestock Prod. Sci.*, 22, 49, 1989.
129. Lewin, H. A., Clamp, P. A., Beever, J. E., and Schook, L. B., Mapping genes for resistance to infectious diseases, in *Gene Mapping: Techniques and Applications*, Schook L. B., Lewin H. A., and McLaren D. G., Eds., Marcel Dekker, Inc., New York, 1991, 283.
130. Lundén, A., Sigurdardóttir, S., Edfors-Lilja, I., Danell, B., Rendel, J., and Andersson, L., The relationship between bovine major histocompatibility complex class II polymorphism and disease studied by use of bull breeding values, *Anim. Genet.*, 21, 221, 1990.
131. Våge, D. I., Lingaas, F., Spooner, R. L., Arnet, E. F., and Lie, Ø., A study on association between mastitis and serologically defined class I bovine lymphocyte antigens (BoLA-A) in Norwegian cows, *Anim. Genet.*, 23, 533, 1992.
132. Våge, D. I., Olsaker, I., Lingaas, F., Spooner, R. L., Van Der Beek, S., Sørensen, A., Arnet, E. F., and Lie, Ø., High levels of linkage disequilibria between serologically defined class I bovine lymphocyte antigens (BoLA-A) and class II DQB restriction fragment length polymorphism (RFLP) in Norwegian cows, *Anim. Genet.*, 23, 125, 1992.
133. Weigel, K. A., Freeman, A. E., Kehrli, M. E., Jr., Stear, M. J., and Kelley, D. H., Association of class I bovine lymphocyte antigen complex alleles with health and production traits in dairy cattle, *J. Dairy Sci.*, 72, 2538, 1990.
134. Arriëns, M. A., Ruff, G., Schällibaum, M., and Lazary, S., Possible association between a serologically detected haplotype of the bovine major histocompatibility complex and subclinical mastitis, *J. Anim. Breed. Genet.*, 111, 152, 1994.
135. Oddgiersson, O., Simpson, S. P., Morgan, A. L. G., Ross, D. S., and Spooner, R. L., Relationship between the bovine major histocompatibility complex (BoLA), erythrocyte markers and susceptibility to mastitis in Icelandic cattle, *Anim. Genet.*, 19, 11, 1988.

136. Da, Y., Shanks, R. D., Stewart, J. A., and Lewin, H. A., Milk and fat yields decline in bovine leukemia virus-infected Holstein cattle with persistent lymphocytosis, *Proc. Natl. Acad. Sci. U.S.A.*, 90, 6538, 1993.
137. Lewin, H. A., Host genetic mechanism of resistance and susceptibility to a bovine retroviral infection, *Anim. Biotechnol.*, 5, 183, 1994.
138. Palmer, C., Thurmond, M., Picanso, J., Brewer, A. W., and Bernoco, D., Susceptibility of cattle to bovine leukemia virus infection associated with BoLA type, *Proc. 91st Ann. Meeting of U.S. Anim. Health Assoc.*, 218, 1987.
139. Lewin, H. A., Wu, M. C., Stewart, J. A., and Nolan, T. J., Association between *BoLA* and subclinical bovine leukemia virus infection in a herd of Holstein-Friesian cows, *Immunogenetics*, 27, 338, 1988.
140. Stear, M. J., Dimmock, C. K., Newman, M. J., and Nicholas, F. W., BoLA antigens are associated with increased frequency of persistent lymphocytosis in bovine leukaemia virus infected cattle and with increased incidence of antibodies to bovine leukemia virus, *Anim. Genet.*, 19, 151, 1988.
141. Solbu, H., Spooner, R. L., and Lie, Ø., A possible influence of the bovine major histocompatibility complex (BoLA) on mastitis, *Proceedings of the Second World Congress on Genetics Applied to Livestock Production*, 7, 368, 1982.
142. Solbu, H. and Lie, Ø., Selection for disease resistance in dairy cattle, *Proceedings of the Fourth World Congress on Genetics Applied to Livestock Production*, 16, 445, 1990.
143. Mejdell, C. M., Lie, Ø., Solbu, H., Arnet, E. F., and Spooner, R. L., Association of major histocompatibility complex antigens (BoLA-A) with AI bull progeny test results for mastitis, ketosis and fertility in Norwegian Cattle, *Anim. Genet.*, 25, 99, 1994.
144. Larsen, B., Jensen, N. E., Madsen, P., Nielsen, S. M., Klastrup, O., and Madsen, F. S., Association of the M blood group system with bovine mastitis, *Anim. Blood Grps. Biochem. Genet.*, 16, 165, 1985.
145. Stear, M. J., Hetzel, D. J. S., Brown, S. C., Gershwin, L. J., Mackinnon, M. J., and Nicholas, F. W., The relationships among ecto- and endoparasite levels, class I antigens of the bovine major histocompatibility system, immunoglobulin E levels and weight gain, *Vet. Parasitol.*, 34, 303, 1990.
146. Stear, M. J., Tierney, T. J., Baldock, F. C., Brown, S. C., Nicholas, F. W., and Rudder, T. H., Class I antigens of the bovine major histocompatibility system are weakly associated with variation in faecal worm egg counts in naturally infected cattle, *Anim. Genet.*, 19, 115, 1988.
147. Stear, M. J., Nicholas, F. W., Brown, S. C., and Holroyd, R. G., Class I antigens of the bovine major histocompatibility system and resistance to the cattle tick (*Boophilus microplus*) assessed in three different seasons, *Vet. Parasitol.*, 31, 305, 1989.
148. Park, C. A., Hines, H. C., Monke, D. R., and Threlfall, W. T., Association between the bovine major histocompatibility complex and chronic posterior spinal paresis — a form of ankylosing spondylitis — in Holstein bulls, *Anim. Genet.*, 24, 53, 1993.
149. Joosten, I., Sanders, M. F., and Hensen, E. J., Involvement of major histocompatibility complex class I compatibility between dam and calf in the aetiology of bovine retained placenta, *Anim. Genet.*, 22, 455, 1991.
150. Weigel, K. A., Kehrli, M. E., Jr., Freeman, A. E., Thurston, J. R., Stear, M. J., and Kelley, D. H., Association of class I bovine lymphocyte antigen complex alleles with *in vitro* blood neutrophil functions, lymphocyte blastogenesis, serum complement and conglutinin levels in dairy cattle, *Vet. Immunol. Immunopathol.*, 27, 321, 1991.
151. Lie, Ø., Solbu, H., Larsen, H. J., and Spooner, R. L., Possible association of antibody responses to human serum albumin and (T,G)-A — L with the bovine major histocompatibility complex (BoLA), *Vet. Immunol. Immunopathol.*, 11, 333, 1986.

152. Stear, M. J., Pokorny, T. S., Muggli, N. E., and Stone, R. T., The relationships of birth weight, preweaning gain and postweaning gain with the bovine major histocompatibility system, *J. Anim. Sci.*, 67, 641, 1989.
153. Beever, J. E., George, P. D., Fernando, R. L., Stormont, C. J., and Lewin, H. A., Associations between genetic markers and growth and carcass traits in a paternal half-sib family of Angus cattle, *J. Anim. Sci.*, 68, 337, 1990.
154. Batra, T. R., Lee, A. J., and Gavora, J. S., Class I alleles of the bovine major histocompatibility system and their association with economic traits, *J. Dairy Sci.*, 72, 2115, 1989.
155. Hines, H. C., Allaire, F. R., and Michalak, M. M., Association of bovine lymphocyte antigens with milk fat percentage differences, *J. Dairy Sci.*, 69, 3148, 1986.
156. Stear, M. J., Pokorny, T. S., Echternkamp, S. E., and Lunstra, D. D., The influence of the *BoLA-A* locus on reproductive traits in cattle, *J. Immunogenet.*, 16, 77, 1989.
157. Lewin, H. A., Beever, J. E., Da, Y., Hines, H. C., and Faulkner, D. B., The bovine B and C blood group systems are not likely to be the orthologues of human RH: an interesting twist in the comparative map, *Anim. Genet.*, 25, 13, 1994.
158. Zonotti, M., Poli, G., Ponti, W., Polli, M., Rocchi, M., Bolzani, E., Longeri, M., Russo, S., Lewin, H. A., and van Eijk, M. J. T., Association of BoLA class II haplotypes with subclinical progression of bovine leukemia virus in Holstein-Friesian cattle, *Anim. Genet.*, in press.

CHAPTER 5

The Caprine Major Histocompatibility Complex

Gabriela Obexer-Ruff, Irma Joosten, and Franz-Werner Schwaiger

CONTENTS

0-8493-8032-4/96/$0.00+$.50

I. INTRODUCTION

The goat (*Capra hircus*) represents an important socio-economic impact in developing countries. In many of these countries its products (meat, milk, fibers) are the main income basis of the local population. The goat is also very well adapted to various extreme environmental and nutritional conditions (climate, food resources, etc.). Furthermore, it is a relatively long-living, multiparous animal with a short generation interval, though often used in various experimental medical studies (serum production, etc.) for different industrial and scientific purposes. In intensive breeding areas goats are kept more individually with breeding and disease records available for the single animal. This enables various studies of different parameters influencing production and fitness traits of this species. The caprine major histocompatibility complex (Mhc) is one of these genetic systems with a major influence on the individual immunocompetence and therefore of primary interest for such investigations in the goat.

The characterization of the caprine Mhc has a relatively young history with a limited number of groups working on it. The first studies on this system have been carried out by the Dutch group of van Dam[1-5] in 1976, who investigated the structure and function of the caprine Mhc genes and gene products in the experimental dwarf goat herd of the University of Utrecht.

In the year 1982, the group of Ruff and Lazary of Berne, Switzerland, initiated a project on the study of the caprine Mhc in Swiss goat breeds with special emphasis on its influence on caprine arthritis-encephalitis (CAE) disease manifestation. The following year, collaboration with the Norwegian group of Larsen and Nesse was taken up to establish a typing system for the local goat breed and a possible comparison of the detected gene products in both countries. In 1988, the first DNA-based investigations on the goat Mhc were also performed in a collaborative project among Nesse et al.[6]

The work carried out initially by van Dam was continued by Hensen and Joosten, who started to define the caprine Mhc system by biochemical means in 1989. Epplen and Schwaiger took up collaboration with the Swiss researchers in 1990 to characterize the system by DNA-based techniques with a special interest in evolutionary aspects across the species barrier.

Matheron et al., working on a disease resistance project of cowdriosis[7] in Creole goats established a collaboration in 1990 with the Swiss laboratory to investigate the role of the Mhc in this tropical caprine disease model.

Finally, Cameron et al.[8] first reported physical mapping data of the caprine Mhc-region in 1990 as well as molecular genetics analysis of a part of the class II region in 1993.[9]

II. RECOGNITION OF THE CAPRINE Mhc COMPLEX

The caprine Mhc complex is called the caprine leukocyte antigen (CLA) system according to the nomenclature used for the human complex and other farm animal species. Its products were named following the rules of the human analogs.

A. Serological Determination of CLA-Antigens

1. Detection of Class I Gene Products

To detect the gene products of the class I type of the caprine Mhc by serological means, alloantisera were produced using different techniques. Van Dam et al.[1-5] described the production of a total of 93 reagents by cross-immunization procedure of related and unrelated animals as well as by the skin graft method. Nesse and Ruff [10-13] reported the production of 363 alloantisera by means of cross-immunization procedures, the collection of sera from primiparous goats 2 to 3 weeks post-partum, as well as by purification of colostrum to obtain immunoglobulins.

No official, internationally organized workshop has been organized for comparison of the obtained goat alloantisera. Unofficially, characterization and nomenclature of the detected specificities was agreed upon in a collaborative study[14] between the Norwegian and the Swiss group. The local as well as the European designations of the CLAs detected are given in the following relevant sections.

The primary investigations of the caprine Mhc carried out by van Dam described a two-loci model with nine serologically defined antigens of the class I type (SD1-9), whereby he regarded only three of the first locus products as truly reliable. He named the characterized products goat lymphocyte or GLA-antigens.[4,5] Using transplantation studies he could demonstrate that skin grafts in Mhc-matched donors resulted in prolonged graft survival indicating that the detected GLA-antigens were coded for by the Mhc genes.[2]

The serological characterization of the caprine Mhc in Norwegian goats was combined with an already existing project studying immune response to diphtheria toxoid in selected lines of the faculty herd. The Norwegian investigations resulted in the definition of 15 CLA antigen specificities (N1-N15) belonging to to at least two closely linked loci. The specificity N15 was not coded by genes located in the same chromosomal region.[12]

We[13] are presently able to define 31 different membrane structures expressed on goat lymphocytes by serological means (Be1-Be30 and CLY). Of these CLAs, 27 are confined to 1 locus, 1 belongs to a second closely linked locus (Be19), and 3 (Be5, Be16, and CLY) are membrane antigens not coded by the Mhc region. The nature of these products has been demonstrated to be of class I type by cosegregation of products defined by IEF techniques[15] and the fact that they did not influence mixed lymphocyte reactions (MLR).[13]

In a comparison test between Norwegian and Swiss serum reagents, 12 European lymphocyte antigen specificities (Eu1-12), assigned to two closely linked loci as well as two non-Mhc products (CLY1.1 and 2.1), were defined.[14] The gene frequencies of the 12 designated Eu-specificities calculated for 500 Norwegian and 914 Swiss dairy goats are given in Table 1.

Of the 15 characterized Norwegian specificities, 2 (N2 and N3) did not occur in the Swiss breeds, whereas a total of 14 further local Swiss specificities were not detected in the Norwegian dairy breed. Three recently detected Swiss specificities (Be28 to 30) were not included in the comparison test.

Using the Swiss and Norwegian reagents, typing of exotic breeds such as Angoras and African dwarf goats is possible. In total, 13 of the Swiss CLA antigens could also be detected in the Creole goats typed for the cowdriosis disease association studies.[16] An additional local specificity (Gu1) characterized by a split of the Swiss specificities Be7 and Be10 could be found for the Creole goats (Ruff et al., personal communications). All known non-Mhc gene products occurred in the Creole goats.

In a comparison test with W. C. Davis from the Washington State University (WSU) of Pullman, anti-H2 mouse mAbs kindly provided by Drs. H. Stimpfling and K. J. Hasenkrug et al.[17] (McLaughlin Research Institute, Great Falls, MT) as well as anti-caprine mAbs produced at the WSU were screened on Swiss goat cells. The following results could be obtained (Ruff and Davis, personal communications): 160 mAbs (20 WSU, 140 McLaughlin Research Institute) recognizing class I antigens were screened and compared to the results obtained by the Swiss alloantisera. Out of the 20 WSU mAbs, 9 showed a discriminative reaction pattern reacting only with some of the 40 animals tested, whereas 11 recognized conserved parts of the class I molecules, reacting with all animals. The test animals have been selected for their different class I specificities detected by the Swiss alloantisera. All 30 different known CLA antigens were represented.

Of the 140 mAbs directed against murine class I products, 15 crossreacted with the 40 typed animals. One was reacting with more than 90% of all the screened animals; the others showed a discriminative reaction pattern. Cosegregation (Fisher's exact test $p > 10^{-4}$) with our local specificities Be2 (mAbs 205 and 270), Be4 (mAbs 4, 12, 137, 248, 262, 288), Be7 (mAb 249), and Be10 (4 and 14) could be observed, indicating conservation of specific determinants across the species barrier. Furthermore, we found selective reactivity patterns of the mAbs when screening with sheep and horse cells (Ruff et al., in preparation).

2. *Detection of Class II Gene Products*

Van Dam et al.[3] defined a class II locus with an unknown number of alleles by means of the mixed lymphocyte culture (MLC) technique.

To produce alloantisera recognizing class II antigens in the Swiss goats, animals were matched for their class I types and selected for immunization when positive

Table 1 Gene Frequency Comparison Test of the 12 Designated European (Eu) Lymphocyte Antigen Specificities Occurring Simultaneously in the Swiss (local designition: Be) and Norwegian (local designition: N) Dairy Goat Breeds

CLA Class I specificities	Saanen N = 370	Chamois colored N = 210	Toggenburg N = 192	Appenzell N = 142	Norwegian dairy goat N = 500
Eu1 (= Be25, N7)	Not detected	0.005	Not detected	0.040	0.021
Eu2 (= Be2, N8)	0.203	0.007	0.183	0.014	0.191
Eu3 (= Be7, N9)	0.136	0.041	Not detected	0.029	0.080
Eu4 (= Be8, N10)	0.033	0.026	0.024	0.073	0.069
Eu5 (= Be11, N14)	0.003	0.057	0.037	Not detected	0.021
Eu6 (= Be13, N11)	Not detected	0.074	0.016	0.077	0.049
Eu7 (= Be15)	0.011	0.026	0.016	0.007	0.006
Eu8 (= Be1, N6)	0.135	0.130	0.029	0.293	0.082
Eu9 (= Be19, N13)	0.040	0.046	0.037	0.073	0.152
Eu10 (= Be22, N4)	0.022	0.005	Not detected	Not detected	0.044
Eu11 (= Be24, N5)	Not detected	0.034	0.065	Not detected	0.015
Eu12 (= Be27, N1)	0.002	0.085	0.008	Not detected	0.133
Non-Mhc					
GLY-1.1 (CLY, N15)	0.049	0.068	0.144	0.004	0.264
GLY2.1 (Be16, N12)	0.031	0.019	0.131	0.004	0.003

in MLRs. These immunizations resulted in alloantisera, which recognized six allelic membrane structures expressed only on B lymphocytes.[18] The serologically defined class II products showed cosegregation with six of the seven DR-like products detected with IEF techniques. The latter were assigned to the cells typed in a comparison test between the Dutch and the Swiss laboratories on the basis of our CLA class I serology data. With agreement on nomenclature between the two groups, the six caprine B cell specificities were designated BeB1-BeB6.[15,18]

In addition, 47 mAbs directed against caprine (N = 7) and murine (N = 40) class II products also provided by Dr. W. Davis of the WSU, Pullman, were screened on Swiss goat cells. A comparison to the Swiss reagents has been carried out in our laboratory by separate typing of B and T lymphocytes of 40 animals. Two of the WSU reagents showed cosegregation with one of the serologically defined class II products.[18] The remaining monoclonals seemed to detect structures showing different segregation pattern of the Mhc. The majority of the mAbs reacted with all the animals tested recognizing conserved parts of the class II molecules. Several of them have been described elsewhere by Davis et al.[19]

a. Comparison of Serologically Detected Class II Specificities and Products Detected by Oligotyping

In a comparison between serology and DNA-based techniques, 34 goats were typed by the two methods. Cosegregation for all the six serologically detected specificities with one of the products defined by oligotyping could be found. The oligotyping method enabled a finer analysis with 14 products found[20] compared to the six serologically[18] detected products generating in a more powerful tool for various association studies (Ruff and Schwaiger, personal communications).

3. Linkage Analysis of CLA Class I and II Gene Products

In studying linkage analysis between the caprine class I and II gene products, one offspring showing a crossing over between the parental class I and class II specificities could be found in a total of 78 informative offspring investigated. The recombinant animal was also confirmed by DNA analysis.[18] In the same study, the presence of linkage disequilibrium between the observed haplotype combinations in 90 typed unrelated offspring resulted in five defined gametic associations with a positive delta value indicating the same phenomenon as observed in other species.

4. Detection of Class III Gene Products

To clarify whether class III antigens which influence complement levels are also located in the goat Mhc region, van Dam et al.[4] looked for associations between GLA-specificities and complement levels. He found a sex linkage where females had significant lower complement levels in combination with a decreased level of animals expressing the antigens SD1, SD2, and SD4.

B. Biochemical Typing of Caprine Class I and Class II Products

Characterization of Mhc antigens can best be achieved by a combination of techniques, from which each single technique should add to the more precise definition of the polymorphism. From the begining, serology has been the main method in use for the typing of caprine lymphocyte antigen (CLA) products. The class I specificities have been assigned to two closely linked loci. However, at the time there was hardly any information on the polymorphism of class II products, and the serological class I typing left a number of questions unaccounted for, notably those with regard to disease association and immune response. Since recent studies provided the necessary evidence for the presence of multiple polymorphic CLA class I and class II genes, it was a logical step to try to optimize and develop additional typing methods for both class I and class II products.

Therefore, we started out by using biochemical methods to add to the more precise definition of CLA class I products, looking for the presence of biochemical subtypes of serologically defined specificities and further evidence of the expression of second locus products. At the same time we used biochemical techniques to provide a framework for the typing of class II products. In other livestock species this approach had already proven its worth.

We studied a panel of Swiss goats of the Saanen breed, compromising dam-offspring combinations from six half-sib sire families and several unrelated animals. All animals were typed for their class I products using the standard serological techniques. Peripheral blood lymphocytes of these animals were ^{35}S-methionine labeled and lysed using TX-114 or NP40 lysis buffer. Class I and class II antigens were precipitated with the mAb B1.1G6, HC10, TH16B, and a polyclonal rabbit anti-human class II antiserum. The antisera are cross-species reactive and are directed against monomorphic determinants on either class I or class II antigens. Immunoprecipitates were analyzed by SDS-PAGE and 1D-IEF (isoelectric focusing).

With the help of segregation analysis and serological class I typing it was possible to establish 18 class I haplotype-specific banding patterns. Direct comparison with serological data showed a good correlation between the two methods: with a few exceptions it was possible to assign characteristic banding patterns to each serologically defined haplotype. For each haplotype the banding pattern comprised two to four bands, a phenomenon that is also seen in other species, e.g., man, mouse, and cattle; it can be explained by the fact that there can be additional locus products expressed which cannot yet be distinguished by the typing methods available and that single products can yield a number of bands representing posttranslational modifications of that particular product. B1.1G6 precipitates class I heavy chains only when complexed to beta-2 microglobulin, and thus the reactivity of the mAb for the different allelic products is dependent on the degree to which complex formation takes place. In general this mAb appears to give very satisfying results with regard to CLA class I typing. The use of the mAb HC10 which recognizes denatured noncomplexed class I heavy chains and thus also additional biosynthetic intermediates revealed extra bands or indicated a higher affinity of the mAb for particular variants or modifications thereof as compared with B1.1G6, thus facilitating the definition of certain haplotypes.

For class II, using a polyclonal anti-human class II antiserum, we could establish seven class II haplotypes. Six of these could be distinguished by five recently defined serological clusters and a single serum. The apparent similarity in focusing pattern of the main products precipitated with the pattern obtained with this antiserum in cattle as well as in humans made it likely that these products are of DR nature. Furthermore, using an anti-class II mAb, TH16B, the presence of at least two expressed loci was suggested, based on differences in molecular weight between the products precipitated by TH16B and the polyclonal antiserum. As a consequence of the relative ease with which class II haplotypes could now be distinguished, the IEF method was particularly useful again in the selection of immunization pairs for the production of class II alloantisera.

Summarizing, the added value of the IEF technique lies first in the possibility to perform selective immunizations to fill in class I serological blanks, as in the case of the Be28-Be30 specificities, and to produce CLA class II-specific antisera. Furthermore, the technique enabled the detection of subtypes of serologically defined specificities as observed for the Be17 serotype. This is of particular importance for functional studies. In general, IEF is valuable for the detection of multiple locus products, but benefits from serological support. A combined use of IEF and serology in developing immunization schemes for the purpose of generating locus-specific antisera could well speed up the full characterization of caprine Mhc class I and II products.

C. DNA-Based Analysis of the Goat Mhc

1. *Mapping of the Goat Mhc*

Only a few attempts were made to characterize the genomic goat Mhc region, which is in the following denominated *Caae* (from *Capra aegagrus hircus*, according to the suggestions of Klein et al.[21]). Cameron et al.[8] investigated the polymorphism and the number of genes in the *Caae* region of 20 goat from four breeds (Saanen, British Alpine, Anglo-Nubian, and Toggenburg) and presented a physical map of class I, class II, and class III regions. They found conservation of the location of class III genes between class I and class II as was expected. A high degree of polymorphism was described for RFLPs of the class I and class II region. They suggest up to 10 to 13 class I genes which is more comparable to mouse[22] than to human. Unfortunately they did not say anything about the number of class II genes. As in human and mouse the region haboring the C4 and CYP21 genes seem to be duplicated but, as Cameron et al.[8] suggest, it is likely that the two copies of the C4 and CYP21 region are unlike all other investigated species orientated in a head-to-tail fashion (as is now also supposed for sheep by Groth and Wetherall; submitted). In the class III region only the C4 and C2 genes revealed limited polymorphism. A comparative map of the Mhc regions for goat, mouse, and human was presented by Cameron et al.[8] where class II could not be linked to class III or class I.

2. *The* DYA *Gene*

RFLP studies of cattle DNA revealed two class II genes presumably unique to the family Bovidae.[23] These genes, designated *DYA* and *DYB* (*DIB*) were shown to cosegregate with the *DOB* gene separated by a recombination fraction of 17 cM from the DQ/DR and the class III region. Additionally, Wright et al.[24,25] described a cosmid clone containing an Mhc class II A and B gene pair that proved to have high sequence similarity to the known *DYA* and *DIB* genes. Sequencing of PCR products using primers derived from the *Bota-DYA*[26] (Acc. Nos. m30119 and m30118) uncovered the third exon of a similar gene in the goat[9] (Acc. No. m94325). A high degree of homology to other species (96% amino acid identity in exon 3 between goat and sheep) and no observed polymorphism between the *DYA* sequences of three goats from the Anglo-Nubian breed suggest a strong sequence conservation. RFLP studies revealed that *Caae-DYA* and *Caae-DRB* are located in the same 250- or 340-kb fragment after *Not*I digestion.

The cloning of a potential B gene partner for the *DYA* gene, the *DYB*, or *DIB*, was reported.[27] It was shown to be a single copy gene of limited polymorphism distinct from other known bovine class II genes with a species distribution of restricted to Cervidae, Giraffidae and Bovidae.[28]

3. *The* DRB *Region of the Goat*

In human, 8 *DRB* genes are known within the *HLA-DRB* region, with four of them being pseudogenes.[29] There are six clearly defined haplotypes which harbor different pseudogenes and numbers of expressed loci (one or two). No observations about the number of *DRB* loci in goats were performed yet. As is described for cattle and sheep a single expressed *DRB* locus is suggested to be present in sheep.[30]

a. The Caae-DRB1 *Locus*

Twenty-two *Caae-DRB* exon 2 alleles amplified from genomic or cDNA derived from five different goat breeds were cloned and sequenced[31] (Table 2). The sequences were cloned and analyzed using a PCR system amplifying *DRB* exon 2[20,31] (Figure 1) and part of the second intron including a simple tandemly arrayed dinucleotide repeat. Electroblot hybridizations of genomically amplified PCR products of several halfsib families (Saanen breed) showed that only one *DRB* locus was amplified. Similarity of the *DRB* sequences presented in Table 2 suggests that these are all members of this locus. On the other hand, from about 3% of the investigated animals three different PCR products representing *DRB* sequences (identified by length differences and specific hybridization) were obtained, two of them shown to be expressed in an individual of the breed "Bunte Edelziege". Therefore, we suggest a single expressed *DRB* locus in goats. In analogy to the polymorphic expressed *DRB* locus of human (*HLA-DRB1*), we have called the expressed *DRB* locus of the goat *Caae-DRB1*.

Table 2 Translated Exons 2 of *Caae-DRB* Genes Compared to Consensus *HLA-DRB*

Amino acid #		11	21	31	41	51	61	71	81	91	
#Ind/Breed											
HLA-DRB Cons	EQ	SKSECHFFNG	TERVRFLDRY	FYNQEEYVRF	DSDVGEYRAV	TELGRPDAEY	WNSQKDLLEQ	RRAAVDTYCR	HNYGVGESFT	VQRR	
*Caae-DRB1*0801*	-y	y-r-------	----------	-h-g--f---	---w--f---	----------	------f--s	--t-------	y---------	----	2/z,a
*Caae-DRB1*0802*	-y	y-r-------	----------	yt-g--tl--	---w--f---	----------	------f--s	--t-------	y---------	----	3/z,e
*Caae-DRB1*0803*	-y	c-r-------	----gl----	-h-g--i---	---w--f---	----------	-----ei--s	--t-------	----v-----	----	2/s,w
*Caae-DRB1*0804*	-y	c-r----s--	-----l----	-h-g--i---	---w--f---	----------	-----ei--s	--t-------	y---------	----	1 / e
*Caae-DRB1*0310*	-y	h---------	----wy----	---g------	-n-w------	a-------k-	-----ei---	--t---v---	----------	----	2/s
*Caae-DRB1*0301*	-y	h---------	----------	---g------	-n-w--f---	a----qt-k-	-----e---r	--te---f--	-----f---a	----	1/a
*Caae-DRB1*0302*	-y	h---------	----------	---g------	-n-w--f---	a-----s-k-	-----e---r	--te---f--	-----f---a	----	2/s
*Caae-DRB1*0303*	-y	y-g-------	----------	---g------	-n-w--f---	a-----s-k-	-----e---r	--te---f--	-----f----	----	2/s
*Caae-DRB1*0304*	-y	h---------	----xx----	---g--xx-x	---w--f---	x----xxxx-	-----xx--x	xxxx---x--	x----f---x	----	/s
*Caae-DRB1*0401*	-y	y-g-------	----------	---g------	-n-w--f---	a----qt-k-	------f--d	s------f--	-----i---a	----	4/s
*Caae-DRB1*0402*	-y	y-g-------	----------	---g------	-n-w--f---	a-------k-	------f--d	s------f--	-----i----	----	1/e
*Caae-DRB1*0101*	-y	--k----s--	-----y----	---g-----y	---r--f---	a-----e---	-----ei---	k--ea--v--	-----i---s	----	4/s
*Caae-DRB1*0102*	-y	--k----s--	----------	-h-g--f---	---w------	a-------k-	-----ei--r	k---------	y---------	----	1/s
*Caae-DRB1*0103*	-y	t-k----s--	----------	-h-g--f-l-	---w------	a-------k-	-----ei--r	k---------	----------	----	1/s
*Caae-DRB1*0104*	-y	t-k----s--	----------	---g-----y	---w--f---	a---------	-----ei--d	s---------	----------	----	1/s
*Caae-DRB1*0105*	-y	t-k--r-s--	----------	---g-----y	---w--f---	a---------	-----ei--d	s---------	----------	----	1/s
*Caae-DRB1*0106*	-y	--k--r-s--	----------	---g-----y	---w--f---	a---------	-----ei--d	s---------	----------	----	1/s
*Caae-DRB1*0107*	-y	t-k--r-s--	----------	---g-----y	---w--f---	a---------	-----ei---	k--e---c--	-----i---s	----	1/s
*Caae-DRB1*0201*	-y	-t-----s--	----------	---g-----y	---w--f---	a-----s-k-	-----ei--d	s---------	-----i---s	----	1/e
*Caae-DRB1*0202*	-y	-t-----s--	-----l----	---g--tl-y	---w------	a---------	-----e----	--te------	----------	----	1/s
*Caae DRB1*0203*	y	-t-----s--	----------	---y-----y	---w------	a---------	-----e----	--te------	----------	----	1/s
*Caae-DRB1*0204*	-y	-t-----s--	----------	---g-----y	---w------	a-----s-k-	-----e----	--te------	-----i---s	----	1/w
*Caae-DRB1*0205*	-y	-t-----x--	----xx----	---g--xx-x	-x-w------	x----xxxx-	-----xx--x	xxxx---x--	---------x	----	/s
*Caae-DRB1*0701*	-y	y-g----s--	-g---l-h-f	yt-g--nl--	---w--f---	--q-qe----	------f---	k--e---v--	-----l----	----	4/s
*Caae-DRB1*0901*	-y	xxx----x--	----xx----	-h-g--f---	-x-w--f---	x---------	------f--x	xxxx---x--	x--------x	----	/s
*Caae-DRB1*0902*	-y	xxx----x--	----xx----	-h-g--xx-x	-x-w--f---	x---------	-----xx--x	xxxx---x--	x----i---x	----	/s
*Caae-DRB1*0903*	-y	xxx----x--	----xx----	-h-g--xx-x	-x-w--f---	x---------	-----xx--x	xxxx---x--	x----x---x	----	/s

Note: Protein sequences deduced from the second exon of *Caae-DRB1* alleles derived from genomic or cDNA or from typing patterns revealed by hybridization of polymorphism-specific oligonucleotides. Grouping is based on the similarities in the β-pleated sheets and the adjacent simple repeat derivative structure (group specific organisaton: GSO, see text and References 31 and 35). For each cloned allele at least two independent clones were sequenced. Double recombination or gene conversion-like events as documented in the resulting exon 2 sequences of goat *DRB1* alleles are obvious. An appearence of several independent point mutations is unlikely, as the same phenomenom was also found for silent substitutions and some of the exchanges are species specific. The number of individuals and the breed in which an allele was found are indicated on the right. Abbreviations (goat breeds) s: Saanen; a: Angora; b: Buren; e: Bunte Edel; w: Weiße Edel; z: African dwarf; x: for this sequence motif there was either no signal or no specific probe constructed.

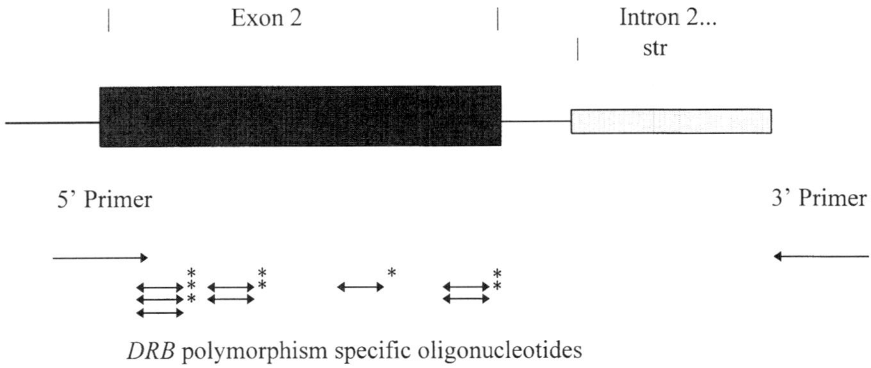

Figure 1 PCR system for the amplification of the second *Caae-DRB1* exon and part of the adjacent intron 2.[20] Approximately 30 bp 3′ the exon 2/intron 2 boundary is a str stretch hypervariable in length but correlated to the exonic polymorphism. Beneath the length variability a set of oligonucleotides specific for the polymorphism found in the bovine *DRB* orthologs complements the typing of caprine *DRB1* alleles.

b. Polymorphism in the Second Exon of DRB1

The deduced amino acid (aa) sequences of all known *DRB1* sequences were used to analyze *Caae-DRB1* polymorphisms and were compared to the three-dimensional crystal structure published by Brown et al.[32] In a comparison of polymorphic aa of human, cattle, and goat *DRB*, each species essentially has polymorphism occurring at the same aa positions. Bovine- or *Pecora*-specific aa were found for several aa positions of *Caae-DRB1* alleles (11, 18, 32, 34, 37, 42, 66, 67, 73) which include not only positions thought to be part of the antigen-binding-site. As expected, the most striking similarities are between the *DRB1* alleles of goat and sheep. As well, the similarity in the number of polymorphisms at most aa sites as the number of *Caprinae*-specific substitutions (e.g., aa position 42 or 73) shows the close relationship of the two animal species.

As was found in other species (mouse,[33] human,[34] artiodactyl species[31]) small regions of *Caae-DRB1* exon 2 seem to have been exchanged between different alleles. Only small regions were exchanged predominantly involving sequences downstream of the 5′ end of exon 2 but not including the simple tandem repeat (str).This was shown by closer phylogenetic relationship between the β-pleated sheet coding sequences and the adjacent str revealed than between the β-pleated sheet and the α-helix coding region. Additionally the C-terminus of *DRB1* first domain seems to have a different evolutionary history compared to all other regions. Hence, goat *DRB1* alleles, as is suggested also for the other bovid species, generate their polymorphism at least in part by "motif shuffling".

c. The Variability of the Intronic Microsatellite Adjacent to the DRB1 *Exon 2*

The *DRB1* exon 2 and a str situated 33 bp 3′ the exon 2/intron 2 boundary was coamplified by PCR (Figure 1). This simple tandemly arrayed repeat (str) has a basic structure (gt)n(ga)m (Table 3) and is preserved in almost the same location between species of primates, rodents, bovids, and cervids. In more than 150 *DRB* alleles examined, none contained a str stretch which is shorter than 44 bp. The length of strs found in different alleles of each species revealed high variability. Both (gt) and (ga) dinucleotide stretches are not only characterized by variations in their lengths but also by "degeneration" involving point mutations disrupting the perfect (gt)n or(ga)m or reamplifation of secondarily derived derivatives (di-, tetra-, hexa-, and octameres). These patterns of **r**epeat **d**erivative **s**tructures (RDS) have species specific characteristics as well as specific *DRB* allelic group characteristics within a species.[31,35] A defined *DRB1* exon 2 sequence is highly correlated with a str of characteristic RDS and in most cases also of a certain length. Hence, polymorphism of the antigen-binding first domain of the DRβ-chain in sheep is correlated with str variability as was shown in human,[36] cattle,[35] and goat.[31] This was called the group specific organisation (GSO) of *DRB* alleles. Yet it seems that the RDS of goat *DRB1* intronic strs reveal the highest divergence in bovids.

The mutation rate of str length variations from *Caae-DRB1* alleles was investigated. In 87 meioses, no differences in the length between the parental alleles and those of the progeny were found.

d. Characteristics of Ovar-DRB *Alleles*

By the computation of similarity (relative silent and replacement substitutions and tree building methods of VOSTORG and PHYLIP phylogenetic analysis programs), two populations of *Caae-DRB1* alleles (A and B) can be distinguished. Members of group A are closely related to the expressed *DRB* genes of cattle and sheep (*DRB3* and *DRB1*, respectively). Group B (*Caae-DRB1*08*) was found in about 5% of the investigated animals and characterized by differences in exon 2 at polymorphic aa sites unlike all the other investigated bovid alleles. A str similar to that found in the goat *DRB* allele group B with also high similarity in the exonic sequence was also present in *Oreamnus americanus* (Mountain goat, member of another mammalian genus), which lead to the assumption that this allele group exists at least 107 years.

Similarly to sheep where some *Ovar-DRB1* alleles were found exhibiting an increase of more than 250 bp in length (see Chapter 6), one allele was present in goat with about an additional 100 bp.The difference in length of this goat *DRB1* allele (*Caae-DRB1*0310*) was generated by an unequal crossing over event resulting in the addition of another complete str basic structure [intron$(gt)_n(ga)_m$ intron$(gt)_n(ga)_m$].

All of these *DRB* sequences appear to be inherited as alleles from one locus. There is the possibility that the different populations of *Caae-DRB* sequences are from separate *DRB* loci, with every Mhc haplotype containing only one of them. However, this seems to be very unlikely.

e. The DRB1 *Nomenclature*

In Table 2, nomenclature is suggested based on the polymorphism of the second exon and of the variability of the intronic str called GSO. Additionally, such *Caae-*

Table 3 Simple Repetitive Sequences in the Second Intron of Sequenced *Caae-DRB1* Alleles

Alleles	Simple tandem repeat sequence									
*Caae-DRB1*0801*		at $(gt)_5$	$(ga)_5$		$(gt)_2gg$			$(ga)_3(caga)_8$	$(gaca)_4(ca)_4cg(ca)_6$	
*Caae-DRB1*0802*		at $(gt)_5$	$(ga)_5$		$(gt)_2gg$			$(ga)_3(caga)_4$	$(gaca)_6(ca)_8$	
*Caae-DRB1*0803*		$(gt)_6$	$(ga)_5$		$(gt)_2gg$			$(ga)_3(caga)_5$	$(gaca)_5(ca)_8$	
*Caae-DRB1*0310*		$(gt)_{12}$	$(ga)_{24}(gc)_2(ggggtgggcggccagtgtggagca)^*$							
		$(gt)_{20}$	$(ga)_{14}gg(ga)_2ggga(gc)_2(ga)_2gc[(ga)_3ca]_2$	$(ga)_4ca$ $(ga)_3$	gtgg	$(ga)_6$	$ca(ga)_6$	$(caga)_2$		
*Caae-DRB1*0301*		$(gt)_{13}$	$(ga)_{10}$	ca		$(ga)_4$				
*Caae-DRB1*0302*		$(gt)_{10}$	$(ga)_{13}$	$aa(ga)_5ca$		$(ga)_6$		$(caga)_2$		
*Caae-DRB1*0303*		$(gt)_{10}$	$(ga)_{13}$	$aa(ga)_5ca$		$(ga)_6$		$(caga)_2$		
*Caae-DRB1*0401*		$(gt)_{27}$	$(ga)_{12}$	$(gc)_2(ga)_2gc[(ga)_3ca]_2[(ga)_4ca]_2(ga)_3$	gtgg	$(ga)_5$	$ca(ga)_6$	$(caga)_2$		
*Caae-DRB1*0402*		$(gt)_{26}$	$(ga)_{12}$	$(gc)_2(ga)_2gc[(ga)_3ca]_2[(ga)_4ca]_2(ga)_3$	gtgg	$(ga)_6$	$ca(ga)_5$	$(caga)_2$		
*Caae-DRB1*0101*	$(gt)_3(gc)_2$	$(gt)_{15}$	$aa(ga)_5$		gtgggaaa		$(ga)_5$	$(caga)_2$	aaca	ga
*Caae-DRB1*0102*	$(gt)_3$ gc	$(gt)_{13}$	$aa(ga)_3caga$		gtgggaaa	$(ga)_5$	$ca(ga)_5$	$(caga)_3$	aaca	ga
*Caae-DRB1*0103*	$(gt)_3$ gc	$(gt)_{13}$	$aa(ga)_3caga$		gtgggaaa	$(ga)_5$	$ca(ga)_5$	$(caga)_3$	aaca	ga
*Caae-DRB1*0104*	$(gt)_3$ gc	$(gt)_{21}$	$aa(ga)_3caga$		gtaggaaa	$(ga)_5$	$ca(ga)_7$	$(caga)_3$	aaca	ga
*Caae-DRB1*0105*	$(gt)_3(gc)_2$	$(gt)_{14}$	$aa(ga)_5$		gtgggaaa	$(ga)_5$	$ca(ga)_7$	$(caga)_3$	aaca	ga
*Caae-DRB1*0106*	$(gt)_3$ gc	$(gt)_{24}$	$aa(ga)_3caga$		gtaggaaa	$(ga)_5$	$ca(ga)_7$	$(caga)_3$	caca	ga
*Caae-DRB1*0107*	$(gt)_3(gc)_2$	$(gt)_{20}$	$aa(ga)_3caga$		gtaggaaa	$(ga)_5$	$ca(ga)_7$	$(caga)_3$	gaca	ga
*Caae-DRB1*0201*	$(gt)_3$ gc	$(gt)_{14}$	$aa(ga)_3caga$		gtgggaaa	$(ga)_5$	$ca(ga)_7$	$(caga)_2$	aaca	ga
*Caae-DRB1*0202*	$(gt)_3$ gc	$(gt)_{20}$	$aa(ga)_3caga$		gtgggaaa	$(ga)_5$	$ca(ga)_7$	$(caga)_3$	aaca	ga
*Caae-DRB1*0203*	$(gt)_3$ gc	$(gt)_{19}$	$aa(ga)_3caga$		gtgggaaa	$(ga)_5$	$ca(ga)_7$	$(caga)_3$	aaca	ga
*Caae-DRB1*0204*		$(gt)_{18}$	$aa(ga)_3caaa$		gtgggaaa		$(ga)_5$	$(caga)_4$	aaca	ga
*Caae-DRB1*0701*		$(gt)_{31}$	$aa(ga)_5$		gtgggaaa		$(ga)_3$	$(caga)_3$	aaca	ga

DRB1 exon 2 sequences were included which were only typed. We suppose that caprine, ovine[31,37] and bovine[38] *3* expressed *DRB* loci are all orthologs. The group numbers (e.g., *Caae-DRB1*01*) were given according to the sequence of the β-pleated sheet. The alleles within a group were numbered in the order of their detection. *DRB1* alleles which are only typed by hybridization and definition of length should have an H suffix. Such typed alleles which cannot be differentiated by their hybridization pattern from sequenced alleles should have the same number.

4. *Aspects of* DRB *Evolution*

Using established tree building methods[31] some *DRB1* alleles of goat and sheep seem to be more closely related to each other than to alleles of the same species. Hence, there is also evidence for a transspecies mode of evolution[39] for the caprine *DRB1* alleles.

Concerning the obvious appearance of genetic exchange between alleles of *Ovar-DRB1*[30,31] and the proposed selective pressure for polymorphism generation,[40] the present *DRB1* alleles may be the result of a "motif-shuffling" process. Therefore, we calculated the number of relative synonymous (d_s) and relative nonsynonymous (d_n) substitutions using the NAG program[41] (Table 4). The relative number of synonymous (nonsynonymous) substitution is calculated by counting all found silent (replacement) substitution divided by all possible synonymous (nonsynonymous) substitutions between two *DRB* alleles. The average value for relative nonsynonymous substitutions between *Caae-DRB1* alleles is comparable to those of other species (man 0.071, cattle: 0.091). For sheep clearly two different point-clusters can be distinguished. This is paralleled in goat but less obvious by overlapping point-clusters. Fine analysis revealed that the two distinct clusters in each species formed by differences in the values of d_s/d_n was generated by the comparison of allele group A with group B. The values for d_s in this comparison (Table 4) is for both species significantly higher as even that of an interspecies comparison of the expressed *DRB* loci from sheep and cattle (0.075 ± 0.031) or goat and cattle (0.072 ± 0.029).Therefore, members of the *Caae-DRB1* allele population B may be preserved in goat since a time before cattle and goat have separated. As there is also only limited evidence of recent sequence exchange between these two groups of *Caae-DRB1* alleles (data not shown), we suggest that exchange of sequence motifs will not occur between alleles of the two *DRB1* populations. As also the str basic structures of both allele groups differ, involvement of str in sequence exchange mechanism seems to be possible.

5. *Informative Markers in the Goat* DRB *Region*

Little work was done to generate informative markers (e.g., length polymorphic str markers) for the goat Mhc. The *Caae-DRB1* intron 2 str[31] provides therefore a first approach to investigate associations between disease and Mhc with molecular biological methods. On the other hand until now eight additional str markers have been established for sheep Mhc (some of them adapted from cattle) and there is a good chance that they are also useful for an application in goat. A panel of informative

Table 4 Mean Numbers of Relative Silent (d_s) and Relative Replacement (d_n) Substitutions with the Respective Standard Deviations (in parentheses)

	All DRB1 alleles	Group A	Group A vs. group B
d_s	0.066 (0.036)	0.051 (0.028)	0.104 (0.027)
d_n	0.092 (0.032)	0.090 (0.034)	0.099 (0.024)

Note: Group A is composed of *CAAe, DRB09* alleles, whereas group B containes all the remaining alleles. Significant differences of the distributions analyzed by the KS-test: relative silentsubstitutions < relative replacement substitutions are only found when all *DRB1* alleles are included or members of the group A are compared with each other ($p < 0.001$). For the comparison of relative silent and relative replacement substitutions 20*Caae-DRB1* sequences were available. The values for the mean numbers of relative substitutions could be distorted if many closely related alleles are biased in the analyzed clusters.

markers covering the whole Mhc region seems to be advantageous as there are hints that we will find a relatively high number of recombinations within the bovid Mhc as was found for sheep (A. Crawford, in preparation).

III. FUNCTION

A. Immune Response Studies

Immune response capacity in relation to Mhc in goats was studied also by van Dam.[5] An influence of certain CLA antigens could be demonstrated.

The Norwegian group[42] has studied the distribution of CLA antigens in diphtheria toxoid-immunized goats. They could not find any association between presence of certain CLA antigens in the animals and their capacity to respond with low or high antibody production.

B. Disease Association Studies

van Dam[5] investigated a possible association of the GLA-system and Jone's disease as well as its influence on trypanosomiasis. He could not find any statistically significant results due to the incompleteness of the characterized system and the low number of animals.

1. CLA and Resistance/Susceptibility to Caprine Arthritis-Encephalitis

As immunopathological mechanisms are involved in the pathogenesis of caprine arthritis-encephalitis (CAE),[43] Ruff et al. studied the possible role of the caprine Mhc in disease manifestation. The influence of its gene products on manifestation of virus-induced arthritis has been investigated in family and population studies.[44] The results clearly demonstrated that susceptibility/resistance to disease onset was linked to the Mhc.[45] The determined CLA products themselves were not directly involved in pathogenesis, but rather in close genetic linkage to the eventual disease genes.

To confirm the above mentioned data, a follow-up study with goats of different genetic background was carried out; 526 Saanen goats, affected (198) and non-affected (328) by clinical arthritis, have been CLA typed to confirm the results mentioned above. The formerly reported association with CAE resistance of the Be7 antigen in the Saanen could be confirmed (corrected $p < 0.008$). Furthermore, an association of the CLA antigens Be1 and Be14 with disease susceptibility could be demonstrated when combining the data for all 807 Saanen goats as demonstrated in Table 5.

Table 5 Relative Risk and Corrected *p* Value of the Three Class I Antigens Showing a Statistically Significant Different Frequency in the CAE-affected and Control Groups in Saanen Goats in Two Subsequent Studies

Class I antigen (year of study)	Affected n/phenotype frequency	Control n/phenotype frequency	Relative risk (odds ratio)	*p* value χ^2	P corrected = P × no. of tests (Bonferroni)
Be1 (1988)	53/0.36	30/0.22	1.99	0.0098	—
Be1 (1992)	43/0.22	45/0.14	1.74	0.0172	—
Combined data			1.85	0.0007	0.003
Be7 (1988)	35/0.24	56/0.41	0.45	0.0017	—
Be7 (1992)	51/0.26	106/0.32	0.73	0.1112	—
Combined data			0.62	0.0020	0.008
Be14 (1988)	15/0.10	05/0.04	2.80	0.0323	—
Be14 (1992)	16/0.08	12/0.04	2.29	0.0286	—
Combined data			2.55	0.0037	0.015

Note: 1988 affected N = 146, controls N = 135, total N = 281; 1992 affected N = 198, controls N = 328, total N = 526, combined total N = 807.

Among the investigated goats, 16 families were included. Three of the families had already been studied in the former investigations, but the pedigrees were extended. The segregation analysis of the inherited CLA types in diseased and nondiseased offspring in multiple-case families still showed a distortion from the expected ratio of 1:1 as reported in previous studies.[45] The CLA class I antigens segregating in association with disease still changed from family to family, indicating a linkage to the Mhc region but no direct involvement of the determined class I products.

From a total of 800 class I typed animals, 357 were also typed for their B-cell specificities on separated T and B cells. Of those, 130 animals were affected and 227 showed no clinical manifestation of arthritis. The BeD2 and BeD5 specificities with a corrected p-value of <0.02 and 0.04 showed a weak association in the Saanen with susceptibility to disease manifestation.[46]

As linkage disequilibrium of the haplotype combination Be1-D5 could be demonstrated for the Saanen goat population, the effect of this haplotype in the case/control study was studied in 60 affected and 76 control goats with known phases of inheritance deduced from segregation analysis data. The relative risk of the combination of the combined antigens Be1-D5 was significantly increased compared to the antigens alone or in combination with other alleles indicating a "haplotype effect".[46] No statistically significant association was found with any of the eight determined oligotypes in an additional study of this material testing only a limited number of animals (Ruff and Schwaiger, personal communication).

Cameron et al.[8] did not find any variation in length around the complement gene cluster when probing C4 of DNA from 21 goats, including 12 with clinically evident arthritis.

2. CLA and Resistance/Susceptibility to Cowdriosis

A genetic component in the resistance/susceptibilty to heartwater disease has been reported for Creole goats.[47] In a joint project with the Swiss group, 177 Creole goats including 4 halfsib families with known resistance/susceptibility to the disease in question have been serologically typed for their CLA class I products. A statistically significant association for some of the determined class I specificities and disease manifestation could be detected.[16] A positive association with resistance to cowdriosis was found for the pan-T-cell marker characterizing the CLY antigen described earlier. Whether the observed differences represent regional sire effects or indicate pathogen-induced selection pressure remains to be clarified. Pépin et al.[48,49] commenced a disease association study at the DNA level using different bovine microsatellites located within the Mhc region or scattered throughout the genome on goat DNA. Further analysis on additional animals is needed to detect any association.

REFERENCES

1. van Dam, R. H., Borst-van Werkhofen, C., van der Donk, J. A., and Goudswaard, J., The production and evaluation of allo-antibodies for GLA typing in goats, *J. Immunogenet.*, 3, 237, 1976.
2. van Dam, R. H., Boot, R., van der Donk, J. A., and Goudswaard, J., Skin grafting and graft rejection in goats, *Am. J. Vet. Res.*, 39, 1359, 1978a.
3. van Dam, R. H., van Kooten, P. J. S., and van der Donk, J. A., *In vitro* stimulation of goat peripheral blood lymphocytes: optimization and kinetics of the response to mitogens and to allogenetic lymphocytes, *J. Immunol. Meth.*, 21, 217, 1978b.
4. van Dam, R. H., d'Amaro, J., van Kooten, P. J. S., van der Donk, J. A., and Goudswaard, J., The histocompatibility complex GLA in the goat, *Anim. Blood Grps. Biochem. Genet.,* 10, 121, 1979.
5. van Dam, R. H., The Major Histocompatibility Complex of the Goat, Definition and some Aspects of its Biological Function, Thesis, Vet. Med. Faculty Utrecht, The Netherlands, 1981.
6. Nesse, L. L., Paulsen, G., Syed, M., and Ruff, G., A human major histocompatibility complex (Mhc) DNA probe recognizes goat genes, *Acta Vet. Scand.*, 29, 193, 1988.
7. Uilenberg, G., Heartwater (*Cowdria ruminantium* infection): current status, *Adv. Vet. Sci. Comp. Med.*, 27, 427, 1983.
8. Cameron, P. U., Tabarias, H. A., Pulendran, B., Robinson, W., and Dawkins, R. L., Conservation of the central Mhc genome: PFGE mapping and RFLP analysis of complement, HSP70, and TNF genes in the goat, *Immunogenetics*, 31, 253, 1990.
9. Mann, A. J., Abraham, L. J., Cameron, P. U., Robinson, W., Giphart, M. J., and Dawkins, R. L., The caprine Mhc contains *DYA* genes, *Immunogenetics*, 37, 292, 1993.

10. Nesse, L. L., Studies on Caprine Lymphocyte Antigens, Thesis, Norwegian Veterinary High School, Oslo, 1985.
11. Nesse, L. L. and Larsen, H. J., Production of alloantibodies against caprine lymphocyte antigens, *Anim. Genet.*, 18, 159, 1987a.
12. Nesse, L. L. and Larsen, H. H., Lymphocyte antigens in Norwegian goats: serological and genetic studies, *Anim. Genet.*, 18, 261, 1987b.
13. Ruff, G., Investigations on the Caprine Leucocyte Antigen (CLA) system, Thesis No. 8468, Swiss Federal Institute of Technology, Zurich, 1987.
14. Nesse, L.L. and Ruff, G., A comparison of lymphocyte antigen specificities in Norwegian and Swiss goats, *Anim. Genet.*, 20, 71, 1989.
15. Joosten, I., Ruff, G., Sanders, M. F., and Hensen, E., Use of isoelectric focusing to define major histocompatibility complex class I polymorphism in goats, *Anim. Genet.*, 24, 47, 1993.
16. Ruff, G., Maillard, J. C., Camus, E., Déprés, E., and Matheron, G., Occurrence of caprine leucocyte antigens (CLA) in Creole goats susceptible/resistant to Heartwater, *Rév. d'Elev. Méd. Vét. Pays Trop* ., 46(1–2), 205, 1993.
17. Hasenkrug, K. J., Cory, J. M., and Stimpfling, J. H., Monoclonal antibodies defining mouse tissue antigens encoded by the H-2 region, *Immunogenetics*, 25, 136, 1987.
18. Ruff, G., Joosten, I., Howald, M., Davis, W.C., and Lazary, S., Serological and biochemical detection of B cell antigens in goats, *Eur. J. Immunogenet.*, 20, 273, 1993.
19. Davis, W. C. and Ellis, J. A., Individual antigens of goats, in *Vet. Immunol. and Immunopathol.*, Vol. 27, Howard, C. J., and Morrison, W. I., Eds., Elsevier, Amsterdam, 1991, 121.
20. Schwaiger, F.-W., Buitkamp, J., Weyers, E., and Epplen, J. T., Typing of artiodactyl Mhc-DRB genes with the help of intronic simple repeated sequences, *Mol. Ecol.,* 2, 55, 1993b.
21. Klein, J., Bontrop, R. E., Dawkins, R. L., Erlich, H. A., Gyllensten,U. B., Heise, E. R., Jones, P. P., Parahm P., Wakeland, E. K., and Watkins, D. I., Nomenclature for the major histocompatibility complexes of different species: a proposal, *Immunogenetics*, 31, 217, 1990.
22. Yuhki, N. and O'Brien, S. J., Molecular characterization and genetic mapping of class I and class II genes for the domestic cat, *Immunogenetics*, 27, 414, 1988.
23. Andersson, L., Lunden, A., Sigurdardottir, S., Davies, C. J., and Rask, L., Linkage relationships in the bovine Mhc region. High recombination frequency between class II subregions, *Immunogenetics*, 27, 273, 1988.
24. Wright, H. and Ballingall, K. T., Mapping and characterisation of the DQ subregion of the ovine Mhc, *Anim. Genet.*, 25, 243, 1994.
25. Wright, H., Ballingall, K. T., and Redmond, J., The DY subregion of the sheep Mhc contains an A/B gene pair, *Immunogenetics,* 41, 131, 1995.
26. van der Poel, J. J., Groenen, M. A. M., Dijkhof, R. J. M., Ruyter, D., and Giphart, M. J., The nucleotide sequence of the bovine Mhc class II alpha genes: DRA, DQA and DYA, *Immunogenetics,* 31, 29, 1990.
27. Stone, R. T. and Muggli-Cockett, N. E., Partial nucleotide sequence of a novel bovine major histocompatibility complex class II beta chain gene, BoLA DIB, *Anim. Genet.,* 21, 352, 1990.
28. Stone, R. T. and Muggli-Cockett, N. E., BoLA-DIB: species distribution, linkage with DOB and Northern analysis, *Anim. Genet.*, 24, 41, 1993.
29. Marsh, S. G. E., and Bodmer, J., HLA class II nucleotide sequences, 1992, *Immunogenetics,* 37, 79, 1993.

30. Schwaiger, F.-W., Weyers, E., Buitkamp, J., Ede, A. J., Crawford, A., and Epplen, J. T., Interdependent Mhc-*DRB* exon-plus-intron evolution in artiodactyls, *Mol. Biol. Evol.,* 11, 239, 1994a.
31. Schwaiger, F.-W., Weyers, E., Epplen, C., Ruff, G., Crawford, A., and Epplen, J. T., On the paradox of Mhc-DRB evolution: α -helix and β-sheet encoding regions diverge while hypervariable intronic simple repeats co-evolve with β-sheet codons, *J. Mol. Evol.*, 37, 260, 1993a.
32. Brown, J. H., Jardetzky, T. S., Gorga, J. C., Stern, L. J., Urban, R. G., Strominger, J. L., and Wiley, D. C., Three-dimensional structure of the human class II histocompatibility antigen HLA-DR1, *Nature,* 364, 3, 1993.
33. Wakeland, E.K., Boehme, S., and She, J. X., The generation and maintainence of Mhc class II gene polymorphism in rodents, *Immunol. Rev.*, 113, 39, 1990.
34. Gyllensten, U., Sundvall, M., and Erlich, H. A., Allelic diversity is generated by intraexon sequence exchange at the *DRB1* locus of primates, *Proc. Natl. Acad. Sci. U.S.A.,* 88, 3686, 1991.
35. Ammer, H., Schwaiger, F.-W., Kammerbauer, C., Gomolka, M., Arriens, A., Lazary, S., and Epplen, J. T., Exonic polymorphism vs. intronic simple repeat hypervariability in Mhc-DRB-genes, *Immunogenetics*, 32, 330, 1992.
36. Rieß, O., Kammerbauer, C., Roewer, L., Steimle, V., Andreas, A., Albert, E., Nagai, T., and Epplen, J. T., Hypervariability of intronic simple (gt)n(ga)m repeats in HLA-DRB genes, *Immunogenetics,* 32, 110, 1990.
37. Ballingall, K. T., Wright, H., Redmond, J., Dutia, B. M., Hopkins, J., Lang, J., Deverson, E. V., Howard, J. C., Puri, N., and Haig, D. M., Expression and characterisation of ovine major histocompatibility complex class II (*OLA-DR*) genes, *Anim. Genet.,* 23, 347, 1992
38. Andersson, L., Sigurdardottir, S., Borsch, C., and Gustafsson, K., Evolution of Mhc polymorphism: extensive sharing of polymorphic sequence motifs between human and bovine DRB alleles, *Immunogenetics,* 33, 188, 1991.
39. Klein, J., Origin of the major histocompatibility complex polymorphism: the trans-species hypothesis, *Hum. Immunol.,* 19, 155, 1987
40. Hughes, A.L. and Nei, M., Nucleotide substitution at major histocompatibility complex class II loci: evidence for overdominant selection, *Proc. Natl. Acad. Sci. U.S.A.,* 86, 958, 1989.
41. Nei, M. and Gojobori, T., Simple methods for estimating the numbers of synonymous and nonsynonymous nucleotide substitutions, *Mol. Biol. Evol.,* 3, 418, 1986.
42. Eide, D. M., Adnoy, T., Klemetsdal, G., Nesse, L. L., and Larsen, H. J., Selection for immune response in goats: the antibody response to diphtheria toxoid after 12 years of selection, *J. Anim. Sci.*, 69, 3967, 1991.
43. Peterhans, E., Zanoni, R., Ruff, G., and Lazary, S., Caprine arthritis-encephalitis, in *Farm Animals in Biomedical Research,* (Adv. in Animal Breeding and Genet., Issue 5), Pliska, V. and Stranzinger, G., Eds., Verlag Paul Parey, Hamburg, 1990, 147.
44. Ruff, G. and Lazary, S., Possible influence of the caprine leucocyte antigen (CLA) system on development of caprine arthritis encephalitis (CAE) in family and population studies, in *Improving Genetic Disease Resistance in Farm Animals,* van der Zijpp, A. J. and Sybesma, W., Eds., Kluwer Academic, Dordrecht, 1989, 107.
45. Ruff, G. and Lazary, S., Evidence for linkage between the caprine leucocyte antigen (CLA) system and susceptibility to CAE virus-induced arthritis in goats, *Immunogenetics*, 28, 303, 1988.

46. Ruff, G., Regli, J., and Lazary, S., Occurrence of caprine leucocyte class I and II antigens in Saanen goats affected by caprine arthritis (CAE), *Eur. J. Immunogenet.*, 20, 285, 1993.
47. Matheron, G., Barré, N., Camus, E., and Gogue, J., Genetic resistance of Guadeloupe native goats to heartwater, *Onderstepoort J. Vet. Res.*, 54, 337, 1987.
48. Pépin, L., Camus, E., Matheron, G., and Bensaid, A., Use of microsatellites as genomic markers to study resistance to cowdriosis, *Rév. d'Elev. Méd. Vét. Pays Trop.*, 46 (1-2), 209, 1993.
49. Pépin, L., Recherche des Polymorphismes Génétiques Chez les Caprins: Application à l'Étude Comparative du Génome des Ruminants et à la Résistance à la Cowdriose, Thesis, Jouy-en-Josas, France, 1994.

CHAPTER 6

The Ovine Major Histocompatibility Complex

Franz-Werner Schwaiger, Jill Maddox, Keith Ballingall, Johannes Buitkamp, Allan M. Crawford, Bernadette M. Dutia, Joerg Thomas Epplen, Elaine D. Ferguson, David Groth, John Hopkins, Susan M. Rhind, David Sargan, John Wetherall, and Harry Wright

CONTENTS

0-8493-8032-4/96/$0.00+$.50

ABSTRACT

The sheep major histocompatibility complex (Mhc) was first identified 15 years ago by serological studies of sheep lymphocyte antigens.[1] Since then both classical class I and class II molecules have been identified as membrane-associated molecules (antigens) on sheep peripheral blood lymphocytes by serological and immunochemical methods. The relative paucity of international resources has meant that serological characterization of the sheep Mhc has been slow. In recent years the emphasis has moved to the use of the molecular techniques to characterize the ovine Mhc region. Relative to other domesticated species the ovine Mhc is poorly characterized. To date the class II region has been the focus for most investigations while the class I region has received little attention since the early flurry of serological effort. The presence of a class III region is still imputed by analogy with other ungulate species.

The sheep Mhc was initially termed OLA (ovine lymphocyte antigens) and is still referred to occasionally by this name, as a reflection of its serological history. However, to strengthen efforts for a systematic nomenclature of the Mhc in all vertebrates,[2] the sheep Mhc will be referred to as *Ovar* (from ***Ovis aries***) in this chapter. The ovine Mhc has been mapped to chromosome 20q15-q23.[3,4] Although no overall physical map exists so far, information from overlapping cosmids and linkage analysis studies indicate similarities of the *Ovar* complex to cattle and human[5,6] (Groth and Wetherall, unpublished data; Crawford et al., unpublished data).

I. THE *OVAR* CLASS I REGION

A. Classical Sheep Class I Genes and Molecules

Class I sheep Mhc molecules were first detected on lymphocytes using the well-established microlymphocytotoxicity assay which had been developed initially for the serological detection of HLA antigens.[1,7-9]

As expected, serological investigations confirmed the complexity of the *Ovar* system and the existence of multiple loci encoding these molecules. Thus Millot in 1978 obtained evidence for two closely linked main *Ovar* loci which he designated A and B, and two minor loci which were not closely linked and which may not be true Mhc loci. The *Ovar-A* and *Ovar-B* loci exhibited linkage disequilibrium. Millot later confirmed the existence of these loci in 1984 and extended to 16 the number of specificities which could be distinguished with the available typing sera. Other laboratories obtained similar results but extensive serological characterization similar to that achieved for human and mouse Mhcs has been difficult due to the lack of resources, interlaboratory collaboration, and standardization.

1. Class I Molecules

There has only been one study elucidating the biochemical nature of the *Ovar* class I lymphocyte antigens. Gogolin-Ewens et al.[10] produced three distinct monoclonal antibodies specific for a monomorphic determinant on *Ovar* class I molecules and used these antibodies to immunoprecipitate sheep lymphocyte membrane proteins which had been surface-labeled with radioactive iodine. These antibodies precipitated a 44,000-daltons polypeptide together with a 12,000-daltons polypeptide similar to β_2-microglobulin. These data indicate that the structure of sheep class I molecules is similar to those of other mammalian species.

2. Molecular Studies of Class I Genes

There have been several studies of sheep class I genes using molecular techniques. Thus Chardon et al.[11] used RFLP analysis based upon cross hybridization with human class I and class II Mhc probes to show that the sheep Mhc was of similar complexity to that of humans. Grossberger et al.[12] screened a sheep thymus cDNA library with a cDNA probe from the HLA-B27 gene at lowered stringency and subsequently isolated 13 cDNA clones for which partial nucleotide sequences were determined. On the basis of DNA sequence the 13 clones represented five distinct groups of sheep class I molecules, requiring the expression of at least 3 loci. One of the sheep class I genes retained an intron between the cytoplasmic domain exons which could lead to a long cytoplasmic tail on the expressed p44 glycoprotein. The sequences of the sheep class I genes were more similar to bovine class I genes than to murine class I genes. A highly polymorphic microsatellite locus has been recently identified in a cosmid clone containing sheep genomic DNA which hybridized with a sheep Mhc Class I cDNA probe.[13]

Garber and colleagues have recently compared the sequence of three PCR amplified sheep class I cDNA fragments with the corresponding sequences of cattle and red deer. They observed insertions and deletions in exon 5 that could be aligned suggesting that these were present in the ancestral artiodactyl lineages (personal communication).

B. Nonclassical Class I Region Genes and Molecules

1. CD1 Genes and Molecules

CD1 molecules belong to a family of nonpolymorphic β_2-microglobulin associated glycoproteins with strong structural homology, but weaker sequence homology, to Mhc class I genes.[14] Five CD1 genes have been identified in humans (CD1A, B, C, D and E),[15] three of which encode the serologically defined CD1a, b, and c antigens.[16] The expression of the three CD1 molecules is restricted to cortical thymocytes and dendritic cells, while skin Langerhans cells and a subset of B cells also express CD1a and CD1c, respectively.[14] In addition, human CD1d has been observed on intestinal epithelial cells.[17] In the mouse two very similar genes, both related to the human CD1D gene, have been described.[18,19] The highest level of murine CD1 tissue expression is in the gastrointestinal tract[20] and the level is comparable to that observed for the human CD1d molecule. While Southern blot analysis has indicated the existence of up to eight rabbit CD1 genes, only two incomplete rabbit CD1 genes have been cloned to date, one CD1B-like and the other CD1D-like.[21] Sequence analysis of the leader, $\alpha 1$ and $\alpha 2$ domains in several species, has permitted the CD1 genes to be divided into two distinct classes. The first class consists of the human CD1A, B, C, and rabbit CD1B genes and the second of the human, rabbit, and mouse CD1D genes with the human CD1E gene being of an intermediate type.[22] Differing patterns of tissue expression and intraclass relatedness suggest that the two subclasses may have different functions. The cellular expression of CD1 molecules also has been detected in cattle, sheep, and pig by species-specific monoclonal antibodies but the relationship of these molecules to the two classes has not yet been defined.[23-25]

2. The Function of CD1

Although the structure and expression of CD1 has been extensively studied in various species, the function of these molecules remains elusive. The Mhc-like nature and expression on classical antigen-presenting cells suggests a role for CD1 in antigen presentation. There has been increasing evidence to indicate that CD4⁻8⁻ T cells are able to recognize nonclassical presentation elements including Mhc class Ib molecules such as Qa and TLa, as well as CD1. The TL and Qa antigens have been implicated as presentational elements for both ab and gd T cells.[26-28] Similarly, a variety of peripheral blood and intestinal T cell lines and clones specifically recognize CD1 molecules.[20,29-31] The expression of CD1 on immature cortical thymocytes at a time when these cells are undergoing selection suggests that CD1 may play a role in T-cell development. CD1 is covalently associated with the CD8 molecule on immature thymocytes, and there is some evidence to suggest that CD1 may be linked to a cellular activation pathway.[14]

3. Ovar-CD1 *Genes and Molecules*

The study of CD1 expression and function in ruminants has utilized specific anti-CD1 monoclonal antibodies.[24] Expression studies show that these antibodies fall into two main groups — group 1 antibodies[24] have a tissue distribution similar to that of the human CD1b and group 2[32] appear analogous to human CD1c (Table 1). Immunochemical analysis indicates that group 2 antibodies also react with group 1 molecules.[24] Group 1 antibodies stain cortical thymocytes and dendritic cells in a variety of tissues as well as skin Langerhans cells. Group 2 antibodies have a more extensive tissue distribution with staining additionally observed on B cells, peripheral blood monocytes, and intestinal dendritic cells. Different antibodies within group 2 show minor differences in tissue distribution. A major difference between species is that sheep peripheral blood monocytes constitutively express CD1c (group 2) whereas the expression of CD1 by human monocytes is only observed in response to the cytokines GM-CSF and IL4.[31]

Table 1 Tissue Distribution of CD1 in the Sheep

Tissue		Group 1[a]	Group 2[b]
Skin	Dermis	+	+
	Epidermis	−	+
Thymus	Cortex	+	+
	Medullary DC[b]	+	+
Spleen	White pulp DC	+	+
	Mz[2] and red pulp	−	+
Intestine	Lamina propria	−	+
Liver	Kupffer cells	−	+
Blood	B cells	−	+
	Monocytes	−	+
Afferent lymph	B cells	−	+
	Dendritic cells	+	+
Efferent lymph	B cells	−	+

[a] Group 1 antibodies: VPM5, CC13, CC14, CC20, CC409. Group 2 antibodies: SBU-T6 (20.27), CC43, CC11819.
[b] DC, dendritic cells; mz, marginal zone.

Studies on the molecular biology of *Ovar-CD1* have utilized the human CD1C α3 domain probe to screen sheep thymocyte cDNA libraries. Initially an ovine CD1B-like clone (SCD1A25) was isolated and the α3 region from this clone was used to identify an additional three ovine clones, SCD1B-42, SCD1B-52, and SCD1T10. Three of the four CD1 clones are truncated at the 5′ end, with sequences beginning near the end of the α1 domain. The fourth clone, SCD1B-52, is truncated by having its α3 domain precisely spliced out. The lack of full-length, correctly spliced sequences suggests a complex alternative splicing situation which parallels the observations of alternative splicing events reported for the human[33] and mouse.[34] The missing 5′ ends from two of the truncated sequences were obtained by polymerase chain reactions, generating full-length coding sequence for two complete *Ovar-CD1* genes. These sequences show 81 to 96% identity at the nucleotide level and 79 to 90% identity at the aa level, suggesting that they represent distinct gene products rather than allelic variants. The *Ovar-CD1* sequences have also been compared to the human, mouse, and rabbit sequences (Table 2). Unexpectedly, given the existence of five different human CD1 genes, all of the ovine CD1 genes are most homologous to human CD1B at both the nucleotide and amino acid (aa) levels. The *Ovar-CD1* sequences also show a high percentage sequence identity to the cottontail rabbit sequence, which is most similar to CD1B. Southern blot analysis of sheep genomic DNA digested with a variety of enzymes and probed with an ovine α3 probe (Figure 1) indicates the possible existence of up to seven ovine CD1 genes.

Table 2 Percent Sequence Identity Among CD1 Genes and Proteins

	SCD1A25[a]	SCD1B-42		SCD1B-52		SCD1T10	
HCD1A	70.80	69.12	**58.84**	63.30	**52.63**	71.78	**64.38**
HCD1B	81.02	79.64	**73.35**	78.99	**68.05**	79.38	**75.00**
HCD1C	70.54	70.17	**58.86**	66.94	**48.35**	70.94	**59.19**
HCD1D	64.24	64.37	**50.30**	56.29	**37.32**	68.84	**53.81**
HCD1E	65.81	64.10	**51.71**	55.98	**32.66**	68.32	**56.54**
MCD1.1	60.80	59.78	**44.31**	50.41	**30.84**	62.89	**48.66**
MCD1.2	60.00	59.48	**43.71**	46.42	**29.60**	62.44	**47.77**
CTRAB	65.41	65.33	**52.92**	55.68	**33.15**	69.71	**57.30**

Note: Plain type refers to nucleic acid sequence, bold type to amino acid sequence.

[a] No possible derived amino acid sequence because of a single nucleotide deletion.

II. THE *OVAR* CLASS II REGION

Most of the Mhc class II gene families that have been described for other species were also found in sheep. Currently the existence of at least 14 distinct loci have been shown (Table 3). As for other species, these genes tend to be clustered in family groups and in *AB* pairs within families.[6] The latter form the classical protein heterodimers on the cell surface. The number of class II genes may vary in different haplotypes, with some sheep lacking *DQA1* and possibly one of the *DQB* genes.

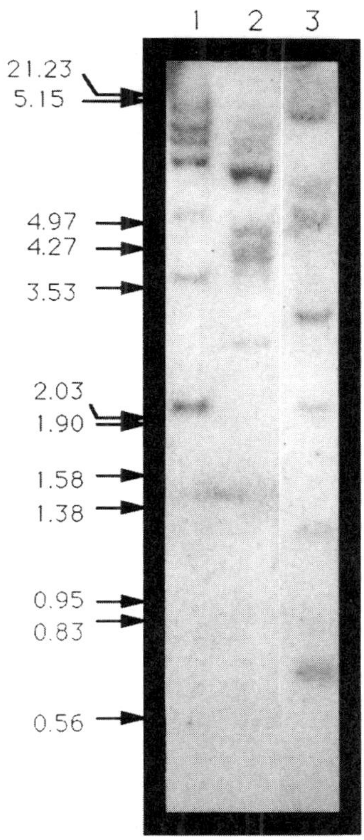

Figure 1 Southern blot analysis of sheep genomic DNA digested with a variety of enzymes and probed with an ovine a3 probe.

A. The Expression of *Ovar* Class II Gene Products

1. *Cloning and Expression of the* Ovar *Class II Genes*

A large number of cosmid clones have been isolated [5,35] (Table 4), which, on the basis of differential hybridization to human and mouse class II probes, contained either a single *B* gene, a single *A* gene, both an *A* and a *B* gene, or in some cases two *B* genes. Restriction mapping showed that some of the clones overlapped, some constituted alleles at a given locus, while others appeared to be unrelated to the rest. Many of the genes in the cosmids have now been sequenced and assigned to a subregion on the basis of their similarity to published human, sheep, and cattle sequences. To identify the genes that encode the *Ovar* class II molecules expressed on the surface of sheep cells, *A/B* gene pairs were cotransfected into mouse L-cells.[35] Expression at the L-cell surface was detected by the use of an indirect fluorescence antibody assay and FACS analysis with a panel of sheep class II-specific *m*onoclonal *a*nti*b*odies (***mAbs***) as first-stage reagents.

Table 3 Genes of the Ovine Mhc Class II Region

Gene	Copy #	Evidence for existence/absence	Polypeptide Length Total	sp	Mat	Current number of peptide alleles	Current number of PCR amplified alleles	Typing systems	Level of RFLP	PCR
DRA	1	cDNA and genomic sequence	253	24	229	2	—	S, R	Low	
DQA1	0 or 1	cDNA and genomic sequence	255	23	232	7 + null	7	S, R	Moderate	Exon 2
DQA2	1	cDNA and genomic sequence	255	23	232	11	11	S, SSCP, R	High	Exon 2
DPA	0	Not apparent on Southern blots								
DNA	1	Genomic sequence								
DYA	1	Genomic sequence						S, R		Exon 2
DMA	1	PCR fragment								Exon 2
DRB1	1	cDNA and genomic sequence	266	29	237	56	75	S+M	Moderate	Exon 2, Exon2/intron2
DRB2	1	Genomic sequence						M	Moderate	Intron 5
DRB3	?	Microsatellite?								
DQB1/DQB2 ~	1 or 2#	cDNA and genomic sequence	261/262			16		S, SSCP	Moderate, complex	Exon 2
DQB3	?	RFLP?								
DPB	1	Southern blot and PCR fragment								Exon 2
DOB	1	Southern blot								
DYB/DIB	1	Genomic sequence								Exon 3/4
DMB	1	Sequenced PCR fragments							Low	Exon 2, exon 3

Table 4 Genomic Cosmid Clones Containing *Ovar* Class II Genes

Cosmid clone	Sheep	Loci	Seq. accn number	Expressed in mouse L-cells	Ref.
22	A	DRB1	Z11520	Yes	6
9.5	B	DRB1		Yes	6
302	C	DRB1	X76905	Yes	36
9.1	B	DRB3/6		No	5
23.1	B	DRB4		No	5
C5.2*	D	DRB2#	M33307	No	80
46	A	DRA/DRB5#	Z11600	Yes	6
61	A	DRA		Yes	6
13/14/41	C	DQA1/DQB1	Z28418/Z28423	Yes	35
07/43/81	C	DQA1/DQB1	Z28422	Yes	35
C8.1	D	DQA1	M33304		57
4	A	DQA1/DQB1 (T)	Z28422	No	5
39.1	B	DQA1/DQB1	Z28518/Z28424	Yes	5
22.1	B	DQA2/DQB1 (T)	Z28421	No	5
2	A	DQA2	Z28419	No	5
9.2	B	DQA2		No	5
C17.2	D	DQA2	M33305		57
62	A	DQB2	Z28425	No	5
09+79	C	DQB2	Z28523	No	35
B10-1	D	DQB	M33306		80
365	C	DYA/DYB (T)	Z27398/Z27401	No	92
12	A	DNA	Z29533	No	92

Note: The cosmid clones containing *Ovar* class II genes isolated from three unrelated heterozygous sheep. * indicates lambda phage genomic clones isolated from a fourth sheep. *Ovar* class II genes within these clones have been characterized by differential hybridization to human and murine subtype-specific probes and by nucleotide sequence analysis. EMBL genebank accession numbers are indicated where available. The Mhc class II genes that are expressed following transfection of mouse L-cells are also indicated. (T), truncated gene, probably due to the cloning process. (#), pseudogene, i.e., a gene unable to transcribe a complete message due to either loss of an exon, or the presence of a premature stop codon.

Three L-cell lines expressing *Ovar-DR* molecules have been generated following the transfection of the nonpolymorphic *DRA* gene with three alleles of the polymorphic *DRB1* locus.[6,36] This has confirmed that the *DRB1* locus is expressed. All transfections involving genes from additional *DRB* loci (*DRB3* and *DRB4*, Table 4) cotransfected with the expressed *DRA* gene failed to result in a detectable expression of class II molecules at the L-cell surface.

L-cell lines expressing *Ovar-DQ* molecules also have been generated.[35] These each express the products of a pair of *DQA1* and *DQB1* genes linked on individual cosmids and isolated from two unrelated heterozygous sheep (Table 4). Despite extensive efforts no transfectants have been generated that express DQ2 molecules. This is despite the evidence for the transcription of these genes, both *in vivo* and following transient transfection of L-cells.

Table 5 Specificity of *Ovar* Monoclonal Antibodies

Monoclonal antibody	DQ1-1 transfectant	DQ1-2 transfectant	DR1-1 transfectant	DR1-2 transfectant	Subtype specificity	Chain specificity
SW 73.2	+	+	+	+	Pan	Beta
VPM 36	+	+	−	−	DQ	Alpha
VPM16	+	+	−	−	DQ	Beta
VPM 40	−	+	−	−	DQ	Beta
VPM 54	−	−	+	+	DR	Alpha
VPM 57	−	−	+	+	DR	Beta
SBU 28.1	+	+	−	+		Conf
SBU 37.68	−	−	+	+	DR	Conf
SBU 38.27	+	+	−	−	DQ	Conf
SBU 42.20	−	+	+	+		Alpha
SBU 49.1	+	−	+	+		Beta

Note: The specificitiy of two panels of monoclonal antibodies for L-cell transfectants each expressing a single Ovar-DQ or DR subtype, using a fluorescence antibody assay and FACS analysis. Positive antibodies were identified as those that increased the fluorescence relative to cells transfected with the thymidine kinase gene only. Transfectants DQ1-1 and DQ1-2 express both alleles at the DQ1 locus isolated from a single heterozygous sheep polymorphic at both *DQA1* and *DQB1*.[35] Transfectants DR1-1 and DR1-2 express the product of the non-polymorphic *Ovar-DRA* gene with two alleles at the *Ovar-DRB1* locus.[36] The chain specificity of mAbs was determined by separating the chains by SDS-PAGE followed by immunoblotting.[38,39]

2. *Monoclonal Antibodies for* Ovar *Class II Gene Products*

As L-cells do not express endogenous *H-2* class II genes, the established transfectants are able to be used to assign subtype specificity to sheep class II specific mAbs as well as being used as immunogens in the development of novel mAb reagents. To date the *Ovar* class II transfectants have been used to determine the fine specificity for two panels of *Ovar* class II-specific mAbs. The results are summarized in Table 5. The VPM series of antibodies[37,38] and the SBUII series[39] have all previously been characterized using immunochemical techniques. The use of transfectants complements these techniques and often identifies cross-locus specificities (DQ and DR) as well as gaps in the specificity of mAbs previously considered to recognize monomorphic determinants that would otherwise remain undetected.

3. *Patterns of* Ovar *Class II Expression*

The exploitation of the cannulated lymphatic model in sheep permits access to individual cell populations in distinct lymphoid environments[40,41] which are not readily accessible in other species. Studies on the expression of class II genes by these cell populations have utilized a panel of anti-sheep class II mAbs including antibodies specific to total sheep class II[42] and to the α and β chains of the sheep homologs of the DQ and DR subregions.[43,44]

Expression studies using the anti-framework mAb, SW73.2, showed that, in contrast to man and mouse, in the blood and lymph of sheep the proportion of class II-positive lymphoid cells is greater than the sum of the B cells, dendritic cells (DC), and monocytes.[44,45] Furthermore, the expression of class II differed between the different compartments.[44] In blood, where the percentage of B cells and monocytes is no more than 45% at least 60% of cells express class II molecules. The discrepancy in afferent (peripheral) lymph is even greater; up to 90% of cells are positive for SW73.2, whereas the sum of B cells and DC is less than 25%. Similarly, efferent lymph, which is comprised entirely of B cells (<25%) and T cells (<75%), has between 45 and 55% class II-positive cells. Two-color flow cytometry using T-cell- and B-cell-specific mAbs showed that the residual class II-positive cells represent T cells of all three subpopulations.[44] The sheep is the only species so far examined where only a proportion of T cells express class II. In man and mice nonactivated T cells do not express class II[46,47] whereas equine[48] T cells are all class II positive.

Use of the subregion-specific mAbs shows that T cells in the different lymphoid compartments have distinct phenotypes. T cells in efferent lymph and blood expressed almost exclusively DR, whereas DQ molecules are essentially absent from T cells in these compartments. In contrast afferent lymph T cells, almost all of which are class II positive, express approximately equal quantities of both DQ and DR. Examination of the afferent T cells with mAbs of different specificities indicates that they are chiefly activated cells.[49] This implies that, as in humans,[50] the expression of DQ by sheep T cells is a marker of recent activation. These data also illustrate the plasticity of T-cell phenotype. The migration of lymphocytes across endothelia and into tissues (e.g., skin) requires adhesion via phosphorylated β2 integrins. This occurs by activation of the leukocytes which is mediated by contact between leukocytes and endothelia and by endothelia-derived factors.[51] The high level of DQ and DR expression by afferent lymph T cells is likely to be a consequence of this activation. Activated cells in the blood express adhesive integrins and would extravasate preferentially to nonactivated cells.[52] These data also imply that the actual act of endothelial interaction also leads to activation and a high level of class II expression. The low expression of DQ in efferent lymph may reflect the fact that these cells do not appear in the efferent lymph until 30 h after they have crossed the endothelium and by this time they may have reverted to a DQ^-DR^+ phenotype.[53] This hypothesis is supported by the fact that T cells isolated from peripheral lymph nodes have a phenotype intermediate between afferent and efferent lymph.

4. *Quantitative Modulation of* Ovar *Class II Surface Expression*

This pattern of class II expression is not fixed in most cells but is modulated during *in vivo* immune responses.[46,54] Different cell populations express different quantities of class II molecules. Afferent dendritic cells (DC) express approximately 3×10^5, afferent B and T cells express 1×10^5, and efferent lymph T cells express 2×10^4 molecules per cell although the level of expression within each of these populations is rather heterogeneous. During primary immune responses *in vivo* the levels of class II expression remain largely unaltered but during secondary responses DC and T-cell expression is rapidly upregulated. By day 2 of the secondary response there is a twofold increase and by day 4 to 5 a four- to sevenfold increase over resting levels by both DC and T cells. B-cell expression of class II remains unaffected.

The functional consequence of modulated class II expression on the antigen-presenting function is that DCs, taken during a secondary response and expressing an increased level of class II, are up to fivefold more active in their ability to present soluble antigen to T cell lines and to stimulate mixed leukocyte reactions (MLR). The relative expression of DC class II correlates temporally with the increased capacity to present antigen. In contrast high class II-expressing efferent lymph T cells cannot present soluble or peptide antigen but are weakly stimulatory in the MLR.[54] These data make clear that the magnitude of T-cell responses to antigen presented by DC is directly related to the quantity of class II expressed by the presenting cell. In contrast the class II molecules expressed by T cells possibly do not simply act as antigen-presenting molecules but may also play a more subtle role in intercellular signaling.

B. A Genomic Map of the *Ovar* Class II Region

The map in Figure 2 is a schematic based on the arrangement of the human class ll region. The shaded boxes represent overlapping cosmid clones, with the *DQ* subregion expanded to show the arrangement of the *DQ* genes. The *DY* genes have been placed close to the *DNA* gene because these genes segregate together in cattle.[55,56] The order of any of the subregions has not been formally proven but is currently under investigation by pulse field gel electrophoresis (PFGE).

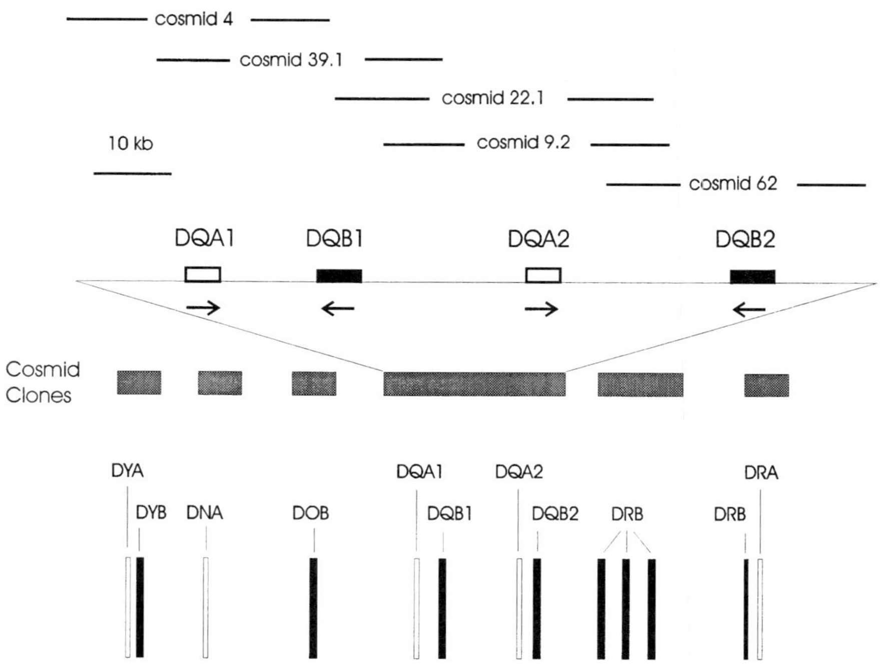

Figure 2 Schematic map based on the arrangement of the human class II region.

C. The *Ovar-DR* Subregion

Genomic regions from four different *Ovar-DRB* loci have been sequenced[5,35,57] and evidence exists for the presence of additional ovine *DRB* genes, both functional and pseudogenes, in sheep. Early work by Puri and colleagues[58] determining N-terminal amino acid sequences identified a number of different B polypeptides from a single sheep and suggested that these provided evidence for at least nine nonallelic *B* genes. Of these, seven B polypeptides were detected in association with *DRA* chains. However, the N-terminal peptide sequences determined in this study were too complex, with multiple residues at most positions, for the accurate determination of the number of polypeptides. At a minimum they provide evidence for the existence of more than one locus whose products are found in association with *DRA* polypeptides. Recently Dutia et al.[59] describe the expression of two distinct DRβ-like molecules. Using immunochemical techniques as well as transfected cell lines, they show that, depending on haplotype, individual animals can express at least two distinct, nonpolymorphic DRb chains each associated with the DRa chain.

The evidence for the existence for extra *DRB* pseudogenes is fairly tentative. Primers designed to amplify microsatellites of two bovine *DRB* pseudogenes,[60,61] distinct from homologues of the *Ovar-DRB1* intron 2 and *Ovar-DRB2* intron 5 microsatellites, amplify distinct fragments in sheep (S. Paterson, unpublished data).

The findings for the sheep *DRB* region reflect the situation found in several other species where multiple *DRB* loci, functional and pseudogenes, occupy a region of several hundred kilobases. Haplotype variation has been found in humans where six different haplotype groups have been identified. These haplotype groups harbor both different numbers of pseudogenes and different numbers of expressed loci (one or two).[62] Differing numbers of *DRB* loci per haplotype have been found also in other primate species.[63,64] The preliminary work described above supports the idea that there is haplotype variation in the number of *DRB* loci in sheep.

1. Ovar-DRA *and* -DRB *Gene Containing Cosmids*

Eleven cosmid clones (Tables 3 and 4) containing *DR*-like genes have been isolated from three libraries.[5,35] Two clones contained the *Ovar-DRA* gene and one of these (clone 46) also possessed a *DRB* pseudogene lacking exons 1 and 2. This represents an analogous situation to that described for *HLA-DRB9*.[62] The products of two other clones, 22/9.5 and 302, have both been expressed on the cell surface of L-cells when cotransfected with the *DRA* gene.[6] Sequencing of the second exons of these *DRB* genes revealed that they were similar to those of the *Ovar-DRB1* locus[65] amplified from genomic DNA by PCR.

The sequences of the second exons from the *B* gene for two other cosmids (9.1 and 32) were shown to be identical[5] and the aa translation of this gene shows only 62% identity when compared with those for the expressed *DRB1* gene. Additionally, the sequence of the *B* gene in cosmid 23.1 (Tables 3 and 4) showed only 65% identity to that for the expressed *DRB1* gene but only 65% identity to the cosmid 9.1/32 sequence. The three genes all contain a simple tandem repeat sequence (str) in intron 2. This intron 2 str is commonly found in *DRB* genes from a number of species (from primates[66] and Artiodactyls[67,68]). The strs in the *DRB* genes from cosmid 9.1/32 and 23.1 are quite distinct from any of the published *Ovar-DRB1* strs or strs from other species and also quite distinct from each other (Wright et al., unpublished observations; see Table 4). The PCR primers described for the amplification for the *Ovar-DRB1* locus[68] do not amplify these divergent sequences even when low-stringency PCR conditions are applied (unpublished observations).

Four different *DRB* genes have been cloned from sheep B (Table 4): the expressed *DRB1* gene in cosmid 9.5, the two genes in cosmid 9.1, one of which has not been sequenced, and the gene in cosmid 23.1. The latter three may be pseudogenes and at present, no homologous cattle sequences exist within the EMBL database. Table 6 shows the homologies of the *Ovar-DRB* loci.

Table 6 Comparison of Polymorphisms in *Ovar-DRB* Genes

Comparison	Range of nucleotide changes Exons 3–6	Range of amino acid changes b2, CP, TM, CY
DRB1 vs. *DRB1**	1–4	0–2
*Ovar-DRB*08* vs. *DRB1*	13–17	2–4
DRB2 vs. *DRB1*	50–51	28–29
DRB2 vs. *Ovar-DRB*08*	55	29

Note: Excluding (*Ovar-DRB1*08* group).

2. *The* Ovar-DRA *Locus*

Sheep possess a single *DRA* gene[69] which encodes a mature polypeptide of 229 aa.[70] Sequencing and RFLP studies of the *Ovar-DRA* gene has revealed that this gene has only a low level of polymorphism.[6,70,71] There are two deduced peptide alleles which show conservative differences at two aa sites (H^{50}/R^{50}, T^{109}/I^{109}).

3. *The* Ovar-DRB1 *Locus*

Most of the *Ovar-DRB1* sequences were generated by using a PCR system amplifying the second exon of the *DRB1* gene (Figure 3, Table 7) together with part of the second intron which includes a simple tandemly arrayed dinucleotide repeat (Table 8). Electroblot hybridizations of genomically amplified PCR products of several large full-sibling (IMF families, AgResearch, A. Crawford) and half-sibling families[72] (Scottish Blackface, M. Stear) showed that only one *DRB* locus was amplified. The identity of some of the genomic *Ovar-DRB1* alleles with those identified by cDNA analysis (Fabb et al., Maddox et al., unpublished data) and with the expressed *DRB* from cosmids 9.5 and 32 (see Table 7) suggests that the PCR amplified genomic *Ovar-DRB1* locus is expressed.

a. *Polymorphisms in the Second Exon of the* Ovar-DRB1 *Genes*

The deduced aa sequences of all known *DRB1* alleles were used to analyze *Ovar-DRB1* polymorphisms and compared to the three-dimensional crystal structure.[73]

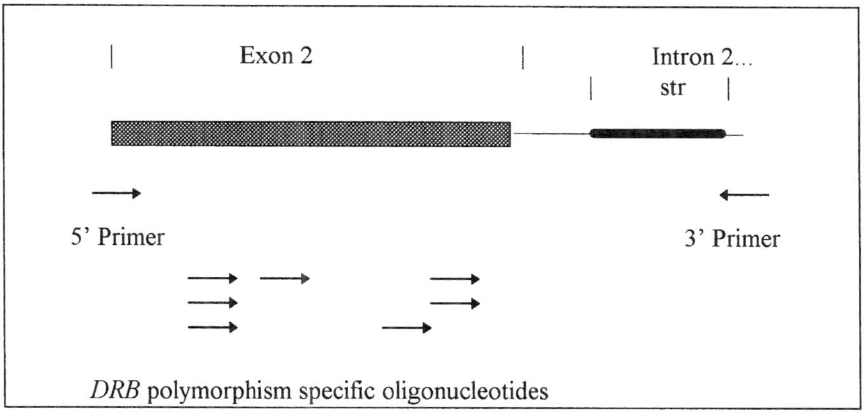

Figure 3 Schematic presentation of the PCR system for the amplification of *Ovar-DRB1* exon 2 with adjacent intronic str. Approximately 30 bp 3′ the exon 2/intron 2 boundary is a str stretch hypervariable in length but correlated to the exonic polymorphism. Beneath the length variability of the PCR products, a set of oligonucleotides specific for the polymorphism found in the bovine *DRB* orthologs complements the typing of ovine *DRB1* alleles.

Table 9 presents a comparison of polymorphic aa of human, cattle, and sheep *DRB* sequences. While each species shows polymorphisms in the same aa positions, bovine or *Pecora* (Cervidoidae + Bovidoidae) specific aa were found for several positions of *Ovar-DRB1* alleles (11, 13, 18, 32, 34, 37, 42, 57, 60, 66, 73). As expected the most striking similarities are found between the *DRB1* loci of goat and sheep. Both the similarity in the number of polymorphisms at most aa sites and the number of Caprinae-specific substitutions (e.g., aa position 42 or 73) show the close relationship of the two species. Interestingly some highly polymorphic aa positions in the first domain of the *DRB* chain (13, 70, and 74) revealed only two or three different aa in *Ovar-DRB1*. They are predominantly occupied by only one of these aa.

As was proposed for other species (e.g., mouse,[74] human,[75] artiodactyl species[68]) small polymorphic DNA motifs within the *Ovar-DRB1* exon 2 seem to have been exchanged between different alleles (Figure 4). Sequence exchange has been observed predominantly downstream of the 5′ end of exon 2 but does not include the str. This was shown by phylogenetic analysis. The β-pleated sheet coding sequences and the adjacent str revealed more similarity in their phylogenetic relationships than the β-pleated sheet and the α-helix coding region. Additionally the C-terminus of *DRB1* first domain seems to have a different evolutionary history compared to all other regions. Hence, *DRB1* alleles generate their polymorphism at least in part by "motif shuffling".

b. Variability of the Intronic Microsatellite Adjacent to the Ovar-DRB1 *Exon 2*

Almost all second introns of primate, rodent, bovid, and cervid *DRB* genes contain a str with the basic structure $(gt)_n(ga)_m$ (for sheep see Table 8) found in approximately the same position.[131] Although the strs are highly variable in length

Table 7 Exon 2 Sequences of *Ovar-DRB* Genes Compared to *HLA-DRB* Consensus Sequence

Aminoacid # / HLA-DRB Cons.	EQ	11 SKSECHFFNG	21 TERVRFLDRY	31 FYNQEEYVRF	41 DSDVGEYRAV	51 TELGRPDAEY	61 WNSQKDLLEQ	71 RRAAVDTYCR	81 HNYGVGESFT	91 VQRR	#Ind/Breed
*Ovar-DRB1*0101L526*	-y	t-k--r-s--	----------	-h-g--tl--	---w------	a-------k-	------f--r	a---------	-----i----	----	8/P,R
*Ovar-DRB1*0101L530*	-y	t-k--r-s--	----------	-h-g--tl--	---w------	a-------k-	------f--r	a---------	-----i----	----	1/P
*Ovar-DRB1*0101H522*	-y	t-k--r-s--	----------	-h-g--tl--	---w------	a-------k-	------f--r	a---------	-----i----	----	1/B
*Ovar-DRB1*0101H526*	-y	t-k--x-x--	----------	-h-g--xx--	---w------	x-------k-	------f--x	xxxxx-----	-----i---x	----	/B,R
*Ovar-DRB1*0101H534*	-y	t-k--x-x--	----------	-h-g--xx--	---w------	x-------k-	------f--x	xxxxx-----	-----i---x	----	/B
*Ovar-DRB1*0110H568*	-y	t-k--x-x--	----------	-h-g--xx--	---w------	x-------k-	-----xx--x	xxxxx-----	-----i---x	----	/B
*Ovar-DRB1*0110H572*	-y	t-k--x-x--	----------	-h-g--xx--	---w------	x-------k-	-----xx--x	xxxxx-----	-----i---x	----	/B
*Ovar-DRB1*0102*	-y	t-k--r-s--	----------	-h-g--tl--	-- w------	a-------k-	------f--r	a--t------	-----i----	----	1/A
*Ovar-DRB1*01201542*	-y	t-k--r-s--	----------	---g------	---w--f---	a-------k-	-----e---?	????------	----???--?	----	1/Mu
*Ovar-DRB1*01031524*	-y	t-k--r-s--	----------	-h-g--tl--	---w------	a-------k-	------f--r	a---------	---------s	----	4/P,T
*Ovar-DRB1*0103H520*	-y	t-k--x-x--	----------	-h-g--xx--	---w------	x-------k-	------f--x	xxxxx-----	---------x	----	/B,T
*Ovar-DRB1*0104*	-y	t-k--r-s--	----------	-h-g--tl--	---w------	a---------	------f--r	a---------	---------s	----	1/A
*Ovar-DRB1*01051514*	-y	t-k--r-s--	----------	---g---a--	---w------	a---------	-----ei---	t---------	----------	----	1/F
*Ovar-DRB1*0105H512*	-y	t-k--x-x--	----------	---g--xx--	---w------	x-------x-	-----xx--x	xxxxx-----	---------x	----	/B
*Ovar-DRB1*0112H522*	-y	t-k--x-x--	----------	---g--xx--	---w------	x-------x-	-----xx--x	xxxxx-----	-----x---x	----	/B
*Ovar-DRB1*01071542*	-y	t-k--r-s--	----------	---g------	---w------	a-------k-	-----ei--r	--te------	-----i---s	----	1/F,S**
*Ovar-DRB1*0111H546*	-y	t-k--x-x--	----------	---g--xx--	---w------	x---- --k-	-----xx--x	xxxxx-----	-----i---x	----	/B,R
*Ovar-DRB1*01081476*	-y	t-k--r-s--	----------	---g---a--	---w------	a-------k-	-----e---r	k--n------	----------	----	1/F
*Ovar-DRB1*01091484*	-y	t-k--r----	-----y----	---g------	-n-w------	a-------k-	-----e---r	k--n------	----------	----	1/T
*Ovar-DRB1*0113H482*	-y	t-k--x-x--	-----x-x--	---g--xx--	---w------	x-------k-	-----xx--x	xxxxx-----	---------x	----	/T,R
*Ovar-DRB1*0114*	-y	t-k-------	-----l-e--	---g------	---w--f---	a-------k-	------f--s	--t-------	----------	----	1/M *
*Ovar-DRB1*0201L544*	-y	-t--------	----------	---g------	---w------	a-------k-	-----ei--r	--te------	-----i---s	----	2/C
*Ovar-DRB1*0201L556*	-y	-t--------	----------	---g------	---w------	a-------k-	-----ei--r	--te------	-----i---s	----	/B,C
*Ovar-DRB1*02021518*	-y	-t--------	----------	---g--tl--	---w------	a-------k-	------f---	t-t-------	-----i---s	----	1/R
*Ovar-DRB1*0202H514*	-y	-t---x-x--	----------	---g--xx--	---w------	x-------k-	------f--x	xxxxx-----	-----i---x	----	/C,R
*Ovar-DRB1*0203L516*	-y	-t--------	----------	---g--tl--	---w------	a-------k-	-----ei--r	k---------	----------	----	1/T
*Ovar-DRB1*0203L512*	-y	-t--------	----------	---g--tl--	---w------	a-------k-	-----ei--r	k---------	----------	----	/B
*Ovar-DRB1*04111486*	-y	a----r----	-------e--	---g--tl--	---w------	a-------k-	-----e---r	k--n------	-----d----	----	1/P
*Ovar-DRB1*04121486*	-y	a----r----	-------e--	---g--tl--	---w------	a-------k-	-----e---r	k--n------	----------	----	1/F
*Ovar-DRB1*04011478*	-y	a----r----	-------e--	---g------	---w------	a---------	------f--r	k--n------	----------	----	1/C
Ovar-DRB1+0401H476	-y	xxx--x-x--	-----x-x--	---g--xx--	---w------	x----xxxx-	------f--x	xxxxx-----	---------x	----	/C
*Ovar-DRB1*04021480*	-y	a----r----	-------e--	---g------	---w------	a----rs---	------f--r	k--n------	----------	----	1/P
*Ovar-DRB1*04031480*	-y	a----r----	-------e--	---g------	---w------	a-------k-	------f--r	k--n------	-----f---s	----	5/M,A
*Ovar-DRB1*0403H476*	-y	xxx--x-x--	-----x-x--	---g--xx--	---w------	x-------k-	------f--x	xxxxx-----	-----f---x	----	/M
*Ovar-DRB1*04041490*	-y	a----r----	-------e--	---g------	---w------	a-------k-	------f--r	k--n------	-----i---s	----	1/F
*Ovar-DRB1*04051476*	-y	a----r----	-------e--	---g------	---w------	a-------k-	------f--r	k--n------	----------	----	1/C
*Ovar-DRB1*0408H470*	-y	xxx--x-x--	-----x-x--	---g--xx--	---w------	x----xxxx-	-----xx--x	xxxxx-----	---------x	----	/R
*Ovar-DRB1*0409H476*	-y	xxx--x-x--	-----x-x--	-h-g--xx--	---w------	x-------k-	-----xx--x	xxxxx-----	-----i---x	----	/R,C
*Ovar-DRB1*0410H484*	-y	xxx--x-x--	-----x-x--	---g--xx--	---w------	x-------k-	-----xx--x	xxxxx-----	---------x	----	/B
*Ovar-DRB1*0501L490*	-y	a-r-------	----------	---g------	---ws-f---	a---------	------f--r	a---------	----------	----	1/F

Ovar-DRB1*0301L478	-y	h---------	----------	-h-g--f---	---w--f---	a----rs---	-----e---r	--te------	-----f---s	----	1/R
Ovar-DRB1*0301L476	-y	h---------	----------	-h-g--f---	---w--f---	a----rs---	-----e---r	--te------	-----f---s	----	3/P,R
Ovar-DRB1*0311H498	-y	h----x-x--	----------	-h-g--f---	---w------	x----xxxxx	-----xx--x	xxxxx-----	-----f---x	----	/R
Ovar-DRB1*0312H498	-y	h----x-x--	----------	-h-g--f---	---w------	x-------k-	-----xx--x	xxxxx-----	-----f---x	----	/R
Ovar-DRB1*0313H500	-y	h----x-x--	-----x-x--	-h-g--f---	---w------	x----xxxxx	-----xx--x	xxxxx-----	-----f---x	----	/R
Ovar-DRB1*0302L500	-y	h---------	----------	---g------	---w--f---	a--------h	-----e---r	---e------	-----i---s	----	1/P
Ovar-DRB1*0303L502	-y	h---------	----------	---g------	---w--f---	a----qs--h	-----e---r	---e------	-----i---s	----	1/M
Ovar-DRB1*0304L494	-y	h---------	----------	---g------	---w--f---	a----qs---	------f--r	a---------	-----f---s	----	1/M
Ovar-DRB1*0305L476	-y	h---------	----------	---g------	---w------	a-------k-	------f--r	k--n------	----------	----	1/F
Ovar-DRB1*0314H476	-y	h----x-x--	----------	xx-g--xx--	---w------	x----xxxxx	------f--x	xxxxx-----	-----f---x	----	/B
Ovar-DRB1*0315H498	-y	h----x-x--	----------	xx-g--xx--	---w------	x----xxxxx	------f--x	xxxxx-----	---------x	----	/B
Ovar-DRB1*0316H480	-y	h----x-x--	----------	---g--xx--	---w------	x----xxxxx	------f--x	xxxxx-----	-----i---x	----	/B
Ovar-DRB1*0317H492	-y	h----x-x--	----------	---g--xx--	---w------	x----xxxxx	------f--x	xxxxx-----	---------x	----	/C
Ovar-DRB1*0306L476	-y	h----r-s--	-----y----	---g------	---w------	a---------	------f--r	k--n------	----------	----	1/C
Ovar-DRB1*0307	-y	h----r-s--	-----y----	---g------	-n-w--f---	a---------	------f--r	k--n------	-----i---s	----	1/K
Ovar-DRB1*0308L480	-y	h----r-s--	-----y----	---g------	-n-w------	a-------k-	-----ei--r	k---------	-----i---s	----	1/M
Ovar-DRB1*0321	-y	h----r-s--	-----y----	---g------	-n-w------	a-------k-	------f---	t-te------	-----i---s	----	1/S **
Ovar-DRB1*0318H492	-y	h----x-x--	----------	---g--xx--	---w------	x----xxxxx	-----xx--x	xxxxx-----	-----f---x	----	/R
Ovar-DRB1*0318H500	-y	h----x-x--	----------	---g--xx--	---w------	x----xxxxx	-----xx--x	xxxxx-----	-----f---x	----	/B
Ovar-DRB1*0319H498	-y	h----x-x--	-----x-x--	---g--f---	---w------	x----xxxxx	-----xx--x	xxxxx-----	-----f---x	----	/C
Ovar-DRB1*0320H500	-y	h----x-x--	----------	---g--xx--	---w------	x----xxxxx	-----xx--x	xxxxx-----	-----i---x	----	/R
Ovar-DRB1*0309L490	-y	h---------	----------	---g------	---w--f---	a-----e-k-	-----ei--r	--te------	-----i---s	----	1/T
Ovar-DRB1*0601L480	-y	r----r-s--	-----y----	---g------	---w------	a-------k-	-----e---r	k--n------	-----i---s	----	1/M
Ovar-DRB1*0310	-y	h---------	----wy----	---g------	-n-w------	a-----s-k-	-----ei--e	----------	-----i----	----	1/R
Ovar-DRB1*0701	-y	----------	----------	yt-g--n---	---w--f---	---------q	------f--r	k-t-------	-----i---s	----	1/Mu
Ovar-DRB1*0801	-y	yr--------	-----l-e--	-h-g--fa--	---w--f---	------a--q	-----ni---	k--e-n-v--	-----f---a	----	1/F
Ovar-DRB1*0801H416	-y	yr---x-x--	-----x-x--	-h-g--xx--	-x-w------	x-----xxxx	-----xx--x	xxxxxxxx--	-----x---x	----	/B,R
Ovar-DRB1*0802	-y	yr--------	-----l-e--	-h-g--fa--	---w--f---	-------v-q	-----ni---	k--e---v--	---------a	----	1/A
Ovar-DRB1*0902	-y	t-r-------	--w--l-e--	---g--r---	---w--f---	a----r--k-	------f--s	k-????????.??????????.????..			
Ovar-DRB1*0903H1>800	-y	xxx--x-x--	-----x-x--	---g--x---	---w------	x-------k-	-----xx--x	xxxxx-----	---------x	----	/C
Ovar-DRB1*0903H2>800	-y	xxx--x-x--	-----x-x--	---g--x---	---w------	x-------k-	-----xx--x	xxxxx-----	---------x	----	/R
Ovar-DRB1*0904H1>800	-y	xxx--x-x--	-----x-x--	---g--x---	---w------	x-------x-	-----xx--x	xxxxx-----	---------x	----	/B
Ovar-DRB1*0904H2>800	-y	xxx--x-x--	-----x-x--	---g--x---	---w------	x-------x-	-----xx--x	xxxxx-----	---------x	----	/B
Ovar-DRB1*0904H3>800	-y	xxx--x-x--	-----x-x--	---g--x---	---w------	x-------x-	-----xx--x	xxxxx-----	---------x	----	/B
Cosmid 23.1	v-	g------s--	-qq-------	i--r--q---	--l------r	--m---agqr	--rwp-a-qr	a----hdf-a	s--effa-r-	----	
Cosmid 9.1	h-	f-g--r-s--	l--m--fa--	i--tk-d-h-	------ft-l	----------	--q---fm--	m--kv--v--	s--qgig--l	r---	

Note: Protein sequences deduced from the second exon of *Ovar-DRB1* alleles derived from genomic or cDNA or from typing patterns revealed by hybridization of polymorphism specific oligonucleotides. Grouping is based on the similarities in the β-pleated sheets and the adjacent simple repeat derivative structure (group specific organization: GSO, see text and References 67 and 68). The number of individuals carrying a given allele, the breeds in which an allele was found, and the frequency of typed alleles in the investigated sheep population is indicated on the righthand side. For each cloned allele at least two independent clones were sequenced. Abbreviations: A: Argali; B: Scottish Blackface; C: Coopworth; F: Finish Landrace; K: Kasach Sheep; M: Spanisch Merino Sheep; P: Perendale; R: Romney; S: Suffolk cross; T: Texel; Mu: Mufflon" x Sequence was not determined by specific hybridization. ? Sequence could not determined by cloning and sequencing. * Sequence from EMBL databank [accession number: M73984; Fabb et al., unpublished.] ** Sequence from Reference 6.

Table 8 Intronic Simple Repeat Sequences in the Second Exon of *Ovar-DRB* Genes

***Ovar-DRB1* alleles** **Allel**	**simple repetitive element**					
*Ovar-DRB1*0101L526/530*		(gt)$_{32}$ (ga)$_{15}$ (gc)$_{5}$	(ga)$_{4}$ (gc)$_{4}$	[(ga)$_{2}$ gc]$_{3}$ gaca (ga)$_{4}$ ca (ga)$_{3}$ gt (ga)$_{7}$	(caga)$_{2}$	(gaca)$_{3}$ ga
*Ovar-DRB1*0102*		(gt)$_{27}$ (ga)$_{15}$ (gc)$_{4}$	(ga)$_{4}$ (gc)$_{4}$	[(ga)$_{2}$ gc]$_{3}$ gaca (ga)$_{4}$ ca (ga)$_{3}$ gtgg (ga)$_{6}$	(caga)$_{2}$	(gaca)$_{4}$ ga
*Ovar-DRB1*01031524*		(gt)$_{40}$ (ga)$_{13}$ (gc)$_{2}$	(ga)$_{4}$ (gc)$_{4}$	[(ga)$_{2}$ gc]$_{3}$ gaca (ga)$_{4}$ ca (ga)$_{3}$ gtgg (ga)$_{6}$	(caga)$_{2}$	(gaca)$_{3}$ ga
*Ovar-DRB1*01201542*		(gt)$_{22}$ (ga)$_{21}$ (gc)$_{2}$	(ga)$_{7}$ (gc)$_{4}$	[(ga)$_{2}$ gc]$_{3}$ (ga)$_{6}$ ca (ga)$_{3}$ gtgg (ga)$_{6}$	(caga)$_{2}$ (ga)$_{2}$	(gaca)$_{2}$ ga
*Ovar-DRB1*0104*		(gt)$_{22}$ (ga)$_{19}$ (gc)$_{2}$	(ga)$_{8}$ (gc)$_{4}$	[(ga)$_{2}$ gc]$_{3}$ (ga)$_{6}$ ca (ga)$_{3}$ gtgg (ga)$_{6}$	(caga)$_{2}$ (ga)$_{2}$	(gaca)$_{2}$ ga
*Ovar-DRB1*01051514*		(gt)$_{13}$ (ga)$_{26}$ (gc)$_{3}$	(ga)$_{5}$ (gc)$_{3}$	[(ga)$_{2}$ gc]$_{3}$ (ga)$_{6}$ ca (ga)$_{3}$ gtgg (ga)$_{6}$	(caga)$_{2}$ (ga)$_{2}$	(gaca)$_{2}$ ga
*Ovar-DRB1*01071542*	[(gt)$_{3}$ gc]$_{2}$ [(gt)$_{2}$ gc]$_{4}$	(gt)$_{19}$ (ga)$_{22}$ gc		[(ga)$_{2}$ gc]$_{4}$ gaca (ga)$_{4}$ ca (ga)$_{3}$ gtgg (ga)$_{6}$	(caga)$_{2}$	(gaca)$_{2}$ ga
*Ovar-DRB1*01081476*	[(gt)$_{3}$ gc]$_{2}$ [(gt)$_{2}$ gc]$_{2}$ gt gc	(gt)$_{9}$ (ga)$_{7}$ (gc)$_{2}$		[(ga)$_{2}$ gc]$_{4}$ gaca (ga)$_{4}$ ca (ga)$_{3}$ gtgg (ga)$_{6}$	(caga)$_{2}$	(gaca)$_{2}$ gg
*Ovar-DRB1*01091484*	[(gt)$_{3}$ gc]$_{2}$ [(gt)$_{2}$ gc]$_{2}$	(gt)$_{13}$ (ga)$_{7}$ (gc)$_{2}$		[(ga)$_{2}$ gc]$_{4}$ gaca (ga)$_{4}$ ca (ga)$_{3}$ gtgg (ga)$_{6}$	(caga)$_{2}$	(gaca)$_{2}$ ga
*Ovar-DRB1*0201L544*	[(gt)$_{3}$ gc]$_{2}$ [(gt)$_{2}$ gc]$_{4}$	(gt)$_{18}$ (ga)$_{24}$ gc		[(ga)$_{2}$ gc]$_{4}$ gaca (ga)$_{4}$ ca (ga)$_{3}$ gtgg (ga)$_{6}$	(caga)$_{2}$	(gaca)$_{3}$ ga
*Ovar-DRB1*02021518*		(gt)$_{21}$ (ga)$_{16}$ (gc)$_{4}$ (ga)$_{2}$ gg	(ga)$_{2}$ (gc)$_{4}$	[(ga)$_{2}$ gc]$_{3}$ (ga)$_{6}$ ca (ga)$_{3}$ gtgg (ga)$_{6}$	(caga)$_{2}$	(gaca)$_{3}$ ga
*Ovar-DRB1*0203L516*		(gt)$_{12}$ (ga)$_{12}$ (gc)$_{3}$ (ga)$_{13}$gc	(ga)$_{5}$ (gc)$_{3}$	[(ga)$_{2}$ gc]$_{3}$ (ga)$_{6}$ ca (ga)$_{3}$ gtgg (ga)$_{6}$	(caga)$_{2}$ (ga)$_{2}$	(gaca)$_{2}$ ga
*Ovar-DRB1*04111486*	[(gt)$_{3}$ gc]$_{2}$ [(gt)$_{2}$ gc]$_{2}$ gt gc	(gt)$_{11}$ (ga)$_{7}$ (gc)$_{2}$		[(ga)$_{2}$ gc]$_{4}$ gaca (ga)$_{4}$ ca (ga)$_{3}$ gtgg (ga)$_{6}$	(caga)$_{2}$	(gaca)$_{2}$ gg
*Ovar-DRB1*04121486*	[(gt)$_{3}$ gc]$_{2}$ [(gt)$_{2}$ gc]$_{2}$ gt gc	(gt)$_{12}$ (ga)$_{7}$ (gc)$_{2}$		[(ga)$_{2}$ gc]$_{4}$ gaca (ga)$_{4}$ ca (ga)$_{3}$ gtgg (ga)$_{6}$	(caga)$_{2}$	(gaca)$_{2}$ gg
*Ovar-DRB1*04011478*	[(gt)$_{3}$ gc]$_{2}$ [(gt)$_{2}$ gc]$_{3}$	(gt)$_{14}$ (ga)$_{7}$ (gc)$_{2}$		[(ga)$_{2}$ gc]$_{2}$ gaca (ga)$_{4}$ ca (ga)$_{3}$ gtgg (ga)$_{6}$	(caga)$_{2}$	(gaca)$_{2}$ gg
*Ovar-DRB1*04021480*	[(gt)$_{3}$ gc]$_{2}$ [(gt)$_{2}$ gc]$_{3}$	(gt)$_{12}$ (ga)$_{7}$ (gc)$_{2}$		[(ga)$_{2}$ gc]$_{2}$ gaca (ga)$_{4}$ ca (ga)$_{3}$ gtgg (ga)$_{6}$	(caga)$_{2}$	(gaca)$_{2}$ gg
*Ovar-DRB1*04031480*	[(gt)$_{3}$ gc]$_{2}$ [(gt)$_{2}$ gc]$_{3}$	(gt)$_{14}$ (ga)$_{7}$ (gc)$_{2}$		[(ga)$_{2}$ gc]$_{2}$ gaca (ga)$_{4}$ ca (ga)$_{3}$ gtgg (ga)$_{6}$	(caga)$_{2}$	(gaca)$_{2}$ gg
*Ovar-DRB1*04041490*	[(gt)$_{3}$ gc]$_{2}$ [(gt)$_{2}$ gc]$_{3}$	(gt)$_{13}$ (ga)$_{7}$ (gc)$_{2}$		[(ga)$_{2}$ gc]$_{4}$ gaca (ga)$_{4}$ ca (ga)$_{3}$ gtgg (ga)$_{6}$	(caga)$_{2}$	(gaca)$_{2}$ gg
*Ovar-DRB1*04051476*	[(gt)$_{3}$ gc]$_{2}$ [(gt)$_{2}$ gc]$_{3}$	(gt)$_{13}$ (ga)$_{7}$ (gc)$_{2}$		[(ga)$_{2}$ gc]$_{2}$ gaca (ga)$_{4}$ ca (ga)$_{3}$ gtgg (ga)$_{6}$	(caga)$_{2}$	(gaca)$_{2}$ gg

*Ovar-DRB1*0501L490*	[(gt)$_{3}$ gc]$_{2}$ [(gt)$_{2}$ gc]$_{2}$ (gt)$_{7}$ (ga)$_{17}$ (gc)$_{2}$	[(ga)$_{2}$ gc]$_{4}$ gaca (ga)$_{4}$ ca (ga)$_{3}$ gtgg (ga)$_{6}$ (caga)$_{2}$ (ga)$_{2}$	(gaca)$_{2}$ ga
*Ovar-DRB1*0301L478*	[(gt)$_{3}$ gc]$_{2}$ [(gt)$_{2}$ gc]$_{2}$ (gt)$_{14}$ (ga)$_{13}$ (gc)$_{2}$	[(ga)$_{2}$ gc]$_{4}$ gaca (ga)$_{4}$ ca (ga)$_{3}$ gtgg (ga)$_{6}$ (caga)$_{2}$	(gaca)$_{2}$ ga
*Ovar-DRB1*0302L500*	[(gt)$_{3}$ gc]$_{2}$ [(gt)$_{2}$ gc]$_{2}$ (gt)$_{13}$ (ga)$_{14}$ (gc)$_{2}$	[(ga)$_{2}$ gc]$_{4}$ gaca (ga)$_{4}$ ca (ga)$_{2}$gcgtgg (ga)$_{6}$ (caga)$_{2}$	(gaca)$_{2}$ ga
*Ovar-DRB1*0303L502*	[(gt)$_{3}$ gc]$_{2}$ [(gt)$_{2}$ gc]$_{2}$ (gt)$_{13}$ (ga)$_{16}$ (gc)$_{2}$	[(ga)$_{2}$ gc]$_{4}$ gaca (ga)$_{4}$ ca (ga)$_{3}$ gtgg (ga)$_{5}$ (caga)$_{2}$	(gaca)$_{2}$ ga
*Ovar-DRB1*0304L494*	[(gt)$_{3}$ gc]$_{2}$ [(gt)$_{2}$ gc]$_{2}$ (gt)$_{14}$ (ga)$_{12}$ (gc)$_{2}$	[(ga)$_{2}$ gc]$_{4}$ gaca (ga)$_{4}$ ca (ga)$_{3}$ gtgg (ga)$_{6}$ (caga)$_{2}$	(gaca)$_{2}$ ga
*Ovar-DRB1*0305L476*	[(gt)$_{3}$ gc]$_{2}$ [(gt)$_{2}$ gc]$_{3}$ (gt)$_{13}$ (ga)$_{7}$ (gc)$_{2}$	[(ga)$_{2}$ gc]$_{3}$ gaca (ga)$_{4}$ ca (ga)$_{3}$ gtgg (ga)$_{6}$ (caga)$_{2}$	(gaca)$_{2}$ gg
*Ovar-DRB1*0306L476*	[(gt)$_{3}$ gc]$_{2}$ [(gt)$_{2}$ gc]$_{3}$ (gt)$_{14}$ (ga)$_{7}$ (gc)$_{2}$	[(ga)$_{2}$ gc]$_{2}$ gaca (ga)$_{4}$ ca (ga)$_{3}$ gtgg (ga)$_{6}$ (caga)$_{2}$	(gaca)$_{2}$ gg
*Ovar-DRB1*0307*	[(gt)$_{3}$ gc]$_{2}$ [(gt)$_{2}$ gc]$_{3}$ (gt)$_{9}$ (ga)$_{7}$ (gc)$_{2}$	[(ga)$_{2}$ gc]$_{4}$ gaca (ga)$_{4}$ ca (ga)$_{3}$ gtgg (ga)$_{6}$ (caga)$_{2}$	(gaca)$_{2}$ gg
*Ovar-DRB1*0308L480*	[(gt)$_{3}$ gc]$_{2}$ [(gt)$_{2}$ gc]$_{2}$ (gt)$_{11}$ (ga)$_{8}$ (gc)$_{2}$	[(ga)$_{2}$ gc]$_{4}$ gaca (ga)$_{4}$ ca (ga)$_{3}$ gtgg (ga)$_{6}$ (caga)$_{2}$	(gaca)$_{2}$ gg
*Ovar-DRB1*0309L490*	(gt)$_{22}$ (ga)$_{4}$ gc [(ga)$_{2}$ gc]$_{3}$(ga)$_{5}$ gc	(ga)$_{2}$ gc gaca (ga)$_{4}$ ca (ga)$_{2}$ gtgg (ga)$_{6}$ ca(ga)$_{4}$ (caga)$_{2}$ (ga)$_{2}$	(gaca)$_{2}$ ga
*Ovar-DRB1*0601L480*	(gt)$_{16}$ (ga)$_{11}$ (gc)$_{3}$ (ga)$_{3}$ (gc)$_{3}$	[(ga)$_{2}$ gc]$_{3}$ gaca (ga)$_{4}$ ca (ga)$_{3}$ gtgg (ga)$_{6}$ (caga)$_{2}$ (ga)$_{2}$	(gaca)$_{2}$ ga
*Ovar-DRB1*0310*	(gt)$_{30}$ (ga)$_{14}$ (gc)$_{3}$ (ga)$_{4}$ (gc)$_{4}$	[(ga)$_{2}$ gc]$_{3}$ (ga)$_{6}$ ca (ga)$_{3}$ gtgg (ga)$_{6}$ (caga)$_{2}$	(gaca)$_{2}$ ga
*Ovar-DRB1*0701*	(gt)$_{23}$ ga gc (ga)$_{2}$ gc (ga)$_{5}$ gc	[(ga)$_{2}$ gc]$_{2}$ gaca (ga)$_{4}$ ca (ga)$_{3}$ gtgg (ga)$_{6}$ (caga)$_{2}$ (ga)$_{2}$	(gaca)$_{2}$ ga
*Ovar-DRB1*0801*	(gt)$_{6}$ at (gt)$_{5}$ ct (gt)$_{12}$ (ga)$_{24}$	gg (caga)$_{2}$	aaca ga
*Ovar-DRB1*0802*	(gt)$_{6}$ at (gt)$_{5}$ ct (gt)$_{9}$ (ga)$_{31}$	gg (caga)$_{2}$	aaca ga

Note: Simple repetitive sequences in the second intron of sequenced *Ovar-DRB1* alleles. The SRD characteristic for the allele group A (see text) is framed.

Table 9 Comparison of Translated *DRB* Exon 2 Polymorphic AS-Positions from the Ag-Binding Domain of Primate and *Pecora* Species

AS pos.	Chimpanzee	Gorilla	Maccace	Man	Cattle	Sheep	Goat
9	E; L; W;/K	E; D; L;/K; W	E;/K	E, K, W	E; Q	E	E
10	Y; Q;/L;	Y; Q; L;/R;	Y; Q;	Y, E; Q, L	Y	Y	Y
11	A; P; S;/H; G; V; L; D; E	S; A; P; R;/L; G; D; Y;	S; A; V; L;/D; I	L; S, D; G; P; V	L; S; Y, A; C; H; T	S; T; A; H; R; Y;	Y; C; H; T; S
12	K; T	K; T	K; T	K; T	K; T	K; R; T	K; T
13	C; S; G; R; Y;/H; F	A; S; G;/Y; H; C	F; S; H;/F; C	G; R; S, F; H; Y	G; R; S; K	S; K	S; R; G; K
14	E; K	E	E	E, K	E	E	E
16	H	H; Y	H; R	H, Y	H	H; R	H; R
18	F;/L	F	F	F	F	S; F	S; F
22	E;/D; Q	E	E	E	E	E	G; E
24	V;/M	V	V	V	V; L	V	V
25	R; Q;/W	R;/P	R	R, Q	R	R	G; W; R
26	F; L; Y;/N	F; L; Y	F; L; Y;/N	F; L; Y	F; L; Y	F; Y; L	F; L; Y
28	D; E; H;/I	D; E;/H	D; E; V; I;/H	D; E; H	D; E; H; N	D; E	D; H
30	R; D; Y; L;/H; G	R; Y;/D; H	R; Y; F; V; H	C; H; Y, G; L; R	C; H; Y; S	Y	F; Y
31	F; I	F; I	F; I; V	F, I; V	F; Y	F; Y	F; Y
32	H; Y	H; Y;/C	H; Y	H; Y	H; Y; T; S	H; T; Y	H; T; Y
33	N	N	N; Q	N, H	N	N	N
34	Q	Q	Q; P; R	Q	G	G; R	G
37	F; Y; D; S;/N	F; Y; L;/N; D;	Y; N; L; F;/D	F; L; N; Y, S	F; L; N; Y; T, R	Y; F; T; N	Y; F; T; N; I
38	V; M; L;/F; A	A; V;/L	A; V; L	V; A, L	V; A	L; A; V	L; V
40	F;/Y	F; Y	F;/Y	F, Y	F	F	F; Y
42	S	S	S	S	S	S; N	S; N
45	G	G;/W	G	G	G; D	G; S	G

47	Y; F	Y;/F	Y; F; H	F; Y	F; Y	Y; F	Y; F
48	R;/Q	R;/Q	R	R	R; Q	R	R
55	R	R	R	R	R; Q; P	R	Q; R
56	P	P	P; R	P	P; Q; R	R; Q; P	Q; P; E
57	D; V;/S	D; G; V; I;/A	D; S; T; E;/V	A; D; S; V	A; D; S; V	D; S; E; A	S; T; E; D
58	A	A	A	A, E	A; R; P; G; V	A; V	A
59	E;/K; V	E	E; Q	E	E; K; V; Q, D	K; E	K; E
60	Y; S	Y; S; N	Y; S	H; Y, S	H; Y; Q; F	Y; H; Q	Y
61	W; C; Y	W	W; L	W	W; C; L	W	W
65	K	K	K	K	K; DEL.	K	K
66	D;/G	D;/G	D; A	D	D; E	E; N; D	E; D
67	I; F; L; Y	I; V; L	I; F; L; Y	F; I; L	F; I; L; T	L; F; I	L; F; I
68	L; V	L	L	L	L	L	L
70	Q; R; H; D	Q; D;/R; E	Q; D; R; E	D; Q; R	D; Q; R; E	R; Q; E	Q; D; S; R
71	R; A; E; S;/T	Q; R; K;/E; N; I; M	K; R; Q; G	A; E; K; R	A; E; K; R; G	R; T; K; A	R; K; S
73	A; G	A; G	A; G	A, G	A	A; T	A; T
74	G; A; E; Q; T; D;/K; R	D; Q; A; S; R; L;/K; P; T	A; E; P; S; R; G; M; Q	A; E, L; Q; R	A; E; N; S; Y	E; A	E; A; N; T
77	T; N	T; N;/I	T; N	T, N	T; R; S	T	T
78	F; Y; V	F; Y;/L; V	F; Y; V	V; Y	V; Y; D	Y; V	Y; V; F; C
85	V; A	V;/I	V; I	V, A	V; G; R	V	V
86	V; G; A; L	V; G;/I	V; G; C; F; A;/F	G; V	G; V; M; F	I; G; F; D	I; G; F; V; L

Note: For the primate species with more than one expressed *DRB*-loci aa found at the polymorphic *HLA-DRB1* locus are listed before the upslash. Polymorphic aa found at the corresponding positions of the other expressed *HLA-DRB* loci are listed after the upslash. Data for human and primate *DRB* sequences are from References 62 and 131.

```
Ovar-DRB1*0101 -y t-k--r-s-- ---------- -h-g--tl-- ---w------ a-------k- ------f--r a--------- -----i---- ----
Ovar-DRB1*0104 -y t-k--r-s-- ---------- -h-g--tl-- ---w------ a--------- ------f--r a--------- ---------s ----
Ovar-DRB1*0302 -y h--------- ---------- ---g------ ---w--f--- a--------h -----e---r ---e------ -----i---s ----
Ovar-DRB1*0303 -y h--------- ---------- ---g------ ---w--f--- a----qs--h -----e---r ---e------ -----i---s ----
Ovar-DRB1*0304 -y h--------- ---------- ---g------ ---w--f--- a----qs--- ------f--r a--------- -----f---s ----
Ovar-DRB1*0305 -y h--------- ---------- ---g------ ---w------ a-------k- ------f--r k--n------ ---------- ----
Ovar-DRB1*0306 -y h----r-s-- -----y---- ---g------ ---w------ a--------- ------f--r k--n------ ---------- ----
Ovar-DRB1*0307 -y h----r-s-- -----y---- ---g------ -n-w--f--- a--------- ------f--r k--n------ -----i---s ----
Ovar-DRB1*0321 -y h----r-s-- -----y---- ---g------ -n-w------ a-------k- ------f--- t-te------ -----i---s ----
```

Figure 4 Putative double recombination or gene conversion-like events as documented in the resulting exon 2 sequences of sheep *DRB1* alleles. An appearance of several independent point mutation is unlikely, as the same phenomenom was also found for silent substitutions and some of the exchanges are species specific. For aa numbers and aa identities see Table 7.

no str shorter than 44 bp was found in more than 150 alleles of expressed *DRB* loci. The perfect $(gt)_n$ or $(ga)_m$ "degenerates" by point mutations to secondarily derived derivative sequences which are often then themselves subjected to the repeat amplification mechanisms (see Table 8). In this way each *DRB* allele generates its own characteristic pattern of ***r***epeat ***d***erivative ***s***tructures (***RDS***). A defined *DRB1* exon 2 sequence is highly correlated with a str of characteristic RDS. The RDS of similar *DRB* alleles also show high similarity within one species.[67,68] Hence, polymorphism of the antigen-binding first domain of the DRB chain in sheep is correlated with str variability as has been shown in human,[66] cattle,[67] and goat.[68] This fact has been annotated as ***g***roup ***s***pecific ***o***rganisation (***GSO***) of *DRB* alleles.[67]

The str length of *Ovar-DRB1* alleles is stable inherited as in 556 meioses no differences in the lengths between the parental alleles and those of the progeny occurred (unpublished observations).

c. *Three Groups of* Ovar-DRB1 *Alleles*

Three distinct groups (A, B, and C) can be defined by sequence analysis of sheep *DRB1* alleles on the basis of the simple repeat sequences (see Tables 7, 8, and 10). The three allele groups produce PCR products of nonoverlapping length ranges (group A: 476 to 572 bp; group B: ~410 bp; group C: ~800 bp) and family segregation analysis reveals that all of these *DRB* sequences appear to be inherited as alleles from one locus. Within group A the RDS exhibits high similarity haboring an almost invariant DNA stretch of about 64 bp in length. Group A alleles are found in about 90% of sheep and are closely related to the expressed *DRB* genes of cattle and goat (*DRB3* and *DRB1*, respectively) as shown by the computation of similarity, relative silent, and replacement substitutions, and tree building using the VOSTORG and PHYLIP phylogenetic analysis programs. Group B (*Ovar-DRB1*0801*/*Ovar-DRB1*0802*) alleles were found in about 5% of investigated animals and are characterized by strong differences in exon 2 at polymorphic aa sites which differ from all the investigated bovid alleles. In addition, the str of group B has a basic structure which differs significantly from all other *Ovar-DRB1* alleles.[65] Group C alleles also represent about 5% of sheep *DRB1* alleles and are characterized by an intronic length variation generating a PCR product of more than 800 bp. While alleles belonging to groups A and B have been completely sequenced, only the first part of the second exon (*Ovar-DRB1*0901*) of a group C allele has been sequenced as group C alleles are yet resistant to all variations of cloning procedures. It remains to be shown if members of group C are expressed. The existence of the longer group C product resembles the situation found in the goat, where some *Caae-DRB1* alleles were found with strs about 100 bp longer than those of other alleles (see Chapter 5). The difference in length of these goat *DRB1* alleles is postulated to have been generated by an unequal crossing over event resulting in the addition of another complete str basic structure [intron $(gt)_n(ga)_m$ intron $(gt)_n(ga)_m$]. Groups A, B, and C alleles have been found in several different sheep breeds, and in addition groups A and B alleles have also been found in the Argali, one of the putative progenitor of the domestic sheep. Whether group C alleles are also found in the Argali is unknown as only two individuals from this breed (representing four alleles) have been investigated.

Table 10 Mean Relative Number of Synonymous and Nonsynonymous Substitutions for Sheep *DRB1* Alleles

	All DRB1 alleles	Group A alleles	Group A vs. Group B alleles
d_s	0.039 (0.052)	0.021 (0.020)	0.172 (0.013)
d_n	0.078 (0.030)	0.071 (0.024)	0.130 (0.015)

Note: The mean numbers of relative silent (d_s) and relative replacement (d_n) substitutions with the respective standard deviations (in parentheses) are shown. Significant differences of the distributions analyzed by the KS-test: relative silent substitutions < relative replacement substitutions are found when all *DRB1* alleles are included ($p < 0.001$) or members of the group A are compared with each other, whereas relative silent substitutions > relative replacement substitutions are found if group A and group B are compared ($p < 0.001$). For the comparison of relative silent and relative replacement substitutions 33 sheep *DRB* sequences were available. The values for the mean numbers of relative substitutions could be distorted if many closely related alleles are biased in the analyzed clusters.

d. Ovar-DRB1 *Allelic Frequencies*

The *Ovar-DRB1* locus has been found to be highly polymorphic. To date, 74 *Ovar-DRB1* alleles have been described by cloning (sequencing) and by oligonucleotide typing (Reference in Table 7). At least 56 *Ovar-DRB1* alleles possess unique second exon sequences which correspond to unique peptide sequences; all other alleles can be distinguished by differences in their str length (Tables 7 and 8). Of these *Ovar-DRB1* sequences, 20 are defined only by oligonucleotide typing in a study of 15 unrelated sheep (4 different breeds; Gostomski, unpublished data) and 100 sheep of a single breed (Scottish blackface).[72] We might assume that many more alleles remained to be discovered for the *Ovar-DRB1* locus, as for expressed *DRB* loci of other bovids (unlike the *HLA-DRB1* locus). A recent investigation of about 100 Scottish blackface sheep uncovered only 19 different alleles by **p**olymorphism-**s**pecific **o**ligonucleotide typing (PSO; see Table 7),[72] 7 of them which had not been described prior to this study. Interestingly, only 13 unique hybridization patterns were found for the 19 alleles, implying that some of the alleles possess identical second exon sequences but different lengths of the strs. Concerning the data for the Scottish Blackface breed, we conclude that about 15 to 20 different *DRB1* alleles will be found per sheep breed. In view of the high recombination rate in the ovine Mhc it will be of interest to determine whether alleles identical in their exon 2 sequence but with different str lengths will be found in different haplotypes. If this were the case str length could then be used as a marker for linkage disequilibria for other neighboring Mhc loci.

e. A Proposal for the DRB1 *Nomenclature*

The following nomenclature is recommend for alleles of the *Ovar-DRB1* gene based on both the polymorphism of the second exon *and* on the variability of the intronic str (see Table 7). This nomenclature attempts to correlate similarities among these genes by giving similar allele names where possible as it seems that the expressed *DRB* loci in goats, sheep[6,68] and cattle[76] (*DRB3*) are orthologs. Group numbers (e.g., **01.*) are based on the sequence of the β-pleated sheet and are similar for both sheep and goat sequences. Alleles within a group are numbered in order of their detection. Alleles with identical coding sequence but different str lengths are distinguished by the addition of the letter "L" followed by the length of the PCR product (including the PCR primer). *DRB1* alleles which have only been typed by PSO hybridization and length of the str have an "H" suffix followed by the length of the PCR product. Typed alleles which can be not differentiated by their hybridization pattern from sequenced alleles should be given the same number.

f. *Aspects of* Ovar-DRB1 *Gene Evolution*

Phylogenetic trees constructed from sheep and goat *DRB1* sequences reveal that some sheep *DRB1* alleles are more similar to goat alleles than to other sheep alleles and vice versa.[68] This agrees with Klein's hypothesis for a *transspecies* mode of evolution of Mhc genes.[77] A reduced level of polymorphism has been found at some of the normally polymorphic *DRB* aa positions in the sheep. This finding suggests that a bottleneck may have occurred in sheep evolution, possibly at the time of domestication. That almost all *Ovar-DRB1* alleles have the same basic str derivative structures further supports this idea. The apparent exchange of segments between *Ovar-DRB1* alleles[65,68] together with the hypothesis of generation of polymorphism by selection pressure[78] has lead to the idea that present-day *Ovar-DRB1* alleles have evolved by an extensive motif-shuffling process. To further examine this possibility the number of relative synonymous (d_s) and relative nonsynonymous (d_n) substitutions among sheep *DRB1* alleles was calculated using the NAG program[79] (Table 10). The relative number of synonymous (nonsynonymous) substitutions is calculated by dividing the number of actual silent (replacement) substitution by all possible synonymous (nonsynonymous) substitutions between two *DRB* alleles. These calculations revealed that the average value for d_n among *Ovar-DRB1* alleles (0.078) is comparable to the averages found in other species (man 0.071; goat 0.092). In contrast, the average value of d_s (0.039) is notably smaller than those of other ungulates (e.g., goat: 0.066). The values for d_s in a comparison of *Ovar-DRB1* allele group A with group B (Table 10) are about twice as high as that for an interspecies comparison between sheep and any other bovid species (e.g., sheep vs. cattle: 0.075 ± 0.029). Hence, members of group B may have arisen even before the divergence of sheep and cattle. The lack of evidence for sequence exchange between the two groups of *Ovar-DRB1* alleles (A and B; data not shown) suggests that exchanges of sequence are group specific. As the basic str structures of both allele groups are extremely different, this may imply that the str may also be involved in the sequence exchange mechanism(s).

4. *The* Ovar-DRB2 *Pseudogene*

The *Ovar-DRB2* pseudogene has several mutations that render it nonfunctional,[80] suggesting that this gene has been nonfunctional for a long period of time. In addition to stop codons in exons 3 and 4, the gene contains frameshift mutations, aberrant splice sites, and may lack exons 1 and 2.[80] In contrast to the *Ovar-DRB1* gene there is no str in the second intron. However, a simple $(AC)_n$ stretch is found in the fifth intron.[80,81] This str has been amplified for all sheep tested from a number of breeds[120] (Crawford et al., in preparation) suggesting that most, if not all, sheep possess this pseudogene.

5. Informative Markers in the Ovar-DRB *Region*

Two different microsatellite markers are known in the *DRB* region. One is located in the second intron of the expressed *DRB1* gene[68] and the other in the fifth intron of the *DRB2* pseudogene.[81] The existence of at least 32 different length variations of the *DRB1* str and thirteen different lengths have been described for the *DRB2* str. Interestingly, typing of these strs on a mapping flock (IMF sheep families; A. Crawford, in preparation) identified five recombination events between the two *DRB* loci in 149 informative meioses. This corresponds to a genetic distance of 2.6 cM between the two loci and is of about the same magnitude as that between *DRB2* and an *Ovar* class I locus. This distance is about ten times larger than that for the most distant *DRB* loci within the human *DRB* subregion. The analysis of haplotypes between unrelated animals also has identified several haplotypes of the ovine *DRB* region combining different *DRB1* and *DRB2* alleles. The same *DRB1* allele was accompanied by up to three different *DRB2* alleles and vice versa. This underscores the genomic instability of the *DRB* subregion.

D. The *Ovar-DQ* Subregion

Restriction mapping of cosmids has revealed that the structure of the *DQ* region is similar to that of the *HLA-DQ* region, with two *A/B* gene pairs in which the genes are orientated tail to tail spanning a distance of 130 kb[5,35] (Figure 5). There is evidence that both sets of *Ovar-DQ* genes encode products that are expressed on the surface of cells (Ballingall et al., unpublished data). In contrast to humans, however, no evidence has been found for the existence of a *DQB3* pseudogene in sheep.

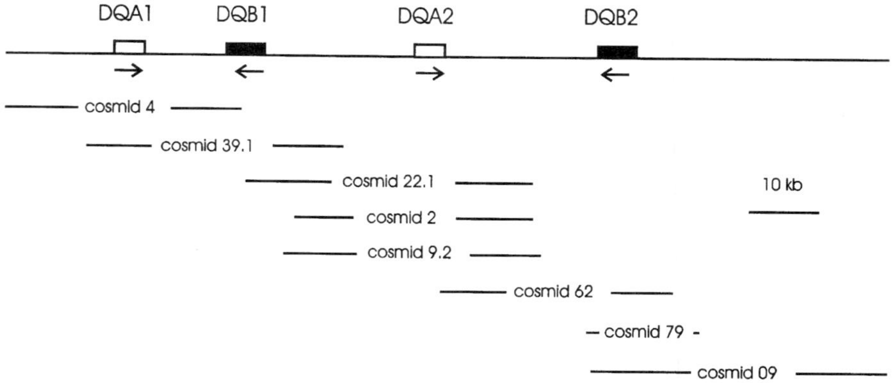

Figure 5 Schematic map of the sheep DQ region.

Ovar-DQA genes can be discriminated between on the basis of sequence information for the second exon of the gene which encodes the α1 domain.[57,70] This region shares less than 81% nucleotide identity, and less than 72% aa identity, whereas exons 3 and 4 share about 92% nucleotide and aa identity. The *DQA* genes in cosmids 4 and 39.1 were assigned to the *DQA1* gene on the basis of sequence analysis, while the *A* genes in cosmids 2 and 22.1 were only 75% similar to the *DQA1* genes and were assigned to the *DQA2* locus.[35] The cosmid *DQB* genes were designated *DQB1* and *DQB2* solely on the basis of their proximity to the *A* genes since they could not be discriminated between on the basis of their nucleotide sequences.

1. **Ovar-DQA1** *and* **DQA2** *Genes*

Sequence and RFLP studies demonstrate that both the *Ovar-DQA* genes are more polymorphic than the *Ovar-DRA* gene, with the *Ovar-DQA2* gene being more polymorphic than the *Ovar-DQA1* gene.[57,70] Specific PCR primers based on genomic sequence have been designed to amplify the second exons of each *Ovar-DQA* gene independently. Currently seven distinct *Ovar-DQA1* and eleven distinct *Ovar-DQA2* alleles have been identified based on sequencing of clones from genomic and cDNA libraries, and PCR-amplified fragments (Figure 6; Ballingall et al., unpublished data; Maddox et al., ISAg 1994 proceedings). Additional *Ovar-DQA2* alleles have been identified by use of the single-strand conformation polymorphism (SSCP) technique (Snibson et al., unpublished data). While all sheep tested possess *Ovar-DRA* and *Ovar-DQA2* genes, a significant proportion of sheep from a number of breeds lack the *Ovar-DQA1* gene.[57,70]

2. Polymorphism in the **Ovar-DQA1** *Gene*

Most of the polymorphism in the *Ovar-DQA1* gene is found in exon 2 although exon 3 also shows a moderate amount of polymorphism.[70] Of the 246 nucleotide sites within the *Ovar-DQA1* second exon, 49 are polymorphic, translating to variation at 27 of 82 aa sites within the α1 domain. Only two alleles have been sequenced for exons 3 and 4. Variation is found at 24 of the 282 nucleotide positions in exon 3 (8.5%) and four of the 152 (2.6%) coding positions in the fourth exon translating to 7 (7.4%) and 3 (2%) variable aa in the α2, and connecting peptide/transmembrane/cytoplasmic regions, respectively. The aa polymorphism within the α1 domain is found mainly in three distinct regions: sites 9 to 18 (I), sites 47 to 56 (II), and sites 67 to 82 (III), corresponding to the first strand of β-sheet and the start and end of the α-helix lining the antigen-binding site (Figure 7). Up to 14% nucleotide and 24% aa differences are found in comparisons between *Ovar-DQA1* alleles.

3. Polymorphism in the **Ovar-DQA2** *Gene*

Polymorphism within the *Ovar-DQA2* gene is also found mainly in the second exon.[70] Of the 246 nucleotide sites within the *Ovar-DQA2* second exon, 79 (32%) are polymorphic, translating to 40 of 82 (49%) variable aa sites within the α1 domain. This contrasts with variation at 9 of the 282 (3.2%) nucleotide positions in exon 3,

(1) nucleotide sequences

```
DQA1.1    ACCCTGACTCAGCTGACCACATTGGCACCTATGGCGTAAACATCTACCAAACATATGGTCCCTCTGGCTACTATACCCATGAATTTGATGGAGATGAAGA 100
DQA1.2    ..................----------------------TG------------------------------T-------------------------C-
DQA1.3    ..............---------------------------G-------------------T------------------------------------C-
DQA1.4    .........------------------T-------------G----------------------------------------------------------
DQA1.5    ....................----C-G-------CGA---TG-------CT-------------------------------------------------
DQA1.6    .........---------------C-G---------A---TG-------CT-------------------------------------------------
DQA1.7    .........---------------C-G---------A---TG-------CT-------------------------------------------------

DQA1.1    GTTCTACGTGGACCTGGAAAAGAGGGAGACTGTCTGGCGTCTGCCTGAGTTTAGTAAATTTACAAGTTTTGACCCTCAGGGTGCACTGAGAAACATAGCT 200
DQA1.2    ----------------------C-----------------------AT-------------GG-GA-------------TT----------------CG-
DQA1.3    A---------------------C-------------------A---AT-------G--G--GG-TA-------G-----TT-------------------
DQA1.4    ----------------------------------------------------------------------------------------------------
DQA1.5    --------------------------------------A--------T-------C------G-------------------------------------
DQA1.6    -----------------------------------------------T-------C------G---G-----T-----------------------C---
DQA1.7    --------------------------------------A--------T-------C------G-------------------------------------

DQA1.1    ACGGTGAAACATAATTTGGAGATCTTGATTCAAAGGTCCAACTCTACTGCTGCTACCAACAGTATG                                  266
DQA1.2    ---C-G--G------------C--A-A----------------------------------.....
DQA1.3    ---T-C---------------C--A-------G----------------.................
DQA1.4    ----C-----------------------------T-----------------------------C-
DQA1.5    ---------------------G--------------------C------.................
DQA1.6    GT--G------GTC--------------------------------------------------A-
DQA1.7    ---------------------G------------------------------------------A-
```

(2) amino acids sequences

```
DQA1.1    DHIGTYGVNI YQTYGPSGYY THEFDGDEEF YVDLEKRETV WRLPEFSKFT SFDPQGALRN IATVKHNLEI LIQRSNSTAA TN
DQA1.2    ---------V ---------F --------Q- ---------- ----M----G D----F---- -R-L-----L I--------- --
DQA1.3      -------V -----S---- --------Q- ---------- ----M--EVG Y--A-F---- ---F-----L M-------.. ..
DQA1.4    ----I----V ---------- ---------- ---------- ---------- ---------- ---A------ ---M------ --
DQA1.5      -AA-AI-V -HS------- ---------- ---------- -H--V--Q-R ---------- ---------V ------P-.. ..
DQA1.6    ---AA--I-V -HS------- ---------- ---------- ----V--Q-R R--------- --VG-QS--- ---------- --
DQA1.7    ---AA--I-V -HS------- ---------- ---------- -H--V--Q-R ---------- ---------V ---------- --
aa/site     23 22 2  22  2   2        2          2  3   323 4  2 2      225 22  3 3  2  2
                 I                                       II                   III
```

Figure 6 The distribution of polymorphism within the second exon of the (A) *Ovar-DQA1* and (B) *Ovar-DQA2* gene. Nucleotides and amino acids within polymorphic regions I, II, III, and IV are underlined.

(1) nucleotide sequences

```
DQA2.1    CTCACTTATCAGCTGACCACTTTGGCTCCTATGGCACAGAGATCTACCAATCTCATGGTCCCTCTGGCCAGTACACCCAGGAATTTGATGGAGACGAGCT
DQA2.2    ............--------G-----AT-------G----CT-------------------------------------C-----------G--------
DQA2.3    --------------------G-----A--------G----AT-----------------------A--G-------------------C-A---------
DQA2.6    .........---------------------------T-C-TG-------------------------------------T--------------------
DQA2.7    ..............------G------------------TT--------------------------------------T---------A----------
DQA2.8    ..............------G-----A--------G----AT-----------------------A--G-------------------C-A---------
DQA2.10   ..............--------------------------------------------------------------------------------------
DQA2.11   ..............------G----------C--------CT------------------------------T-------------------------T-
DQA2.12   ..............------G-----AT-------G----CT-------------------------------------C-----------G--------
DQA2.13   ..............------G-----AT------------CT--------------------------G------T---CTT---------------AGA
DQA2.14   ..............-C----G-----A--------G----AT-----------------------A--G------------------GC-A--------C

DQA2.1    GTTTTATGTGGACCTGGGGAAGAAGGAGACTGTCTGGAGGCTGCCTATGTTTAGCCAGTTTGCAGGTTTTGATCCACAGGGTGCACTGAGTGAAATAGCT
DQA2.2    -------------------------------------C--------------G-TG-A--CA--A-------C--G--A---------------------
DQA2.3    -C---------------A------A------------C--------------G-----------------C-CATT--A-T----------A-C------
DQA2.6    ------------------------------A------C----------------------------------C------C---------T-C--T-----
DQA2.7    -----------------A-------------------C----------------------------------C--T--A------------A-C------
DQA2.8    -C---------------A------A------------C--------------G-----------------C-CATT--A-T----------A-C------
DQA2.10   ----------------------------------------------------------------------C-CATT--A-T----------A-C------
DQA2.11   -C-----------------------------------C---------------------------A------C--T--A-----------GA-C------
DQA2.12   -------------------------------------C--------------G-TG-A--CA--A-------C--G--A---------------------
DQA2.13   -----------------A-------------------C--------------GATG-A--AAG-A-A-----C--G--A--------A-A-A-C------
DQA2.14   -C--------------.A------A------------C--------------G-----------------C-CA-T--A-T----------A-C------

DQA2.1    ACAGCAAAACAAAACTTGGATATCCTGACTAAACGCTCCAACTTTACCCCTGCTATCAATGGTAAGTGTCCACC
DQA2.2    -----G-----C------------A---T------T----------------T--------.............
DQA2.3    ----AG-----C---------G--A--------T-G-A------------A-T---------------------
DQA2.6    ---T-G--G--C---------TA-A---------A--------------A---C-----C--------------
DQA2.7    -----G-----C------------A-----C--T-GCA------C------A-T---..................
DQA2.8    -----G-----C---------G--A--------T-G-A------------A-T---..................
DQA2.10   -----G-----C-G-------G--A--------T-G-A------------------..................
DQA2.11   ---------G-C-C-----------------------------C------A-T---..................
DQA2.12   -A-----------C----------A---T------T----------------T---..................
DQA2.13   -T---G-----C-------------------------A------------A-T---..................
DQA2.14   ----AG-----C---------G--A--------T-C-G------------A-T---..................
```

(2) amino acid sequences

```
DQA2.1     DHFGSYGTEI YQSHGPSGQY TQEFDGDELF YVDLGKKETV WRLPMFSQFA GFDPQGALSE IATAKQNLDI LTKRSNFTPA IN
DQA2.2     --V-I--ADF ---------- -H-------- ---------- ------GE-T S--------- -----H---- MI-------V --
DQA2.3     --V-T--A-F -------SE- -----E---L ----E----- ------G--- --HI-V---N ---E-H---V M--WY----V --
DQA2.6     -------IHV ---------- -H-------- ---------- ---------- -----R-VIQ L--S-H---Y M--H------ --
DQA2.7     ..V-----V- ---------- -H---R---- ----E----- ---------- ---------N -----H---- M-QWH-S--V ..
DQA2.8     ..V-T--A-F -------SE- -----E---L ----E----- ------G--- --HI-V---N -----H---V M--WY----V ..
DQA2.10    ..-------- ---------- ---------- ---------- ---------- --HI-V---N -----HS--V M--WY----- ..
DQA2.11    ..V-----DF ---------- ---------L ---------- ---------- D-------RN -----DT--- ------S--V ..
DQA2.12    ..V-I--ADF ---------- -H-------- ---------- ------GE-T S--------- --K---T--- MI-------V ..
DQA2.13    ..V-I---DF --------E- IHL-----E- ----E----- ------DELR R-------NN --I--H---- ----Y----V ..
DQA2.14    ..V-T--A-F -------SE- ----GE--PL ----?----- ------G--- --HT-V---N ---E-H---V M--WC----V ..
             2 3  343         22  222 23  32       3        3223 4 23 3 233 2 33 33  3 22234 2 22
               I                      II                          III            IV
```

Figure 6 (continued)

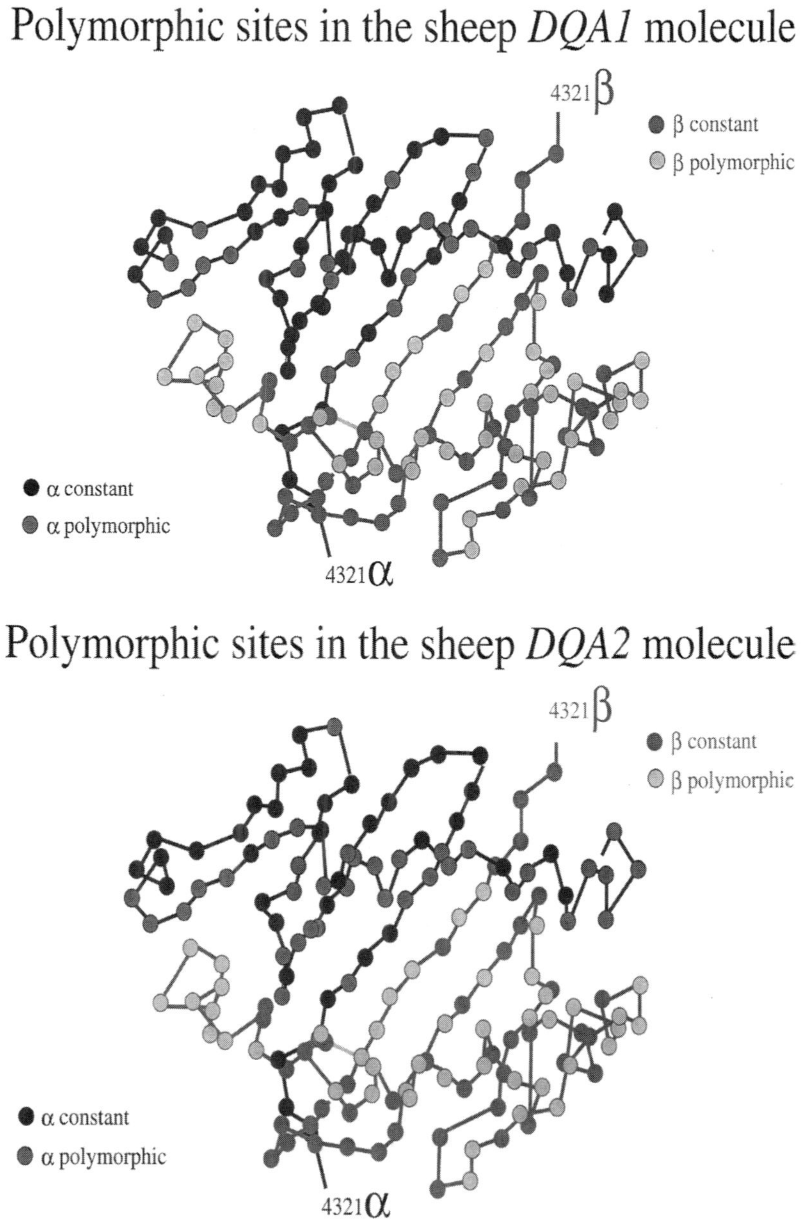

Figure 7 DQ antigen-binding groove.

and 8 of the 152 (5.3%) coding nucleotides in exon 4, translating to two aa changes in the α2 domain (2%) and none in the connecting peptide/transmembrane/cytoplasmic region. The polymorphic regions found within the α1 domain differ from those found for the *Ovar-DQA1* gene. Most of the polymorphism is found in four regions: sites 9 to 15 (I), sites 23 to 35 (II), sites 52 to 66 (III), and sites 68 to 85 (IV)

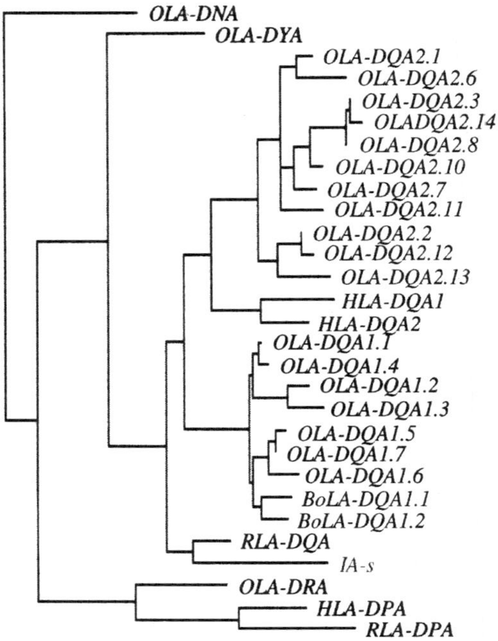

Figure 8 Phylogenetic tree of DQA sequences.

corresponding to the first two strands of β-sheet and a large part of the α-helix within the antigen-binding site (Figure 7). Up to 20% nucleotide and 33% aa differences are found in comparisons between *Ovar-DQA2* alleles.

4. Relationships Among Ovar-DQA1 *and* DQA2 *Alleles*

Allele comparisons between the different *Ovar-DQA* genes show that the two genes are distinguished by 19 to 28% and 28 to 49% nucleotide and aa differences, respectively. A consensus phylogenetic tree constructed from 100 bootstrap replicates of *DQA* second exon nucleotide sequences (Figure 8) showed that the *Ovar-DQA1* sequences group together as do the *Ovar-DQA2* sequences; however, the different genes are found on separate branches of the tree. Both of the *Bota-DQA* sequences are found on the same branch as the *Ovar-DQA1* sequences; hence, both are probably alleles of the same *Bota-DQA* gene rather than being alleles from different *DQA* genes. *DQA* equivalents from non-artiodactyl species are all slightly more similar to the *Ovar-DQA2* gene than to the *Ovar-DQA1* gene.

5. The Ovar-DQB1 *and* Ovar-DQB2 *Loci*

The second exons of *Ovar-DQB* genes have been sequenced following amplification by the polymerase chain reaction or isolation from genomic and cDNA libraries. Sixteen distinct exon 2 sequences, coding for 16 distinct aa sequences, have been characterized from only 18 sheep. Three of these sequences belong to the *Ovar-DQB1* gene and two sequences belong to the *Ovar-DQB2* gene (see Figure 9;

```
                          |    P                                                                                   |
OLA-DQB*1           DFVYQFIGHCYFT NGTERVRHVT RYIYNQEEFV RFDSDWDEHR AVTPLGRPDA EYWNSQKDIM ERTRAEVDTV CRHNYQVEAA FTWQRR
OLA-DQB*2           ---H--K-L---- -------L-- -----R--Y- ------G-Y- ---¥---RQ- ---------T --V------- ------A-LI TSL---
OLA-DQB*3                           -----S-N -----R--Y- ------G-Y- --------S- --------FL -Q----A--
OLA-DQB*4              ---------- ---------- ---------- --------Y- -------RS- ---------- ---------- ----------
OLA-DQB1*5          ---H----Q---- -------L-- ---------- --------Y- -------RH- --F------L -Q-Q-A---- --N-------
OLA-DQB*6              H--K-Q---- -------L-- -----R---- ------G-Y- ---------- ------E--- -Q------R- -----R-Y-P
OLA-DQB*7              V--KC----- -------Y-- -----R--YA --------Y- G---P-QRQ- ---------L ---------- --N--R-Y-P
OLA-DQB*8              V--K-L---- -------S-N -----R--Y- --------Y- --------E- ---------- -Q-------- --N---A-
OLA-DQB*9              V--K-L---- -------S-N --------Y- --------Y- ---S----E- ---------- -Q----A--- ------N-
OLA-DQB*10             ----C----- -------Y-- -N-------- --------Y- -------RS- ---------L ---------- --N--R--
OLA-DQB*11             H--K-L---- -------L-- -----R--YA ------G-Y- -------RQ- ---------- --V------- ------A-
OLA-DQB*12             ----YQ---- -------S-K -Q---RQ-H- -----VN-F- --S----K-- --F--H¥-FL -Q----T--- ------L-   Ê
OLA-DQB1*13         ---FL-MGQ---- -------L-- --------HL ------G-Y- ----P-QRQ- --F------L ---G--A--- ---------- ------
OLA-DQB1*14         ---V--K-L---- -------S-N --------Y- --------Y- ---------- ---------L --V---A--- ------N-LI TSL---
OLA-DQB1*15         ---H--K-R---- ---------- -----R--YA --------YR -------RS- ------E-FL -Q------R- -----R-Y-P ------
OLA-DQB1*16         ---H--KCL---- -------Y-- -----R--YA --------Y- G---P-QRQ- --------FL ---------- S-N--R-Y-P ------
HLA-DRB  Kons.         EQSKSECHFF NGTERVRFLD RYFYNQEEYV RFDSDVGEYR AVTELGRPDA EYWNSQKDLL EQRRAAVDTY CRHNYGVGES FTVQRR
Cosmid 23.1            v-g------s ---qq----- --i--r--q- ----l----- -r--m---ag qr--rwp-a- qra----hdf -as--effa- r-----
Cosmid 9.1             h-f-g--r-s --l--m--fa --i--tk-d- h-------ft -l-------- ----q---fm --m--kv--v --s--qgig- -lr---
# of polymorph. aa     32 334           4 3  3  22 33        23 32 2 232 235   2  23 23  223 23 2  2 2  24223 222
                       HVRI             HVRII                           HVRIII      HVRIV              HVRV
```

Figure 9 *Ovar-DQB* alleles: peptide sequences encoded by exon 2. Horizontal lines (|) indicate the cysteines involved in disulphide bridge formation. P: glycosylation site; ¥: amino acid deletion. Amino acids corresponding to the oligonucleotide primer sequence are bolt. HVR regions[83] are underlined.

Ballingall et al., unpublished data) while the rest cannot be assigned to either gene as yet. The second exons of *Ovar-DQB* genes are highly polymorphic with 29% of nucleotide and 45% of aa sites showing variation.[87] Most of the variation is found within five polymorphic regions: I sites 9 to 15; II sites 26 to 33; III sites 45 to 57; IV 61 to 81, and V sites 84 to 91 . The positions of these regions are similar to a composite of the hypervariable regions described for *DQB* genes of mouse and man.[82,83] Comparison of ovine and bovine *DQB* sequences reveals that some ovine *DQB* sequences are more similar to bovine *DQB* counterparts than to other ovine *DQB* sequences, suggesting that the artiodactyl *DQB* gene and allele lineages predate the separation of the ovine and bovine species 20 million years ago.

Only three sequences are available to investigate polymorphism in exons 3, 4, 5, and 6 of *Ovar-DQB* sequences (Maddox et al., unpublished data). Overall, variation is found at 5 to 8% of nucleotide sites for these exons translating to 6 to 8% of aa sites. It is unknown whether the fifth exon of *Ovar-DQB* genes is expressed. The genomic sequence possesses normal splice sites and the exon is in-frame;[80] however, two distinct *Ovar-DQB* cDNA clones isolated from a mesenteric lymph node lymphocyte library both lack this exon. Exon 5 splice site variants have been found in *HLA-DQB* sequences, and the presence or absence of exon 5 in man has been found to be allele and tissue specific with some alleles expressing both forms of the *HLA-DQB* chain.[84,85] In contrast *Bota-DQB* genes have a two base pair deletion from exon 5 which causes a frame-shift and renders exon 5 incapable of being expressed.[61]

6. **Ovar-DQB** *Genes Are Difficult to Type*

DNA typing systems for the *Ovar-DQB* genes are complicated by the existence of the two highly similar *Ovar-DQB* genes. RFLP studies show complex band patterns for several enzymes and it is difficult to ascribe RFLPs to individual genes. A recent study by Grain and colleagues[86] characterized a number of *DRB*/*DQB* haplotypes. Drawbacks of this study were the cross-hybridization between *Ovar-DRB* and *Ovar-DQB* genes and the failure to examine bands of less than 1 kb as *Ovar-DQB* bands of less than 1 kb are known to be produced by digestion with several enzymes (e.g., *Taq*I). Although the SSCP technique has been shown to produce unique migration patterns for each sequenced allele, difficulties exist with typing *Ovar-DQB* alleles using this method.[87] This is due to the complexity of the SSCP patterns with several bands per allele, depending on the conditions used, and the presence of fragments from two *DQB* genes. However, SSCP should prove a useful tool for screening subcloned PCR fragments for new alleles prior to sequencing them.

7. Is there an **Ovar-DQB3** *Gene?*

An equivalent to the *Bota-DQB3* exon 2 sequence, termed *Ovar-DQB*12*, has been identified in sheep.[87] Whether this sequence represents a separate gene to *Ovar-DQB1* and *Ovar-DQB2* genes or whether this sequence represents a divergent allele of one of the these genes is currently unknown. Probing of genomic Southerns using a *Ovar-DQB*12* probe identifies a set of fragments distinct from those identified by the *Ovar-DQB*1* probe.[87] These fragments are not found in all sheep and are polymorphic. The identification of separate sets of *Ovar-DQB* fragments is not surprising as relatively stringent washing conditions were used to prevent cross-hybridisation with *Ovar-DRB* genes, and the overall similarity between the *Ovar-DQB* probes is only 81%.

8. *Evolution of* Ovar-DQ *Sequences*

It seems likely that the duplication which gave rise to the *Ovar-DQ1* and *Ovar-DQ2* loci preceded the divergence of the ovine and bovine species more than 20 million years ago. In addition, it appears that the development of some of the allele lineages for both the *Ovar-DQA1* and *Ovar-DQB* genes also predated the divergence of ovine and bovine species.

E. Does the Sheep Have *DP* Genes?

Several studies using a human *DPA* probe have failed to identify a homologous *DPA* gene in sheep.[5,69] In contrast, probing of genomic Southern blots using a human *DPB* exon 2 cDNA probe identified fragments which did not cross-hybridize with human *DRB*, *DQB*, or *DOB* probes.[69] In an attempt to study this gene PCR primers were designed based on conserved regions within the second exons of human and rabbit *DPB* sequences. A fragment of the right size was amplified by PCR from ovine genomic DNA; however, sequencing of this fragment revealed that this fragment was derived from the *Ovar-DIB/DYB* gene (Davies and Maddox, unpublished work). A study is currently being undertaken comparing RFLP patterns obtained using the *HLA-DPB* and *Ovar-DIB/DYB* probes to determine whether the unique fragments detected by the *HLA-DPB* probe belong to the *Ovar-DIB/DYB* gene.

F. The *Ovar-DN* Region

The presence of an *Ovar-DNA* gene (formerly called *DZA*) was demonstrated on Southern blots by cross hybridization with an *HLA-DNA* probe.[69] The gene was cloned[5] and sequenced (EMBL Acc. No. z29533, Wright et al., unpublished). Like the *DRA* gene it appears to be monomorphic. The aa translation of exons 2 to 4 of the *Ovar-DNA* gene is shown in Figure 10 where they are compared with the homologous human and mouse *DNA* (*H2-OA*) genes. It has all the salient features of an Mhc class II *A* gene. These include two exons coding for the two extracellular domains and one coding for a proline-rich connecting peptide, a hydrophobic transmembrane region, and a cytoplasmic tail. The gene also has the two conserved N-linked glycosylation sites, NGT and NAT, and the two conserved cysteine residues which form the disulfide bond in the $\alpha 2$ domain. The sequence most resembles that of the *HLA-DNA* gene, with the identities at the aa level being 78, 81, 54, 74, and 60% for the first and second domains, the connecting peptide, and the transmembrane and cytoplasmic domains, respectively. A phylogenetic analysis of the second exons from representative human, mouse, cattle, and sheep *A* genes, together with the three

exon 2

```
Ovar-DNA  DHMGSYGPAFYQSYDGAGQFTYDFDGEQLFSVDLKKREAVWRLPEFGNFAYFDPQNGLVSIAMIKAHLEVLVERSNGTRAPN
HLA-DNA  A---------------GAS----HE--E----------S----------D--R-----G--AG--A------DI-------RS--I-
H-2Oa    ----------------AS----HE-----I------NE-V--------D--HS-F-S--M--S-------DI-------R---VS
```

exon 3

```
Ovar-DNA VPPRVAVLPKSRVQLGEPNVLICIVDNIFPPVINATWLRHGQPVTQGVTQTSFYAQPDHSFRKFHYLTFVPLADDFYDCKVEHWGLDQPLFQHW
HLA-DNA  -----T-------E--Q--I--------------I----N--T--E--A-----S----L-------P---S-E-V---Q-------A--LR--
H-2Oa    -----T----T-----K---------D-------V----NS--I-K--A-----S--N-R-----------S-E-V-----------T--L---
```

exon 4

```
Ovar-DNA EPQVPTVLPDTTETLVCALGLILGLGGFLGGITLIITGTCLSSAPR
HLA-DNA  -L---IPP--AM---------AI--V---V-TV---M--YV--V--
H-2Oa    ----L-PP-------I-G---V---M-C-L-TV-M----RRP-IR-
```

Figure 10 Multiple alignment of the amino acid translations of exons 2, 3, and 4 of the ovine, human, and mouse DNA genes.

DNA sequences, clearly showed that the *HLA-DNA*, *Ovar-DNA*, and the *H2-OA* genes group together on a separate branch.

Whether the *Ovar-DNA* gene is expressed is currently unknown. The *Ovar-DNA* has been shown to be transcribed, as specific RNA has been detected following transient transfection and in Northern blots of a number of tissues (Wright and Ballingall, unpublished observations). The murine homologue to *Ovar-DNA* has been shown to be expressed in partnership with the *H2-OB* gene.[88] Unfortunately, the *Ovar-DOB* gene has not been cloned, and, while attempts were made to express the *Ovar-DNA* gene by transfecting cosmids containing the gene into mouse L-cells in combination with expressible *Ovar-DQB1* and *DRB1,* genes no products were detected.

G. The *Ovar-DY/DI* Region

The presence of two class II genes presumed to be unique to the ruminant was deduced from RFLP studies of cattle DNA.[55] These genes were designated *DYA* and *DY* and were shown to segregate together with the *DOB* gene in one region separated by a recombination distance of 17 cM from the region which contains the *DQA*, *DQB*, *DRB*, *DRA,* and *C4* loci. Subsequently, *Bota-DYA* was sequenced[89] (EMBL acc. # m30119 and m30118). The sequence of part of a similar gene in the goat was obtained by PCR using primers derived from the bovine sequence[90] (EMBL acc. # m94325). However, there has been no report of the cloning of a *B* gene partner for the *DYA* gene. A novel bovine class II *B* gene, designated *Bota-DIB*, was sequenced by Stone and Muggli-Cockett.[91] This was shown to be a single-copy gene of limited polymorphism, which on the basis of RFLP analysis was probably not *Bota-DYB* but did appear to be distinct from other known bovine class II genes. The species distribution of this *B* gene was shown to be restricted to Cervidae, Giraffidae, and Bovidae (*Pecora).*[56] Wright et al.[92] described a cosmid clone containing an Mhc class II *A* and *B* gene pair which proved to be very similar to the *Bota-DYA* and *Bota-DIB* genes. The predicted aa translations of exon 2 and 3 of the sheep *A* gene are shown in Figure 11 together with the those of the cattle and goat sequences. The *A* gene in the cosmid had all the salient features of an Mhc class II *A* gene. It showed a high sequence similarity to the bovine (89 and 92% aa identity in exons 2 and 3, respectively) and caprine (96% aa identity in exon 3) *DYA* genes and much less so to the *Ovar-DRA*, *DQA1*, *DQA2,* and *DNA* genes (54, 67, 66, and 53% identity, respectively, at exon 2). The *A* gene was clearly the sheep homolog of the *Bota-DYA* gene. The aa sequences of the second, third, and fourth exons of the *B* gene in the cosmid clone are shown in Figure 12 together with those of the *Bota-DIB* gene. The presence of a *BamH*I site in exon 2 of the sheep gene caused a truncation at this point during the cloning procedure and so a part of exon 2, the whole of exon 1, and all the upstream regulatory elements were missing. The aa identities between the sheep gene and the *Bota-DIB*, *Ovar-DQB,* and *Ovar-DRB* genes for exon 2 were 98, 49, and 53%, respectively. The authors designated the *B* gene *Ovar-DYB* because of its proximity to the *A* gene which is clearly *Ovar-DYA*. The distance between the

```
exon 2
Bota-DYA DHVGTYGTNVYQTYGTSGQFTFEFDGDELFYVDLGKKETVWRLSEFSNITKFEVQSALRNIVMSKRNLDILIKNSSFTPATS
Ovar-DYA ---------------A-------------------R--------P--N---M--I----------------M---N-----N

exon 3
Bota-DYA EIPEVAVFPKSSVVLGIPNTLICQVDNIFPPVINITWFYNGHFVAEGIAETTFYPKSDHSFLKFSYLTFLPTSEDFYDCRVEHWGLEEPLVKHW
Ovar-DYA D---------------------------------------Q------VT--------------------V-A---------------G------
Cahi-DYA                                            ---------------------Q-----V-----------------------
```

Figure 11 Alignment of the amino acid translation s of exons 2 and 3 of the sheep, cattle, and goat DYA genes.

```
exon 2
Bota-DIB NFVYQFKGMCYFTNGTEHVRLVARQIYNKEEILHFDSDLGEFVAVTELGRVCAEIWNTQKDLLAEFRAYVDTLCRHNYKETAGFTVQRR
Ovar-DYB                                ---------------------------------------E------------------

exon 3
Bota-DIB VEPTVTVSPASTEALNHHNLLVCSVTDFYPRQVKVKWFRNQQEQTAGVGFTPLTQNGDWTYQIHVMLETVPQLGDVYVCHLDHPSLQSPITVEW
Ovar-DYB --------------P-L----------N-------I---------------------------------I----------V-------------

exon 4
Bota-DIB RAQSESAQSKMQSGIGGFVLGLIFLGVGLFVHFWDKR
Ovar-DYB ------V----W---------------------Q---
```

Figure 12 Comparison of the amino acid translations of exons 2 to 4 of *Bota-DIB* and *Ovar-DYB*.

genes is about 11 kb and the transcriptional orientation is tail to tail, a structure similar to that of the *HLA-DQ* subregion,[93] the *Bota-DQ* subregion,[61] and the *Ovar-DQ1* genes.[5,35] There is as yet no estimate of the distance between the *DY* locus and the other class II loci in the sheep. Although there is no estimate of the distance between *Bota-DYA* and *Bota-DRB3* in terms of kb of DNA, the genetic distance in terms of recombination frequency has been estimated at 15 cM,[94] a figure which agrees well with the 17 cM estimated for the distance between the two subregions containing *DY/DO* on the one hand and *DQ/DR* on the other.[55] These frequencies contrast with the estimated recombination distance of only 3 cM between the *HLA-DQ* and *HLA-DP* subregions which are separated by 300 kb on the molecular map.[95] Others have shown that the *Bota-DIB* gene maps to the group containing the *DOB* gene and hence, by inference, the *DYA* gene.[56] Whether an *Ovar-DY* class II molecule is expressed at the cell surface or not remains to be demonstrated. Wright et al.[92] were unable to obtain expression of the *DY* genes following transfection into mouse L-cells. This was perhaps not surprising since the *B* gene in the cosmid was truncated (Figure 12). However, cotransfection of the *DY* cosmid with phage DNA containing the *Bota-DIB* gene kindly supplied by Dr. R. Stone was also unsuccessful (unpublished observations). Evidence for transcription of the *DYA* gene was obtained when total RNA prepared from the transfected L-cells was hybridized with a *DYA* probe. Stone and Muggli-Cockett[56] were unable to detect significant amounts of *Bota-DIB* mRNA in Northern blots of peripheral blood monocytes. A phylogenetic analysis of *Bota-A* genes by Scott et al.[57] placed the *DYA* gene on the same branch as the *DQA* genes but indicated that *DYA* branched off long before the duplication which led to *DQA1* and *DQA2*. The high degree of sequence similarity between sheep *DY* genes and the homologous cattle *DYA* and *DIB* genes provides an argument that these are evolutionary conserved Mhc genes of unknown function.

H. The *Ovar-DM* Region

Sheep possess *Ovar-DMA* and *Ovar-DMB* genes as evidenced by PCR amplification of fragments from exons 2 and 3 of the *Ovar-DMB* gene and exon 2 of the *Ovar-DMA* gene using primers based on conserved sequences between the murine and human genes (Figure 13). Preliminary work using an *Ovar-DMB* exon 3 probe on DNA from seven sheep revealed only 1 EcoRI polymorphism and no TaqI polymorphisms (Maddox and van Oorschot, unpublished data).

III. THE *OVAR* CLASS III REGION

Characterization of the mammalian Mhc has been assisted greatly by the identification of the central region genes, often called class III region genes. In many, if not all, mammalian species class III region genes separate the clusters of class I and class II loci which consist mainly of genes encoding the diverse and highly polymorphic families of leukocyte antigens. Molecules encoded by Mhc class III region genes are heterogeneous with respect to both structure and function and are either

(a) exon 2 nucleotide sequences

```
Consensus   CTTYGTGGCYCATGTGSAAAGCACSTGYSTGYTGGAYGATGMTGGRAMYCCAMARGABTTCACRTAYTGYRTYTCCTTCAACAARGAYYTGCTGRCCTGC
Ovar-DMB    ---T-----C------C-------C--TC--T----C----A---G-AT---A-A--G-----G--T--CG-C-----------G--CT-----A-----
HLA-DMB     ---C-----C------G-------C--TC--T----T----C---G-CT---A-G--T-----A--C--CA-C-----------G--TC-----A-----
H-2IMB      ---T-----T------G-------G--CG--C----T----C---A-CC---C-G--C-----A--C--TG-T-----------A--TC-----G-----

Consensus   TGGGATCCMVWRSWGRVHRRDATRGYYCCYYGYGAATTTGGGGTGCTGWAYVSMTTGGCSRADDWYYTYTCAVRKHHCCTCAACMRWRAMGABRBYCTKM
Ovar-DMB    --------CCTGCA-GCCGGT--G-TT--TC-T---------------A-CGGC-----CA-GTACC-C---GATTC-------AGTG-C--TGTC--GA
HLA-DMB     --------AGAGGA-AATAAG--G-CC--TT-C---------------A-TAGC-----GA-TGTCC-C---CAGCA-------CAAA-A--CACC--GA
H-2IMB      --------AATAGT-GGAAAA--A-TC--CT-T---------------T-TCCA-----CG-AAATT-T---AGGAT-------AAAG-A--GAGT--TC

Consensus   TSCAGCGYTTGYVMARYGGKCTKCMGRAYTGTGCCWCMCACACCCAGCCCTTCTGGR
Ovar-DMB    -C-----C---TCC-GT--T--G-A-G-C------A-C-------------------
HLA-DMB     -G-----C---CGC-AT--G--T-A-A-T------A-A------------------G
H-2IMB      -C-----T---CAA-AC--G--T-C-G-C------T-C------------------A
```

(b) b1 domain amino acid sequences

```
Consensus   FVAHV-STC-LDD-G-P--FTYC-SFNKDLL-CWDP------P-EFGVL--LA---S--LN----L-QRL--GL--CA-HTQPFW
Ovar-DMB    -----Q---L---D-N-KE----V-------T----LQAGMV-R-----NG--KYL-DS--SDDV-I---SS--QD--T------
HLA-DMB     -----E---L---A-T-KD----I-------T----EENKMA-C-----NS--NVL-QH--QKDT-M---RN--QN--T------
H-2IMB      -----E---V---A-T-QD----V-------A----IVGKIV-C-----YP--ENF-RI--KEES-L---QN--PD--S------
```

(c) exon 3 nucleotide sequences

```
Consensus   GCTAYGTSTGGGGCTTCTATCCAGCRGAHGTGRCYATCACRTGGAKGARGAAYGGGMAGSWKGTCMYKYCYCACACGGCAVHRMSBRSAR
Ovar-DMB    ----T--G-----------------G--C---G-C-----G----G--G---C---C--GAG---CT..-C---------CAGCCTGG-G
HLA-DMB     ----T--G-----------------A--A---A-T-----G----G--A---C---A--CTT---ATGC-T----..---GTGCGCAC-A
H-2IMB      ----C--C-----------------G--T---A-C-----A----T--A---T---C--CTT---CCTT-C----..---ACAAGGAG-A

Consensus   GRYBRYYCAGCCCAATGGAGACTGGACATACCAGA
Ovar-DMB    -GTCATC----------------------------
HLA-DMB     -ACTGCC----------------------------
H-2IMB      -ACGGCT----------------------------
```

(d) b2 domain amino acid sequences

```
Consensus     YVWGFYPA-V-ITW--NG--V--H-------QPNGDWTYQ
Ovar-DMB      --------D-A---RR--QE-LP-GTAWRVI---------
HLA-DMB       --------E-T---RK--KL-MP-SSAHKTA---------
H-2IMB        --------D-T---MK--QL-PS-SNKEKTA---------
```

Figure 13 Nucleotide sequence of DMB genes. Regions corresponding to oligonucleotide primer sequences are underlined.

intracellular or extracellular proteins rather than membrane-associated molecules. There has been some debate concerning the relevance of class III region genes within the Mhc. Klein[96] has argued that the close linkage of class III region loci to the lymphocyte antigen-encoding class I and II regions is serendipitous. However, many class III loci are polymorphic and encode molecules which have accessory functions in adaptive immune responses. In addition, Dawkins et al.[97] have shown that class III loci are important markers for highly conserved chromosomal segments (haplotypes) derived from common, though remote, ancestors. These "ancestral haplotypes" provide an explanation for the extensive linkage disequilibrium observed between distinct Mhc loci and are particularly useful in mapping Mhc-associated disease susceptibility loci.[98]

A. C4 and 21-Hydroxylase Genes

One of the most characteristic of the class III region genes is that for complement component C4. Multiple copies of this locus are present in the Mhcs of humans and mice with each copy also manifesting a high degree of polymorphism. In all instances where the C4 gene has been mapped in mammalian species, close linkage to the Mhc has been demonstrated. In contrast to the well-defined class III regions of the human and mouse, the existence of a class III Mhc region in sheep is still based on circumstantial evidence. This evidence derives from comparisons with related species, namely goats and cattle, and synteny between several loci. Thus in both cattle and goats the complement C4 gene has been demonstrated to be linked to either class I or class II loci.[99,100] Further, as will be shown below, the cluster of genes characteristic of the class III region of goats and cattle, and other mammalian species, also occurs in sheep. Chardon et al.[101] were the first to identify a sheep complement C4 gene using cross-hybridization with a human C4 cDNA probe but this group was unable to demonstrate polymorphism of the sheep C4 gene and hence linkage to *Ovar* could not be established. Groth et al.[102,103] described an association (linkage disequilibrium) between RFLPs of the complement C4 gene and a class II *DQB* gene in unrelated sheep. This observation is consistent with linkage between these loci but unfortunately does not settle the question beyond doubt.

Groth and Wetherall (submitted for publication) have recently isolated a family of overlapping cosmid clones containing sequences from C4 genes from a sheep genomic cosmid library prepared from the liver of a male sheep (obtained from Clontech, USA) by cross-hybridization with a 2.1-kb bovine C4 cDNA probe.[104] Partial digestion of cosmid insert DNA with EcoRI and subsequent hybridization with radiolabeled T3 or T4 oligonucleotides permitted the relative ordering of DNA fragments generated from cosmid insert DNA by complete EcoRI digestion. Additional cosmid clones adjacent to the complement C4 genes were obtained by a cosmid walking procedure using DNA probes for several other loci (21-hydroxylase, complement C2 and factor B) known to occur in the human Mhc class III region. With this procedure the relative positions of these loci were mapped within an approximately 150-kb segment of sheep DNA. Clear evidence was also obtained for duplicated complement C4 and 21-hydroxylase loci.

Of particular interest was the observation that in sheep the order of the complement C4 and 21-hydroxylase loci was inverted (...CYP21...C4...C4. CYP21...) relative to that present in the corresponding regions of the human, mouse, and cattle Mhc. In addition, it was shown by using an α-chain-specific cDNA probe that the direction of synthesis of one of the two C4 loci was opposite to that of its partner. These data indicate the presence of a chromosomal inversion in this region of the sheep chromosome. With respect to this finding it is of interest that Cameron[100] also obtained evidence consistent with either a chromosomal inversion in this region of the goat Mhc or a large separation between these loci. The results summarized herein favor the former explanation. In addition to the genes mapped above, a polymorphic microsatellite locus has been detected close to the Bf locus in this region of sheep DNA. This microsatellite locus is being used in conjunction with another polymorphic microsatellite within the class I region to type several sheep families to establish finally whether the typical class III gene cluster described above is Mhc linked as would be expected.

A preliminary map of this region is shown in Figure 14 together with maps for the corresponding human and mouse Mhc regions. These data provide strong circumstantial evidence for linkage of this group of genes to the ovine Mhc.

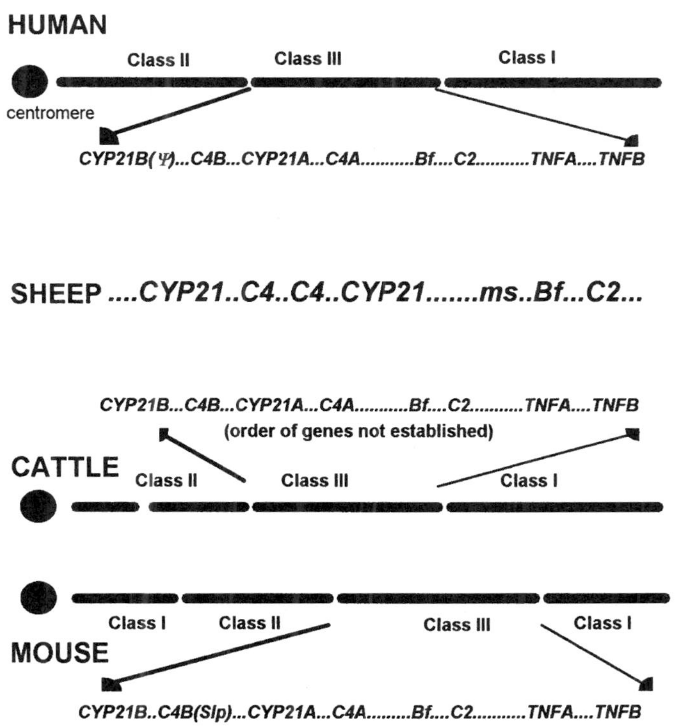

Figure 14 Syntenic group of sheep loci showing similarity with the class III Mhc region loci in the human, cattle, and mouse. The order of genes for the cattle class III region has not been established. Distances are not shown to scale.

The data summarized above are consistent with the earlier demonstration of two functional isotypes of C4 protein present in sheep plasma.[105] More recently Ren et al.[106] cloned and sequenced DNA fragments obtained by PCR amplification of sheep DNA and identified five distinct C4 sequences. This report indicates the existence of at least three C4 loci on one chromosome. Thus the number of sheep C4 genes present per chromosome resembles the situation observed in both humans and mice where there is individual variation in the number (2 to 5) and function of C4 genes per haplotype.

While a polymorphic CA dinucleotide microsatellite is present in at least one of the two cattle 21-hydroxylase genes,[107] neither of the sheep 21-hydroxylase genes contains a (CA) dinucleotide microsatellite sequence as determined by Southern blotting with a radiolabeled oligonucleotide (Groth and Wetherall, unpublished observation). This observation is consistent with the findings of Moore et al.[107] who were unable to demonstrate polymorphism in sheep using the polymerase chain reaction with oligonucleotide primers for the cattle 21-hydroxylase associated microsatellite locus.

B. The Tumor Necrosis Factor

Tumor necrosis factor α (cachectin) and tumor necrosis factor β (lymphotoxin) are related cytokines, the genes for which are closely linked in mouse, human, and cow[108-110] and may have arisen by gene duplication. The genes for the tumor necrosis factors of mouse and man map to within the class III region very close to the class I genes (Figure 14), although no chromosomal location has been assigned in ruminants. While these proteins are highly pleiotropic in function and recognize the same two receptors with high affinity on many cell types, they are produced by different cell lineages. TNFα is produced by appropriately activated cells of the monocyte/macrophage lineage, neutrophils, mast cells, some lymphocytes including natural killer cells, astrocytes, keratinocytes, fibroblasts, and at low levels by a variety of other cell types. In contrast lymphotoxin is produced exclusively by T and B lymphocytes.

A ovine TNFα cDNA has been cloned by three groups, using either PCR based methods[111,112] or conventional screening of a stimulated macrophage library.[113] The published sequences show minor differences in the 3′ untranslated region of the gene including a total of five single base changes and four short (one to three base pair) insertions. One sequence[99] also differs from the other two within the coding sequence, by a three-base deletion leading to the absence of a glutamine residue at position 63 of the preprotein sequence. Assuming that the intron/exon structure of TNFa in the sheep is the same as it is in the cow, this apparent deletion occurs at the 5′ end of exon 2 in the gene structure, where a three-base duplication may give rise to a variation in the choice of splice acceptor site (Figure 15).

The predicted preprotein derived from these sequences is of 234 aa length. By analogy with TNFα maturation in humans, maturation of this preprotein would involve removal of the first 78 aa. There is one N-linked glycosylation site at position

	exon 1	intron 1	exon 2
bovine	AGG GAA GAG	gtg agt ttt tct ccc cag	CAG TCC CC
bovine	AGG GAA GAG		CAG TCC CC
ovine cDNA5,6	AGG GAA GAG		TCC CC
ovine cDNA4	R E E		Q S P

Figure 15 Variants in splicing may give rise to two different messages for TNFa. cDNA sequence of ovine TNFa is aligned with part of the genomic sequence of bovine TNFa, showing the potential alternate splice acceptors at the 5′ end of exon 2. The repeat CAG at the splice acceptor is underlined.

19 of the mature protein, and this site is used in the natural product from LPS-stimulated macrophages[114] (The same site occurs and is used in murine, but not human or bovine TNFα). Comparison of the ovine TNFa sequence with that found in other species shows a very high degree of sequence conservation in ruminants, being 93% similar at the protein level to the bovine sequence,[110] and over 97% similar to the goat sequence[115] (see Figure 15). At the DNA level, this homology extends throughout the 5′ and 3′ untranslated regions of the gene, as well as the structural sequence. In particular a 34 base A + T-rich RNA destabilising sequence typical of cytokine genes is completely conserved across sequenced mammalian TNFa genes. When compared with nonruminant TNFα amino acid sequences the ovine sequence is between 72 and 79% identical.

Recombinant TNFα has been expressed in mammalian cells,[113] in bacteria,[116] and in yeast.[114] The expressed protein is about three orders of magnitude less cytotoxic for murine TNF-sensitive cell lines than is human or murine TNFα (about 10^5 units/mg).[114,116] In contrast the porcine cell line pk1512 is about 100 times more sensitive to ovine TNF, suggesting a species barrier in recognition of the type one TNF receptor between mouse and sheep.

No reports have been published on the ovine lymphotoxin gene or on the presence of sheep loci associated with intracellular processing of antigen or transport of peptides to the cell membrane. It is very likely however that these will be identified in the near future by cross hybridization with DNA probes from the homologous loci of other mammalian species.

IV. A GENETIC LINKAGE MAP OF SHEEP Mhc REGION

A genetic linkage map of the sheep chromosome 20, which contains the sheep Mhc region, was generated from the segregation analysis of nine polymorphic microsatellite markers (derived from cattle or sheep genomes; see Table 11) in the AgResearch International Mapping Flock (IMF). The IMF comprises 9 3-generation pedigrees and consists of 127 animals. All F1 sires had a common Texel sire and one of four Coopworth dams. The F1 dams all had different Perendale-Coopworth cross dams and one of three Merino-Romney cross sires. The number of F_2 offspring varied from 7 to 17 individuals per family.

Table 11 Markers Used in the Linkage Map of Sheep Chromosome 20

Marker	Primers 3′-5′	Annealing temperature (°C)	Informative meioses	Ref.
BM1258	GTATGTATTTTTCCCACCCTGC GAGTCAGACATGACTGAGCCTG	63	141	130
BM1818	AGCTGGGAATATAACCAAAGG AGTGCTTTCAAGGTCCATGC	53	176	130
BM1905	GTCCATGGGTTCACAAAGAG ACGCCTGCTGATGCTGTAG	57	35	130
OarCP73	AAAACTGAGAAATATTCAGATGCAAC TAAACGTCCATCAACAGAGGAAGGG	63	78	Ede et al. (in prep)
OarHH56	GCAACCCACTCATCTCTCCGTGTC GAAAACTTAAGTTCCAGCTATTAAAATAGC	60	156	121
OLADRB	CTGACCCAGA$^{T}/_{G}$TGAGTGAAGTATC TCTCTGCAGCACATTTCCTGG	59	179	68
OLADRBps	CTGCCAATGCAGAGACACAAGA GTCTGTCTCCTGTCTTGTCATC	63	173	120
OMHC1	ATCTGGTGGGCTACAGTCCATG GCAATGCTTTCTAAATTCTGAGGAA	60	197	13

With the exception of the *Ovar-DRB1* DNA locus which was typed on the basis of both size of the microsatellite and the binding of allele-specific oligonucleotides to the polymorphic *DRB1* exon 2 region,[68] all microsatellites where analyzed as previously described.[117]

Genotypes were checked for consistency with the pedigree records. Each marker was also checked for Mendelian segregation using the USERM11 module of MENDEL.[118] Preliminary determination of linkage groups was performed by two-point linkage analyses, ignoring grand parental information. This was supplemented by information from the bovine map, where available.

Multipoint linkage analysis was performed using the BUILD option of CRI-MAP.[119] Genotypes resulting in double recombinants for the most likely order were rechecked. In addition, the genotypes of families, whose proportion of recombinants for a particular pair of markers was atypical, were rechecked as were the genotypes of markers which considerably increased the map length.

The resulting map positions from the BUILD option of CRI-MAP were then checked using the FLIPS options to ensure that the odds in favor of the order presented was at least 1000:1 over the alternative orders. Each marker in each linkage group was checked to make sure that it belonged in its most likely map position against being unlinked to the group at odds of at least 1000:1, and each linkage group map was broken at each interval to ensure that the two resulting groups were linked rather than unlinked, at odds of at least 1000:1.

The resulting genetic linkage map of sheep chromosome 20 covers a total of 58.6 cM. As a result of using three bovine markers on the map, the map of sheep chromosome 20 can be compared with that of cattle chromosome 23 which contains the bovine Mhc. Both the order and the approximate genetic distance between markers are conserved between cattle and sheep. Three markers from the Mhc region form a cluster in the central region of the linkage map. These are microsatellites

Markers	Recombination Fraction between adjacent markers	Distance between adjacent markers (Kosambi cM)	Cumulative map length (Kosambi cM)
BM1905			0.0
	0.14	14.6	
OarHH56			14.6
	0.05	4.5	
BM1818			19.1
	0.10	10.4	
OMHC1			29.5
	0.03	2.9	
OLADRBps			32.4
	0.03	2.5	
OLADRB			34.9
	0.12	12.6	
BM1258			47.5
	0.11	11.1	
OarCP73			58.6

Figure 16 A sex-averaged genetic linkage map of sheep chromosome 20.

from a class one gene (OMHC1), the pseudogene *DRB2* (OLADRBps), and the expressed *DRB1* gene (OLADRB). An unexpected result was the finding of recombinants between the expressed DRB gene and the pseudogene. Five recombinants from a total of 149 co-informative meioses were identified. Multipoint linkage analysis determined a genetic distance between them of 2.6 cM (see Figure 15).

V. ASSOCIATION STUDIES BETWEEN THE *OVAR* COMPLEX AND DISEASES

A. Class I

Disease association studies received considerable impetus from the pioneering studies of Outteridge and colleagues,[122] who showed that a serologically detected class I *Ovar* antigen designated SY1 was associated with resistance of sheep to the economically important nematode parasite, *Trichostrongylus colubriformis*, and that lambs bred for the SY1 phenotype were also resistant to these parasites. A similar association also has been recently found in another breed of sheep, the New Zealand Romney. In 1990 Luffau et al.[123] reported an association between another *Ovar* type (P14) and resistance to a different nematode, the parasite *Haemonchus contortus*, although this association has not been confirmed to date.

Outteridge and colleagues also have reported some evidence for a positive association between resistance to footrot, an infection caused by the bacterium *Bacteroides nodosus*, and the *Ovar* specificities SY1b and SY6.[124] Because of the economic importance of footrot disease, this observation has led to further efforts using molecular techniques to define the nature of the association of this disease with *Ovar*. A major study is presently being undertaken within Australia, by the

CSIRO and a number of collaborating laboratories, in flocks of sheep bred for resistance or susceptibility to footrot using a large number of molecular markers to better define this association (Raadsma and Nicholas, personal communication).

Millot et al.[125,126] reported an association between serologically defined *Ovar* markers and susceptibility or resistance to infection with scrapie virus in the Isle de France breed of sheep. A more recent study by Cullen [127] was not able to confirm these observations.

B. Class II

Only a few studies searching for an association between *Ovar-DRB1* encoded genes and disease have been performed. A serological study by Douch and Outteridge[128] found an association between FEC and SY1, but the numbers of investigated animals was low. Serological and molecular studies failed to find any association between *H. contortus* infection and the ovine Mhc.[81,129] Within these studies extensive RFLP polymorphism was shown for the *DRB* region: Blattman et al.[81] described 21 polymorphic RFLP bands for the *DRB* region with *TaqI* digestion.

In the search for an association between *DRB1* and infection of *Ostertagia circumcincta,* 300 animals (including 100 unrelated) Scottish Blackface sheep were investigated.[72] The lambs were kept together after weaning to ensure equal natural exposure to potentially infective larvae. Fecal samples were taken from 200 lambs at 4-week intervals from the age of 1 to 6 months. The worms were eliminated by anthelminthics in the same intervals. By determination of length variation of intronic simple repetitive sequences and oligonucleotide hybridization of polymorphisms within the *DRB1* exon 2, 19 different *DRB1* alleles were identified in this sheep breed. Multivariate analysis showed that genetic effects on *O. circumcincta* infection were strongest at 6 month of age. The *DRB1* locus accounts for an additional 11% of the variation of fecal egg counts after accounting for the effects of sex, sire, and dam. For the statistical analysis of the data which revealed genetic effects an allele substitution model was applied. The univariate analysis indicated that substitution of the most common allele (*Ovar-DRB1*0101L526*) by allele *Ovar-DRB1*0203L512* would result in a 58-fold reduction in fecal egg counts in 6-month old lambs, suggesting that allele *Ovar-DRB1*0203L512* confers resistance to Ostertagiasis. These results were statistically significant even after Bonferroni correction.

There is no example where polymorphic Mhc alleles show 100% association with a defined disease status in either man or in animals. The probability for finding an association between a multiallele system and parasite infections depends on several factors: (1) The typing system, (2) the kind of statistical model selected for the analysis, (3) the investigated parasite, (4) the sheep breed, and (5) the parameters used to assess disease resistance.

Directional breeding in favor of an increased frequency of the critical allele within a flock could result in animals with superior resistance to Ostertagiasis. This may, however, not be desirable as homozygosity for the *DRB1* allele conferring resistance would restrict the variety of different peptides that can be presented by

Mhc molecules in this population and consequently the immune responsiveness to other pathogens. Association studies including the knowledge of the *DRB1* antigen-binding region will nevertheless give a better understanding of immunological processes leading to susceptibility or resistance.

REFERENCES

1. Millot, P., The major histocompatibility complex of sheep (*OLA*) and two minor loci, *Animal Blood Groups Biochem. Genet.*, 9, 121, 1978.
2. Klein, J., Bontrop, R. E., Dawkins, R. L., Erlich, H. A., Gyllensten, U. B., Heise, E. R., Jones, P. P., Parahm, P., Wakeland, E. K., and Watkins, D. I., Nomenclature for the major histocompatibility complexes of different species: a proposal, *Immunogenetics,* 31, 217, 1990.
3. Mahdy, E. A., Makinen, A., Chowdhary, B. P., Andersson, L., and Gustavsson, I., Chromosomal localization of the ovine *major histocompatibility complex* (OLA) by in situ hybridization, *Hereditas,* 111, 87, 1989.
4. Hediger, R., Ansari, H. A., and Stranzinger, G. F., Chromosome banding and gene localizations support extensive conservation of chromosome structure between cattle and sheep, *Cytogenet. Cell. Genet.,* 57, 127, 1991.
5. Deverson, E. V., Wright, H., Watson, S., Ballingall, K. T., Huskisson, N., Diamond, A. G., and Howard, J. C., Class II major histocompatibility complex genes of the sheep, *Anim. Genet.,* 22, 211, 1991.
6. Ballingall, K. T., Wright, H., Redmond, J., Dutia, B. M., Hopkins, J., Lang, J., Deverson, E. V., Howard, J. C., Puri, N., and Haig, D. M., Expression and characterisation of ovine major histocompatibility complex class II (*OLA-DR*) genes, *Anim. Genet.,* 23, 347, 1992.
7. Stear, M. J. and Spooner R. L., Occurrence of cytotoxic antilymphocyte antibodies in sheep, *Res. Vet. Sci.,* 34, 223, 1983.
8. Cullen, P. R., Bunch, C., Brownlie, J., and Morris, P. J., Sheep lymphocyte antigens: a preliminary study, *Animal Blood Groups Biochem. Genet.,* 13, 149, 1982.
9. Millot P., The OLA major histocompatibility complex of sheep. Study of six new factors and evidence of a third locus of the complex: *OLA-C*, *Exp. Clin. Immunogenet.,* 1, 31, 1984.
10. Gogolin-Ewens, K. J., Mackay, C. R., Mercer, W. R., and Brandon, M. R., Sheep lymphocyte antigens (OLA). I. Major histocompatibility complex class I molecules, *Immunology,* 56, 717, 1985.
11. Chardon, P., Kirszenbaum, M., Cullen, P., Geffrotin, C., Auffray, C., Strominger, J. L., Cohen, D., and Vaiman, M., Analysis of the sheep Mhc using HLA class I, II, and C4 cDNA probes, *Immunogenetics,* 22, 349, 1985.
12. Grossberger, D., Hein, W., and Marcuz, A., Class I major histocompatibility complex cDNA clones from sheep thymus: alternative splicing could make a long cytoplasmic tail, *Immunogenetics,* 32, 77, 1990.
13. Groth, D. M. and Wetherall, J. D., Dinucleotide repeat polymorphisms within the ovine Mhc class I region, *Anim. Genet.,* 25, 61, 1994.
14. Calabi, F., Yu, C. Y., Bilsland, C. A. G., and Milstein, C., CD1: from structure to function. Ch.8, pp215-243, in *Immunogenetics of the Major Histocompatibility Complex*, Srivastava, R., Ram, B. P., and Tyle, P., Eds., VCH Publishers, Cambridge, England, 1991.

15. Martin, L. H., Calabi, F., and Milstein, C., Isolation of CD1 genes: a family of major histocompatibility complex related differentiation antigens, *Proc. Natl. Acad. Sci. U.S.A.,* 83, 9154, 1986.
16. Martin, L. H., Calabi, F., Lefebvre, F.-A., Nilsland, C. A. G., and Milstein, C., Structure and expression of the human thymocyte antigens CD1a, CD1b, and CD1c, *Proc. Natl. Acad. Sci. U.S.A.,* 84, 9189, 1987.
17. Blumberg, R. S., Terhorst, C., Bleicher, P., McDermott, F. V., Allan, C. H., Landau, S. B., Trier, J. S., and Balk, S. P., Expression of a nonpolymorphic Mhc class I-like molecule, CD1D, by human intestinal epithelial cells, *J. Immunol.,* 147, 2518, 1991.
18. Bradbury, A., Tertia Belt, K., Neri, T. M., Milstein, C., and Calabi, F., Mouse CD1 is distinct from and coexists with TL in the same thymus, *EMBO J.,* 7, 3081, 1988.
19. Balk, S. P., Bleicher, P. A., and Terhorst, C., Isolation and expression of cDNA encoding the murine homologues of CD1, *J. Immunol.,* 146, 7, 1991.
20. Balk, S. P., Ebert, E. C., Blumenthal, R. L., McDermott, F. V., Wucherpfennig, K. W., Landau, S. B., and Blumberg, R. S., Oligoclonal expansion and CD1 recognition by human intestinal epithelial lymphocytes, *Science,* 1991.
21. Calabi, F., Tertia Belt, K., Yung Yu, C., Bradbury, A., Mandy, W. J., and Milstein, C., The rabbit CD1 and the evolutionary conservation of the CD1 gene family, *Immunogenetics,* 30, 370, 1989.
22. Calabi, F., Jarvis, J. M., Martin, L., and Milstein, C., Two classes of CD1 genes, *Eur. J. Immunol.*, 19, 285, 1989.
23. Dutia, B. M. and Hopkins, J., Analysis of the CD1 cluster in sheep, *Vet. Immunol. Immunopathol.*, 27, 189, 1991.
24. MacHugh, N. D., Bensaid, A., Davis, W. C., Howard, C. J., Parsons, K. R., Jones, B., and Kaushal, A., Characterisation of a bovine thymic differentiation antigen analogous to CD1 in the human, *Scand. J. Immunol.*, 27, 541, 1988.
25. Pescovitz, M. D., Lunney, J. K., and Sachs, D. H., Preparation and characterisation of monoclonal antibodies reactive with porcine PBL, *J. Immunol.*, 133, 368, 1984.
26. Bonneville, M., Ito, K., Krecko, E. G., Itohara, S., Kappes, D., Ishida, I., Kanagawa, O., Janeway, C. J., Jr., Murphy, D. B., and Tonegawa, S., Recognition of a self major histocompatibility complex TL region product by gd T-cell receptors, *Proc. Natl. Acad. Sci. U.S.A.*, 86, 5928, 1989.
27. Milligan, G. N., Flaherty, L., Braciele, V. L., and Braciele, T. J., Non-conventional (TL-encoded) major histocompatibility complex molecules present processed viral antigen to cytotoxic T lymphocytes, *J. Exp. Med.*, 174, 133, 1991.
28. Vidovic, D., Roglic, M., McKune, K., Guerder, S., MacKay, C., and Dembic, Z., Qa-1 restricted recognition of foreign antigen by a gd T-cell hybridoma, *Nature*, 340, 646, 1989.
29. Porcelli, S., Brenner, M. B., Greenstein, J. L., Balk, S. P., Terhorst, C., and Bleicher, P. A., Recognition of cluster of differentiation 1 antigens by human CD4-8-cytolytic T lymphocytes, *Nature*, 341, 447, 1989.
30. Panja, A., Blumberg, R. S., Balk, S. P., and Mayer, L., CD1d is involved in T cell-intestinal epithelial cell interactions, *J. Exp. Med.*, 178, 1115, 1993.
31. Porcelli, S., Morita, C. T., and Brenner, M. B., CD1b restricts the response of human CD4-8- T lymphocytes to a microbial antigen, *Nature,* 360, 593, 1992.
32. Howard, C. J., Sopp, P., Parsons, K. R., Bembrodge, G. P., and Hall, G., A new bovine leukocyte antigen cluster comprising two monoclonal antibodies, CC43 and CC118, possily related to CD1, *Vet. Immunol. Immunopathol.*, 39, 69, 1993.

33. Calabi, F., Bilsland, C.A.G., Yung Yu, C., Bradbury, A., Tertia-Belt, K., Martin, L. H., and Milstein, C., Recent progress in the molecular study of CD1, in *Leukocyte Typing IV.* Knapp, W. et al., Eds., Oxford University Press, New York, 1989, 254.
34. Bradbury, A., Calabi, F., and Milstein, C., Expression of CD1 in the mouse thymus. *Eur. J. Immunol.*, 20, 1831, 1990.
35. Wright, H. and Ballingall, K. T., Mapping and characterisation of the DQ subregion of the ovine Mhc, *Anim. Genet.*, 25, 243, 1994.
36. Ballingall, K. T. Dutia, B. M., Hopkins, J., and Wright, H., Analysis of the fine specificities of sheep major histocompatibility complex class II-specific monoclonal antibody using mouse L-cell transfectants, *Anim. Genet.*, 26, 79, 1995.
37. Hopkins, J., Dutia, B. M., and McConnell, I., Monoclonal antibodies to sheep lymphocytes. I. Identification of the Mhc class II molecules on lymphoid tissue and changes in the level of class II expression on lymph-borne cells following antigen stimulation in vivo, *Immunology,* 59, 433, 1986.
38. Dutia, B. M., Hopkins, J., Allington, M. R., Bujdoso, R., and McConnell, I., Characterisation of monoclonal antibodies specific for a and b chains of sheep Mhc class II, *Immunology,* 70, 27, 1990.
39. Puri, N. K., Gogolin-Ewens, K. J., and Brandon, M. R., Monoclonal antibodies to sheep Mhc class I and II molecules: biochemical characterisation of three class I gene products and four distinct sub-populations of class II molecules, *Vet. Immunol. Immunopathol.*, 15, 59, 1987.
40. Hall, J. G., Studies of the cells in the afferent and efferent lymph of lymph nodes draining the site of skin homografts, *J. Exp. Med.,* 125, 737, 1967.
41. Hopkins, J., McConnell, I., Bujdoso, R., and Munro, A. J., Studies of Mhc class II products on sheep peripheral and efferent lymph cells, in *Immunology of the Sheep,* Morris, B. and Miyasaka, M., Eds., Editiones (Roche), Basel, 1985, 441.
42. Hopkins, J., Dutia, B. M., and McConnell, I., Monoclonal antibodies to sheep lymphocytes. I. Identification of the Mhc class II molecules on lymphoid tissue and changes in the level of class II expression on lymph-borne cells following antigen stimulation in vivo, *Immunology,* 59, 433, 1986.
43. Dutia, B. M., Hopkins, J., Allington, M. R., Bujdoso, R., and McConnell, I., Characterisation of monoclonal antibodies specific for a and b chains of sheep Mhc class II, *Immunology,* 70, 27, 1990.
44. Dutia, B. M., McConnell, I., Bird, K., Keating, P., and Hopkins, J., Patterns of Mhc class II expression on T cell subsets in different immunological compartments. 1. Expression on resting T cells, *Eur. J. Immunol.,* 23, 2882, 1993.
45. Mackay, C. R., Marston, W. L., and Dudler, L., Naive and memory T cells show distinct pathways of lymphocyte recirculation, *J. Exp. Med.,* 171, 801, 1990.
46. Ko, H. S., Fu, S. M., Winchester, R. J., Yu, D. T. Y., and Kunkel, H. G., Ia determinants on stimulated human T cells. Occurrence on mitogen and antigen activated T cells, *J. Exp. Med.,* 150, 246, 1979.
47. Singh, S. K., Abramson, E. J., Krco, C. J., and David, C. S., Endogenous synthesis of Ia antigens by a cloned T cell line, *J. Mol. Cell. Immunol.,* 1, 147, 1984.
48. Crepaldi, T., Crump, A., Newman, M., Ferrone, S., and Antczak, D. F., Equine T lymphocytes express Mhc class II antigens, *J. Immunogenet.,* 13, 349, 1986.
49. Mackay, C. R., Marston, W. L., and Dudler, L., Naive and memory T cells show distinct pathways of lymphocyte recirculation, *J. Exp. Med.,* 171, 801, 1990.
50. Oshima, S. and Eckels, D. D., Selective expression of class II Mhc isotypes by MLC-activated human T lymphocytes, *Hum. Immunol.,* 27, 208, 1990.

51. Butcher, E. C., Leukocyte-endothelial cell recognition: three (or more) steps to specificity and diversity, *Cell,* 67, 1033, 1991.
52. Chin, W. and Hay, J. B., A comparison of lymphocyte migration through intestinal lymph nodes, subcutaneous lymph nodes and chronic inflammatory sites, *Gastroenterology,* 79, 1231, 1980.
53. Hall, J. G., Scollay, R. G., and Smith, M. E., Studies on lymphocytes of sheep. I. Recirculation of lymphocytes through peripheral lymph nodes and tissues, *Eur. J. Immunol.,* 6, 117, 1976.
54. Hopkins, J., McConnell, I., Dalziel, R. G., and Dutia, B. M., Patterns of Mhc class II expression by T cell subsets in different immunological compartments. 2. Altered expression and cell function following activation in vivo, *Eur. J. Immunol.,* 23, 2889, 1993.
55. Andersson, L., Lunden, A., Sigurdardottir, S., Davies, C. J., and Rask, L., Linkage relationships in the bovine Mhc region. High recombination frequency between class II subregions, *Immunogenetics,* 27, 273, 1988.
56. Stone, R. T. and Muggli-Cockett, N. E., *BoLA DIB*: species distribution, linkage with DOB and Northern analysis, *Anim. Genet.,* 24, 41, 1993.
57. Scott, P. C., Gogolin-Ewens, K. J., Adams, T. E., and Brandon, M. R., Nucleotide sequence, polymorphism and evolution of ovine Mhc class II *DQA* genes, *Immunogenetics,* 34, 69, 1991.
58. Puri, N. K. and Brandon, M. R., Sheep Mhc class II molecules. Identification and characterisation of four distinct subsets of sheep Mhc class II molecules, *Immunology,* 62, 575, 1987.
59. Dutia, B. M., McConnell, I., Ballingall, K., Keating, P., and Hopkins, J., *Anim. Genet.,* in press.
60. Muggli-Cockett, N. E. and Stone, R. T., Partial nucleotide sequence of a bovine major histocompatibility class II DRB-like gene, *Anim. Genet.,* 20, 361, 1989.
61. Groenen, M. A. M., Van der Poel, J. J., Dijkhof, R. J. M., and Giphart, M. J., The nucleotide sequence of the bovine Mhc class II *DQB* and *DRB* genes, *Immunogenetics,* 31, 37, 1990.
62. Marsh, S. G. E. and Bodmer, J., HLA class II nucleotide sequences, 1992, *Immunogenetics,* 37, 79, 1993.
63. Brändle, U., Hideki, O., Vincek, V., Klein, D., Golubic, M., Grahovac, B., and Klein, J., Trans-species evolution of *Mhc-DRB* haplotype polymorphism in primates: organization of *DRB* genes in the chimpanzee, *Immunogenetics,* 36, 39, 1992.
64. Schönbach, C., Vincek, V., Mayer, W. E., Golubic, M., O'hUigin, C., and Klein, J., Multiplication of the *Mhc-DRB5* loci in the orangutan: implications for the evolution of *DRB* haplotypes, *Mam. Gen.,* 4, 159, 1993.
65. Schwaiger, F.-W., Weyers, E., Buitkamp, J., Ede, A. J., Crawford, A., and Epplen, J. T., Interdependent *Mhc-DRB* exon-plus-intron evolution in artiodactyls, *Mol. Biol. Evol.,* 11, 239, 1994.
66. Rieß, O., Kammerbauer, C., Roewer, L., Steimle, V., Andreas, A., Albert, E., Nagai, T., and Epplen, J. T., Hypervariability of intronic simple (gt)(ga) repeats in HLA-DRB genes, *Immunogenetics,* 32, 110, 1990.
67. Ammer, H., Schwaiger, F.-W., Kammerbauer, C., Gomolka, M., Arriens, A., Lazary, S., and Epplen, J. T., Exonic polymorphism versus intronic simple repeat hypervariability in Mhc-DRB-genes, *Immunogenetics,* 32, 330, 1992.
68. Schwaiger, F.-W., Weyers, E., Epplen, C., Ruff, G., Crawford, A., and Epplen, J. T., On the paradox of *Mhc-DRB* evolution: α-helix and β-sheet encoding regions diverge while hypervariable intronic simple repeats co-evolve with β sheet codons, *J. Mol. Evol.,* 37, 260, 1993.

69. Scott, P. C., Choi, C.-L., and Brandon, M. R., Genetic organisation of the ovine Mhc class II region, *Immunogenetics,* 25, 116, 1987.
70. Fabb, S. A., Maddox, J. F., Gogolin-Ewens, K. J., Baker, L., Wu, M. J., and Brandon, M. R., Isolation, characterization and evolution of ovine major histocompatibility complex class II *DRA* and *DQA* genes, *Anim. Genet.,* 24, 249, 1993
71. Escayg, A. P., Montgomery, G. W., Hickford, J. G., and Bullock, D. W., A BglII RFLP at the ovine Mhc class II *DRA* locus, *Anim. Genet.,* 24, 217, 1993.
72. Schwaiger, F.-W., Gostomski, D., Epplen, J. T., Stear, M., and Buitkamp, J., Ovine *Major Histocompatibility Complex DRB1* allele is associated with decrease in faecal egg counts following natural, predominantly *ostertagia circumcincta* infection, submitted.
73. Brown, J. H., Jardetzky, T. S., Gorga, J. C., Stern, L. J., Urban, R. G., Strominger, J. L., and Wiley, D. C., Three-dimensional structure of the human class II histocompatibility antigen HLA-DR1, *Nature,* 364, 33, 1993.
74. Wakeland, E. K., Boehme, S., and She, J. X., The generation and maintainence of Mhc class II gene polymorphism in rodents, *Immunol. Rev.,* 113, 39, 1990.
75. Gyllensten, U., Sundvall, M., and Erlich, H. A., Allelic diversity is generated by intraexon sequence exchange at the *DRB1* locus of primates, *Proc. Natl. Acad. Sci. U.S.A.,* 88, 3686, 1991.
76. Andersson, L., Sigurdardottir, S., Borsch, C., and Gustafsson, K., Evolution of Mhc polymorphism: extensive sharing of polymorphic sequence motifs between human and bovine *DRB* alleles, *Immunogenetics,* 33, 188, 1991.
77. Klein, J., Origin of the major histocompatibility complex polymorphism: the trans-species hypothesis, *Hum. Immunol.,* 19, 155, 1987.
78. Hughes, A. L. and Nei, M., Nucleotide substitution at major histocompatibility complex class II loci: evidence for overdominant selection, *Proc. Natl. Acad. Sci. U.S.A.,* 86, 958, 1989.
79. Nei, M. and Gojobori, T., Simple methods for estimating the numbers of synonymous and nonsynonymous nucleotide substitutions, *Mol. Biol. Evol.,* 3, 418, 1986.
80. Scott, P. C., Maddox, J. F., Gogolin-Ewens, K. J., and Brandon, M. R., The nucleotide sequence and evolution of ovine Mhc class II B genes: *DQB* and *DRB*, *Immunogenetics,* 34, 80, 1991.
81. Blattman, A. N., Hulme, D. J., Kinghorn, B. P., Woolaston, R. R., Gray, G. D., and Beh, K. J., A search for association between major histocompatibility complex restriction fragment length polymorphism bands and resistance to *Haemonchus contortus* infection in sheep, *Anim. Genet.,* 24, 277, 1993.
82. Braunstein, N. S., Germain, R. N., Loney, K., and Berkowitz, N., Structurally interdependent and independent regions of allelic polymorphism in class II Mhc molecules. Implications for Ia function and evolution, *J. Immunol.,* 145, 1635, 1990.
83. Lundberg, A. S. and McDevitt, H. O., Evolution of major histocompatibility complex class II allelic diversity: direct descent in mice and humans, *Proc. Natl. Acad. Sci. U.S.A.,* 89, 6545, 1992.
84. Seidl, C. and Lee, J. S., Expression of alternatively spliced HLA class II transcripts in lymphoid and non-lymphoid tissues, *Immunogenetics,* 35, 385, 1992.
85. Senju, S., Kimura, A., Yasunami, M., Kamikawaji, N., Yoshizumi, H., Nishimura, Y., and Sasazuki, T., Allele-specific expression of the cytoplasmic exon of *HLA-DQB1* gene, *Immunogenetics,* 36, 319, 1992.
86. Grain, F., Main, M. C., Labonne, M. P., Lantier, F., Lechopier, P., Gebuhrer, L., Asso, J., and Maddox, J., Restriction fragment length polymorphism of *DQB* and *DRB* genes of the ovine major histocompatibility complex, *Anim. Genet.,* 24, 377, 1993.

87. van Oorschot, R. A. H., Maddox, J. F., Adams, L. J., and Fabb, S. A., Characterisation and evolution of ovine Mhc class II *DQB* sequence polymorphism, *Anim. Genet.*, 25, 417, 1994.
88. Karlsson, L., Suhr, C. D., Sprent, J., and Peterson, P. A., A novel class II Mhc molecule with unusual tissue distribution, *Nature,* 351, 485, 1991.
89. van der Poel, J. J., Groenen, M. A. M., Dijkhof, R. J. M., Ruyter, D., and Giphart, M. J., The nucleotide sequence of the bovine Mhc class II alpha genes: *DRA*, *DQA* and *DYA*, *Immunogenetics,* 31, 29, 1990.
90. Mann, A. J., Abraham, L. J., Cameron, P. U., Robinson, W., Giphart, M. J., and Dawkins, R. L., The caprine Mhc contains *DYA* genes, *Immunogenetics,* 37, 292, 1993.
91. Stone, R. T. and Muggli-Cockett, N. E., Partial nucleotide sequence of a novel bovine major histocompatibility complex class II beta chain gene, *BoLA DIB, Anim. Genet.,* 21, 352, 1990.
92. Wright, H., Ballingall, K. T., and Redmond, J., The *DY* subregion of the sheep Mhc contains an A/B gene pair, *Immunogenetics,* 40, 230, 1994.
93. Campbell, R. D. and Trowsdale, J., A map of the human Mhc, *Immunol. Today,* 14, 347, 1993.
94. van Eijk, M. J. T., Russ, I., and Lewin, H. A., Order of bovine *DRB3, DYA* and *PRL* determined by sperm typing, *Mamm. Genome,* 4, 113, 1993.
95. Campbell, R. D. and Trowsdale, J., A map of the human Mhc, *Immunol. Today,* 14, 347, 1993.
96. Klein, J. and Figueroa, F., Evolution of the major histocompatibility complex, *Crit. Rev. Immunol.,* 6, 295, 1987.
97. Degli-Esposti, M. A., Andreas, A., Christiansen, F. T., Schalke, B., Albert, E., and Dawkins, R. L., An approach to the localisation of the susceptibility genes for generalised myasthenia gravis by mapping recombinant ancestral haplotypes, *Immunogenetics,* 35, 355, 1991.
98. Dawkins, R. L., Christiansen, F. T., Kay, P. H., Garlepp, M., McCluskey, J., Hollingsworth, P. N., and Zilko, P. J., Disease associations with complotypes, supratypes and haplotypes, *Immunol. Rev.,* 70, 5, 1983.
99. Andersson, L., Lunden, A., Sigurdardottir, S., Davies, C. J., and Rask, L., Linkage relationships in the bovine Mhc region. High recombination frequency between class II subregions, *Immunogenetics,* 27, 273, 1988.
100. Cameron, P. U., Tabarias, H. A., Pulendran, B., Robinson, W., and Dawkins, R. L., Conservation of the central Mhc genome: PFGE mapping and RFLP analysis of complement, HSP70, and TNF genes in the goat, *Immunogenetics,* 31, 253, 1990.
101. Chardon, P., Kirszenbaum, M., Cullen, P., Geffrotin, C., Auffray, C., Strominger, J. L., Cohen, D., and Vaiman, M., Analysis of the sheep Mhc using HLA class I, II, and C4 cDNA probes, *Immunogenetics,* 22, 349, 1985.
102. Groth, D. M., Immunogenetics of Complement Component C4, PhD thesis, University of Western Australia, 1992.
103. Groth, D. M., Lintorn-Terry, E., Outteridge, P., Windon, R. G., and Wetherall, J. D., Taq 1 RFLP polymorphism of the ovine complement component C4 gene, *Nucleic Acid Res.,* 18, 3102, 1990.
104. Groth, D. M., Lintorn-Terry, E., Taylor, L., Carrick, M., and Wetherall, J. D., EcoRI and BamHI RFLP polymorphism of bovine complement C4 gene, *Nucleic Acid Res.* 18, 3107, 1990.
105. Groth, D. M., Wetherall, J. D., Taylor, L., Sparrow, P. R., and Lee, I. R., Purification and characterisation of ovine C4: evidence for two molecular forms in ovine plasma, *Mol. Immunol.,* 25, 577, 1988.

106. Ren, X. D., Dodds, A. W., and Law, S. K. A., Thethiolester and isotypic sites of complement component C4 in sheep and cattle, *Immunogenetics,* 1992.
107. Moore, S. S., Sargeant, L. L., King, T. J., Mattick, J. S., Georges, M., and Hetzel, D. J. S., The conservation of dinucleotide microsatellites among mammalian genomes allows the use of heterologous PCR primer pairs in closely related species, *Genomics*, 10, 654, 1991.
108. Muller, U., Jongeneel, C. V., Nedospasov, S. A., Lindahl, K. F., and Steinmetz, M., Tumor necrosis factor and lymphotoxin genes map close to H-2D in the mouse major histocompatibility complex, *Nature,* 325, 265, 1987.
109. Nedospasov, S. A., Shakov, A. N., Turetskaya, R. L., Mett, V. A., Azizov, M. M., Georgiev, G. P., Korobko, V. G., Dobrynin, V. N., Porilipov, S. A., Bystrov, N. S., Boldreva, E. F., Chuvpilo, S. A., Chumakov, A. M., Shingarova, L. N., and Ovchinnkov, Y. A., Tandem arrangement of genes coding for tumour necrosis factor (TNF-α) and lymphotoxin (TNF-β) in the human genome, *Cold Spring Harbor Symp. Quant. Biol.,* 51, 611, 1986.
110. Cludts, I., Cleuter, Y., Kettmann, R., Arsene, B., and Droogmans, L., Cloning and characterization of the tandemly arranged bovine lymphotoxin and tumour necrosis factor alpha genes, *Cytokine,* 5, 336, 1993.
111. Young, A., Hay, J., and Chan, J., Primary structure of ovine tumor necrosis factor alpha cDNA, *Nucleic Acids Res.,* 18, 6723, 1990.
112. Green, I. R. and Sargan, D. R., Sequence of the cDNA encoding tumor necrosis factor-a: problems with cloning by inverse PCR, *Gene,* 109, 203, 1991.
113. Nash, A., Barcham, G., Brandon, M., and Andrews, A., Molecular cloning, expression and characterization of ovine TNFα, *Immunol. Cell Biol.,* 69, 273, 1991.
114. Green, I. R., Fiskerstrand, C., Giovani, B., Roy, D. J., Peterhans, E., and Sargan, D. R., Expression and characterisation of ovine TNFα: some evidence for species specificity, *Cytokine,* 5, 213, 1993.
115. Goldstein, I. M., Henner, D., and Talhouk, A., EMBL database library, accession no. X14828. n.b. The comparison referred to above is of the area downstream of a frameshift mutation in the leader peptide of this sequence, relative to all other mammalian TNFs. 97% identity with the ovine sequence occurs throughout the remainder of the leader sequence and the mature protein sequence, 1989.
116. Nash, A. D., Lofthouse, S. A., Barcham, G. J., Jacobs, H. J., Ashman, K., Meeusen, E.N.T., Brandon, M. R., and Andrews, A. E., Recombinant cytokines as immunological adjuvants, *Immunol. Cell Biol.,* 71, 367, 1993.
117. Crawford, A. M., Buchanan, F. C., and Swarbrick, P. A., Ovine dinucleotide repeat polymorphism at the MAF18 locus, *Anim. Genet.,* 21, 433, 1990.
118. Lange, K., Weeks, D., and Boehnke, M., Programs for pedigree analysis: MENDEL, FISHER and dGENE, *Genet. Epidemiol.,* 5, 471, 1988.
119. Lander, E. S. and Green, P., Construction of multilocus genetic linkage maps in humans, *Proc. Natl. Acad. Sci. U.S.A.,* 84, 2263, 1987.
120. Blattman, A. N. and Beh, K. J., Dinucleotide repeat polymorphism within the ovine Mhc, *Anim. Genet.,* 23, 392, 1992.
121. Ede, A. J., Pierson, C. A., Henry, H. M., and Crawford, A. M., Ovine microsatellites at the *OarAE64, OarHH22, OarHH56, OarHH62* and *OarVH4* loci, *Anim. Genet.,* 25, 51, 1994.
122. Outteridge, P. M., Windon, R. G., and Dineen, J. K., An association between a lymphocyte antigen in sheep and the response to vaccination against the parasite *Trichostrongylus colubriformis*, *Int. J. Parasitol.,* 15, 121, 1985.

123. Luffau, G., Khang, J. V. T., Bouix, J., Nguyen, T. C., Cullen, P., and Ricordeau, G., Resistance to experimental infections with *Haemonchus contortus* in Romanov sheep, *Genet. Sel. Evol.,* 22, 205, 1990.
124. Outteridge, P. M., Stewart, D. J., Skerman, T. M., Dufty, J. H., Egerton, J. R., Ferrier, G., and Marshall, D. J., A positive association between resistance to ovine footrot and particular lymphocyte antigen types, *Aust. Vet. J.*, 66, 175, 1989.
125. Millot, P., Chatelain, J., Baille, V., and Cathala, F., The OLA major histocompatibility complex of sheep. Genetic frequencies compared in Prelpe sheep with and without scrapie, C.R. Acad. Sci., 298, 305, 1984.
126. Millot, P., Scrapie and the sheep Mhc — claims of linkage refuted, *Immunogenetics*, 27, 1, 1988.
127. Cullen, P. R., Scrapie and the sheep Mhc: claims of linkage refuted, *Immunogenetics,* 29, 414, 1989.
128. Douch, P. G. and Outteridge, P. M., The relationship between ovine lymphocyte antigens and parasitological and production parameters in Romney sheep, *Int. J. Parasitol.,* 19, 35, 1989.
129. Cooper, D. W., van Oorschot, R. A., Piper, L. R., and Le Jambre, L. F., No association between the ovine leucocyte antigen (OLA) system in the Australian merino and susceptibility to *Haemonchus contortus* infection, *Int. J. Parasitol.,* 9, 695, 1989.
130. Bishop, M. D., Kappes, S. M., Keele, J. W., Stone, R. T., Sunden, S. L. F., Hawkins, G. A., Toldo, S. S., Fries, R., Grosz, M. D., Yoo, J., and Beattie, C. W., A genetic linkage map for cattle, *Genetics*, 136, 619, 1994.
131. Schwaiger, F.-W. and Epplen, J. T., Exonic *Mhc-DRB* polymorphisms and intronic simple repeat sequences: *Janus'* faces of DNA sequence evolution, *Immunol. Rev.,* 143, 199, 1995.

CHAPTER 7

The Cervine Major Histocompatibility Complex: Genetic Structure and Function of the Cervine DRB Genes

Peter A. Swarbrick and Allan M. Crawford

CONTENTS

ABSTRACT

The expressed major histocompatibility complex (Mhc) class II DRB genes of 50 unrelated red deer were examined by the polymerase chain reaction (PCR) amplification, cloning, and sequencing of DRB exon 2 from cDNAs. Deer like other mammals have highly polymorphic Mhc class II DRB genes; 48 different sequences were identified. Like other mammals, most of the amino acid variation occurred at positions that have been demonstrated to form the antigen-binding groove. Eighteen deer-specific substitutions were found; 11 occur at antigen-binding positions. Between one and four DRB sequences were found per deer; Mendelian inheritance of these sequences in pedigrees showed that haplotypes may have one or two expressed DRB loci.

0-8493-8032-4/96/$0.00+$.50

I. INTRODUCTION

Mhc class II molecules are heterodimeric cell surface proteins composed of two peptide chains, A and B, which are expressed by cells involved in the immune response.[1] Mhc class II molecules present intracellularly degraded peptides of 8 to 25 residues in length to CD_{4+} T-lymphocytes.[2] In humans there are at least five types of related class II gene groups: DP, DQ, DR, DN, and DO; the cattle homologs to human DQ and DR have been found.

Class II molecules have extremely polymorphic regions, which occur predominantly in exon 2 of B genes and in DQA.[3] In humans 60 alleles from the DRB1 locus have been found.[4] Recently the three-dimensional structure of the human class II HLA-DR1 molecule has been published.[2] This structure shows a majority of the polymorphic sites are positioned in such a way as to form part of the antigen-binding groove and that these amino acids are the ones that interact with the bound peptide.

In humans four expressed DRB loci have been identified (DRB1, 3, 4, and 5); any one haplotype contains a DRB1, and either DRB3, DRB4, DRB5 or no additional DRB loci. Thus in any one haplotype there are up to two expressed DRB loci whereas only one expressed DQB, DRA, and DQA has been detected. Different haplotypes contain a varying number of pseudogenes. Similar patterns of Mhc Class II genetic polymorphism have been shown to exist in bovidae,[5-7] primates,[8] dog,[9] and fish.[10] However, only one transcribed DRB gene has been reported in ruminants. This chapter examines the expressed DRB exon II sequences in red deer.

II. THE EXPRESSED DRB GENES

A. Methods

To examine the expressed DRB genes of deer, we used cDNA as the template for the PCR.[12] The amplified region spans an intron, which in cattle is approximately 5 kbp long; thus any contaminating genomic DNA in the cDNA preparation could not have provided a template for PCR. Using this system we are confident that only transcribed sequences are amplified. The PCR primers were designed from published cattle sequence.[11] A primer sequence was taken from regions of the DRB gene in which the DNA sequence is conserved and hence would not discriminate between loci. PCR primer A, 5′-CAAAGTGACACGTCGCCGCTC-3′, was taken from the 3′ region of exon 2 and primer B, 5′-AGCTCTGACAGTGATACTGATGG-3′, was designed from the 3′ region of exon 1. PCR products were blunt-end ligated into pBluescript (Stratagene) and then electroporated into *Escherichia coli*. Isolated plasmids were sequenced on an Applied Biosystems 373A DNA sequencer and analyzed on ABI Seqed software.

B. Results

Between one and four sequences were found in each deer; most often we isolated either three or four sequences.[12] A total of 48 different sequences were found in the

50 unrelated deer examined (Figure 1). Almost half of these were found in only one or two animals, indicating that there remain a considerable number of undescribed DRB sequences in the deer population. Although our approach did not discriminate between loci amplified, it was likely that not all the sequences contained in the 50 animals sampled were uncovered. There is some evidence (not shown) that the downstream (exon 2) primer's binding region has some variability, which would result in the nonamplification of some sequences. We sequenced an average of 18 clones per deer and hence there is a reasonable chance of "missing" one or more of the sequences, especially if some sequences have lower expression than others, as is the case in the DRB3, 4, and 5 genes of humans.

III. RELATIONSHIPS OF DEER DRB GENES TO THOSE OF OTHER MAMMALS

Cattle are thought to have only one transcribed DRB gene (DRB3);[13] only one expressed DRB has been demonstrated in sheep, although it was expected up to three would be found.[14] There is more than one transcribed DRB gene in most deer; most of the 50 deer we sampled contained 3 or 4 different DRB sequences, indicating the presence of more than one expressed DRB gene.

The expressed DRB sequences were isolated from a small pedigree (Figure 2). Four different sequences were found in the sire; each of the progeny inherited two of the sire sequences and the balance from the dams. This pattern exhibits Mendelian inheritance and confirms the observation that deer contain more than one expressed DRB gene. This finding contrasts with other ruminants where only one expressed DRB gene has been demonstrated.

It is not yet clear whether, in deer, one of the DRB loci is highly polymorphic and the other(s) oligomorphic or whether more than one locus is highly polymorphic. Most of the sequences are found in three or fewer animals; the remaining are found more frequently. Few animals contain more than two sequences, which are rarely found. This may indicate that the situation is more like the human/chimpanzee with one highly polymorphic locus and several oligomorphic loci. This question will be resolved by the characterization of each locus using locus-specific primers.

Comparison of expressed deer DRB sequences with expressed human, chimpanzee, cattle, sheep, and goat DRB genes showed high homology. The majority of amino acid positions of exon 2 which have a high frequency of substitutions occur in three main regions (positions 9 to 13, 25 to 28, and 67 to 74); this is typical of DRB sequences found in other species. Amino acid positions that form the antigen-binding groove were usually very polymorphic in all species, with most of the amino acids shared at each position between each species (Table 1). There were however some sites in the groove that had low rates of polymorphism or were monomorphic in all species, for example, positions 65, 68, and 81. There must be some other constraints on these positions, which even though they have been shown in the three-dimensional crystal structure to form the groove,[2] they are not highly variable. Position 71 was extremely polymorphic in deer; 10 different amino acids substitutions were found. This compares to an average of 4.6 across the other species. There were also

```
                    10         20         30         40         50         60         70         80
HLA-DRB  GDTRPRFLEQ SKSECHFFNG TERVRFLDRY FYNQEEYVRF DSDVGEYRAV TELGRPDAEY WNSQKDLLEQ RRAAVDTYCR HNY
DRB1     RKI--H---Y A-R----P-- -Q--G----- ---G------ ---W------ --------K- -----EYM-R A--------- ---
DRB2     REI--H---Y A-R--Y-S-- -Q--G----- ---G------ ---W------ --------K- -----EYM-R A--------- ---
DRB6     RKI--H---Y A-R----S-- -Q--G----- ---G------ ---W------ --------K- -----EYM-R A--------- ---
DRB3     REI--H---Y A-R----S-- -Q--G----- ---G------ ---W------ --------K- -----EYM-R A--------- ---
DRB4     REI--H---Y A-R----P-- -Q--G----- ---G------ ---W------ --------K- -----EYM-R A--------- ---
DRB28    RKI--H---Y A-R----S-- -Q-----E-- I--R------ ---W------ --------K- Y----E---E K--E------ ---
DRB30    RKI-LH---Y A-R----S-- -Q--G----- I--W------ ---W------ --------K- Y--L-E---E K--E------ ---
DRB32    REI-LH---Y A-R----S-- -Q--G----- I--W------ -C-W------ --------K- Y----E---E E--E------ ---
DRB31    REI--H---Y F------S-- -----L---- V--R--F--Y -------Q-- --------K- L----E---D N------F-- ---
DRB36#   RE---H---Y A-R--P-S-- -Q--QY---- -C-G--F--- ---W------ --------K- -----E---R K--N------ -D-
DRB9 *   RKM--H---Y A-R--P-S-- -Q--QY---- -C-G--F--- ---W------ --------K- -----EYM-- T--E--R--- ---
DRB14    RKM--H---Y A-R--P-S-- -Q--QY---- -C-G--F--- ---W------ --------K- -----EI--R K--N------ -D-
DRB25#   RKI--H---Y A-R--P-S-- -Q--QY---- ---G------ ---W------ --------K- -----EI--E K--E------ ---
DRB33    REI--H--LY --R----S-- -Q--QY---- ---G--F--- ---W------ ------V--- L----E---R K--N------ Y--
DRB16    RKI--H--LY --R----S-- -Q--QY---- ---G--F--- ---W------ --------K- -----EI--R K--N------ ---
DRB17#   REI--H--LY --R----S-- -Q---Y---- ---G--F--- G--W------ --------K- -----EI--R K--N------ ---
DRB7     REI--H--LY A------S-- -Q--G----- ---G--F--- ---W------ ------V--- L----EYM-- T--E------ ---
DRB49    RK---H--LY A------S-- -Q--GL---- ---G--F--- ---W--F--- ---------G --R--E---- ---------- ---
DRB8     RKI--H--LY T------S-- -Q--GL---- ---G------ ---W------ ------S-K- -----EYM-- T--E--R--- ---
DRB5 #   RE---H--LY TT-----S-- -Q--GL---- ---G------ ---W------ ------S-K- -----EYM-- T--E--R--- ---
DRB41    RE---H--MY T-K----S-- -Q--GL---- ---G--F--- ---W--F--- -----R---G --R--E---- ---------- ---
DRB47#   RKI--H--LY TT-----S-- -Q--Q----- ---G--F--- ---W------ ---------G --R--EF--- A--------- ---
DRB11    RKI--H---H L-A------- ---MQ--A-- ---G---A-- ------F--- ---------- -----EI--- HG-E------ ---
DRB18    RKI--H---H L-A------- ---MQF-A-- ---G---A-- ------F--- --------K- -----EI--R K--N------ -D-
DRB12    REI--H---H L-A------- ---MQS-A-- ---G---A-- ------F--- --------K- -----EI--- H--E--R--- ---
DRB38*   REI--H---H L-A-R----- ---MQ--A-- ---G--F--- ---W--F--- ------A--G -----E---- -------V-- ---
DRB13    REI--H---Y A------S-- -Q-------- ---G--F--- ---W--F--- --------K- L--R-EI--- K-----R--- ---
DRB35*   REIQ-H---Y H------S-- -Q--GY-E-- I--R------ ---W------ ------S-K- M----E---R K--N------ Y--
DRB15*   RE---H---Y H------S-- -Q--GY-E-- IH-G--F--- ---W------ --------K- -----EI--R K--N------ ---
DRB40#   REI--H---Y A------S-- -Q--Q--Q-- IH-G--F--- ---W------ ------V--- L----E---- --GE--R--- ---
DRB48    RE---H---Y A------S-- -Q--Q--Q-- I--R--F--- ---W------ --------K- Y----E---- K-----R--- ---
DRB20#   REI--H---Y A------S-- -Q--Q--Q-- I--R---A-- ---W--F--- --------K- -----EI--R V--E---V-- --N
DRB19    REI--H---Y A------S-- -Q-----Q-- I--R---A-- ---W--F -- --------K- -----EI--R A--E---V-- ---
DRB21    REI--H---Y A------S-- -Q-----E-- I-SR-G---- ---W------ --------K- Y----EI--- ---------- ---
DRB42    RKI--H---Y A------S-- -Q-------- ---R------ ---W--F--- ------S-K- ------FM-- K--E---V-- ---
DRB43    REI--H---Y A------S-- -Q-----E-- ---R------ ---W--F--- ------S-K- ------FM-- K------V-- ---
DRB44    RE---H---Y A------S-- -Q--EY---- ---G------ ---W--F--- --------K- ------FM-- K--E---V-- ---
DRB34*   REI--H---Y Y--------- ---------- ---G---L-- ---W--F--- --------K- -----E---R K--N------ ---
DRB46    RKI--H---Y Y--------- ---------- -S-G--TL-- ------F--- ---------L ----R-F--E K------F-- ---
DRB26*   REI--H---Y A------S-- ---------- ---G--T--- ---------- --------K- -----EI--E K--------- ---
DRB27    RKI--H---Y P------S-- -Q-----G-- I--R------ ---W------ --------K- Y----E---E K--E------ ---
DRB23*   RQI--H---Y P------S-- -Q-L---E-- I--R------ ---W------ --------K- Y----E---E K--E--A--- ---
DRB24    RKI--H---Y P------S-- -Q-----E-- I--R------ ---W------ --------K- Y----E---E K--E------ ---
DRB39#   REI--H---Y P------S-- -Q-----E-- I--R------ ---W--F--- --------KG --R--E---- ---------- ---
DRB29*   RKI--H---Y P------S-- -Q-A---E-- I--R------ ---W------ --------K- Y----E---E ---E------ ---
DRB37*   RE---H---H H-C----S-- ----Q--Q-- I--R------ ---W------ ------S-K- Y----E---- K-----R--- ---
DRB45    RKI--H--LH P-A----S-- -D--Q--A-- I--R------ ---------- ----L----L -----EY--R S----N---- ---
DRB10    REI--H---- L-G----S-- -Q--RL-Q-- I--R---A-Y ---W--F--- ---------- -----EYM-E L--------- ---
```

Figure 1 The 48 expressed DRB sequences found in red deer. This table shows an alignment of all the DRB amino acid sequences found in red deer.[12] The top line shows the HLA-DRB consensus sequence and each letter indicates where the amino acid at this position is different from the consensus sequence. - indicates the amino acid is the same as the HLA-DRB consensus sequence; * indicates that one cloned found containing this sequence. # indicates that this sequences was found in only one deer.

some sites outside the groove that had moderate levels of polymorphism (e.g., positions 24, 25, and 26).

Deer DRB sequences contain 25 amino acid substitutions which have not been found in other ruminants or primates (Table 1). The residue arginine at position 15 [R(15)], A(24), E(25), S(33), M(61), V(71), and A(77) were found only in single clones, which were isolated only once. These substitutions may represent Taq polymerase errors, which arose during amplification or perhaps sequencing errors. With the exception of A(77), a single-base substitution would explain these substitutions. R(15) is almost certainly a Taq polymerase error as position 15 is cysteine in all

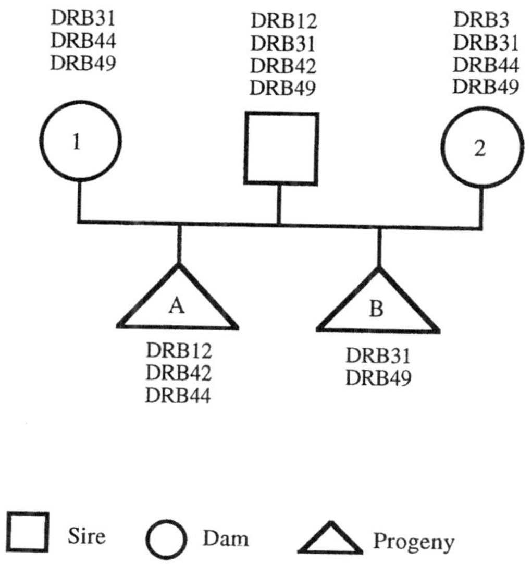

Figure 2 Mendelian inheritance. Mendelian inheritance of DRB sequences in a simple deer pedigree. Progeny A has received sequence DRB44 from dam 1; sequences DRB12 and DRB42 were inherited from the sire. Progency B has inherited sequences DRB31 and DRB49 from both parents, making it homozygous for this haplotype.

other species. This cysteine forms a disulfide bridge with another cysteine at position 79 to give structural integrity of the DRB protein. These DRB sequences should be tagged as requiring confirmation with a second isolation.

The remaining 18 substitutions were found in either more than one sequence or in the same sequence that was isolated from more than one animal. These represent confirmed substitutions which are unique to the red deer. Of the 18 unique substitutions, 11 were found at the peptide binding site (PBS).

Position 34 has amino acid residues that are specific to either primates or ruminants. In primates this position is almost always glutamine with only 3 exceptions (arginine) in the 221 published sequences.[4,8] In ruminants position 34 is never glutamine but glycine in sheep, goat, cattle and half of the deer sequences; arginine and sometimes tryptophan can also occupy this position in deer.

Interestingly, the substitution of methionine at position 68 occurs in over a quarter of the deer sequences. This substitution is not found in any of the other ruminants and is extremely rare in primates.

Like other species the DRB sequences of deer have a rate of nonsynonymous substitution (replacement substitution) significantly higher ($p > 0.001$) than synonymous substitution, especially at PBS. This finding, typical of class II polymorphic sequences, is thought to be caused mainly by overdominant selection (heterozygote advantage) operating in the antigen-binding groove.[15] Interestingly the rate of synonymous substitutions for deer was similar to other ruminants in the range of 3 to 6% (see Chapter 6), which is significantly smaller than primates (7 to 9%). This may suggest higher rates of recombination of these sequences occurring in ruminants.

Table 1 The Amino Acids Found in each Species at each Site

HLA consensus	Human	Chimpanzee	Goat	Sheep	Cattle	Deer	Deer only
G (1)	G					R	
D	D					EKQ	
T	T					IMT	
R	QR					QR	
P	P					LP	
R	CR					H	
F	F					F	
L	L	L				L	
E*	EKQW	EFKLWY	E	E	EQ	ELM	M
Q (10)	ELQY	ELPQY	Y	Y	Y	HQY	(H)
S*	ADGLRPSV	ADEGHPSV	CHSTY	AHRSTY	ACHLST	AFHLPSTY	F
K	KT	KT	KT	KRT	KT	KT	
S*	CFGHRSY	ACGHRSY	GKRS	KRS	GKRS	ACGKRS	
E	EK	EK	E	E	E	E	
C	C	C	C	C	C	CR	R
H	HQY	H	HR	HRS	H	HPY	P
F	F	F	F	F	F	F	
F	FL	FL	FS	FS	F	FPS	P
N	N	N	N	N	NS	N	
G (20)	G	EG	G	G	G	G	
T	T	T	T	T	T	T	
E	E	DEQ	ER	E	E	DEQ	
R	R	RW	R	R	R	R	
V	V	MV	V	V	LV	ALMV	A
R	QRW	QRW	GRW	RW	R	EGQR	E
F	FLNY	FLNY	FLY	FLY	FLY	FLSY	S
L	L	L	L	L	L	L	
D*	DEIH	DEHIK	DH	DE	DEHN	ADEGQ	AG(Q)
R	R	R	R	R	R	R	
Y* (30)	CDGHLRY	DGHLRY	FY	Y	CHSY	Y	
F	FIV	FI	FY	FY	FY	FIV	
Y*	HY	HY	HTY	HTY	HTY	CHSY	(C)S
N	HNW	N	N	N	N	NS	S
Q	Q	Q	G	G	G	GRW	(R)W
E	E	E	E	E	E	E	
E	E	DE	E	E	E	EG	(G)
Y*	DFLNSY	DFNSY	FINTY	FNTY	FLNTY	FTY	
V*	ALV	AFLMV	LV	ALV	V	ALV	
R	R	R	LR	R	R	R	
F (40)	FY	FY	FY	F	F	FY	
D	DN	DN	D	D	D	DG	G
S	S	S	NS	NS	S	CS	C
D	D	D	D	D	D	D	
V	LV	LV	RW	W	W	VW	
G	G	G	G	GS	DG	G	
E	E	E	E	E	E	E	
Y*	FY	FY	FY	FY	FY	FY	
R	QR	QR	R	R	QR	QR	
A	A	A	A	A	A	A	
V (50)	V	V	V	V	V	V	
T	RT	T	AT	A	T	T	

Table 1 (continued) The Amino Acids Found in each Species at each Site

HLA consensus	Human	Chimpanzee	Goat	Sheep	Cattle	Deer	Deer only
E	E	E	E	E	E	E	
L	L	LP	LQ	L	L	L	
G	G	G	G	G	G	G	
R	R	R	QR	R	QR	LR	L
P*	P	P	EPQ	PQR	PQR	PR	
D	ADSV	DSV	DEST	ADES	ADRSV	ADSV	
A	AE	A	A	A	ARV	A	
E	E	EKV	EK	EK	EKV	EK	
Y* (60)	HSY	SY	Y	HQY	HQY	GLY	GL
W*	W	CWY	W	W	CLW	LMWY	M
N	N	N	N	N	N	N	
S	S	S	S	S	GS	RS	(R)
Q	Q	Q	Q	Q	Q	LQ	L
K*	K	K	K	K	K	KR	R
D	D	DG	DE	DEN	DE	ED	
L	FILY	ILY	FIL	FIL	FILT	FILY	
L*	L	LV	L	L	L	LM	(M)
E	E	AE	E	E	EG	E	
Q* (70)	DQR	DEHQR	DQRS	EQRS	DEQR	DEQR	
R*	AERK	AEKRST	KRS	ACKRT	AEGKR	AEHKLNRSTV	HL(N)V
R	R	GR	R	R	R	GR	
A	AG	ADGR	AT	AT	A	AG	
A*	AELQR	ADEGKPQRT	AE	AEN	AENSY	AEN	
V	V	EGV	AV	V	V	V	
D	D	DGR	D	DN	D	DN	
T	NT	NPT	T	T	RT	ART	A
Y*	VY	FVY	CFVY	VY	VY	FVY	
C	C	C	C	C	C	C	
R (80)	R	R	R	R	R	R	
H*	HY	HY	HY	H	H	HY	
N*	N	N	N	N	DNS	DN	
Y	Y	Y	Y	Y	Y	NY	(N)
	n=69	n=64	n=21	n=35	n=35	n=50	

Note: The left column vertically contains the HLA-DRB exon 2 consensus sequence for positions 1 to 83. The other columns show the different amino acids found in each species at each position. ( ) amino acid found in other primates.[8] * is the antigen-binding site. Human sequences were taken from Reference 4. Chimpanzee sequences were taken from Reference 16. Goat and Sheep sequences were taken from Reference 6. Cattle sequences were taken from References 6 and 7.

IV. EVOLUTIONARY ASPECTS

The debate over the genetic mechanisms for the generation of diversity of class I and class II sequences has occupied much journal space. There was evidence of some diversity generating mechanisms operating in deer sequences.

1. Point mutation. Phylogenetic analysis of the deer sequences showed that sequence numbers 1, 2, 3, 4, and 6 formed a distinct branch, they are therefore closely related (data not shown). There are only three amino acid substitutions separating the five

```
                  10         20         30         40         50         60         70         80
                   |          |          |          |          |          |          |          |
HLA-DRB  GDTRPRFLEQ SKSECHFFNG TERVRFLDRY FYNQEEYVRF DSDVGEYRAV TELGRPDAEY WNSQKDLLEQ RRAAVDTYCR HNY

DRB14    RKM--H---Y A-R--P-S-- -Q--QY---- -C-G--F--- ---W------ --------K- -----EI--R K--N------ -D-
DRB9     RKM--H---Y A-R--P-S-- -Q--QY---- -C-G--F--- ---W------ --------K- -----EYM-- T--E--R--- ---
DRB5     RE---H--LY TT-----S-- -Q--GL---- ---G------ ---W------ ------S-K- -----EYM-- T--E--R--- ---

DRB24    RKI--H---Y P------S-- -Q-----E-- I--R------ ---W------ --------K- Y----E---E K--E------ ---
DRB39    REI--H---Y P------S-- -Q-----E-- I--R------ ---W--F--- --------KG --R--E---- ---------- --
DRB49    RK---H--LY A------S-- -Q--GL---- ---G--F--- ---W--F--- ---------G --R--E---- ---------- ---
```

Figure 3 Possible gene conversion in deer sequences. The top line shows the HLA-DRB consensus sequence. For each triplet of sequences the middle sequence has a higher homology with the upper sequence up to about position 54. From about position 61 the middle sequence has greater homology to the lower sequence. Thus the middle sequence appear to be composites of the upper and lower sequences.

sequences, each of which could have been generated by single point mutations, for example, sequences 6 and 3 differ at position 2 by a single amino acid substitution where lysine (GAG) is substituted for glutamic acid (AAG). At position 16 sequences 3 and 2 differ by one amino acid where histidine (CAT) is substituted for tyrosine (TAT). Similarly, other sequences were observed to differ by only a few point mutations. Conversely, they could have arisen by concerted evolution; however, if this had occurred there would be a higher frequency of silent substitution differences between them (these were not seen).

2. Gene conversion. This occurs when there is a sequence exchange between members of a gene family. In class II DRB genes, gene conversion appears to be operating, giving rise to hybrid sequences which, for example, have high homology in the 5′ regions and lower homology in the 3′ regions and vice versa. We have observed some sequences that fit this model of diversity generation (Figure 3). Sequences 14 and 9 have identical 5′ homology up to about position 60, at which point the homology is reduced and is then nearly identical to sequence 49. Sequence 49 and sequence 9 have lower homology below about position 60. At this point we are unable to determine whether what appears to be gene conversion in deer sequences is inter- or intralocus exchange.
3. The trans-species hypothesis proposes that polymorphism is ancient and that polymorphism in contemporary species is inherited from ancestral species. This theory accounts for the fact that often some of the class II sequences of different contemporary species appear to be more closely related than some of the sequences within a species, i.e., that some of the sequences from cattle may appear to be more closely related to some deer sequences than to other cattle sequences. In deer some sequences are more similar to sequences from other ruminants than to other deer sequences; however, this again may be due to convergent evolution.

V. SUMMARY

The expressed DRB sequences of red deer share many features typical of DRB sequences found in other species. The deer DRB locus is extremely polymorphic. We found 48 sequences from 50 unrelated animals. Most of the variation occurs at positions that form the PBS. PBS codons have high levels of nonsynonymous substitution, typical of DRB sequences.

At least 18 substitutions unique to deer were found, some of which were represented in many deer sequences.

There appears to be some of the same mechanisms for the generation of sequence diversity operating in red deer. Examples of possible point mutations and gene conversion were found.

The major difference between red deer and other ruminants is that red deer have more than one transcribed DRB gene.

REFERENCES

1. Klein, J., *Natural History of the Major Histocompatibility Complex,* John Wiley & Sons, New York, 1986.
2. Brown, J. H., Jardetzky, T. S., Gorga, J. C., Stern, L. J., Urban, R. G., Strominger, J. L., and Wiley, D. C., Three-dimensional structure of the human class II histocompatibility antigen HLA-DR1, *Nature (London),* 364, 33, 1993.
3. Rask, I., Gustafsson, K., Larhammer, D., Ronne, H., and Peterson, P. A., Generation of class II antigen polymorphism, *Immunol. Rev.,* 84, 123, 1985.
4. Marsh, S. G. E. and Bodmer, J. G., HLA class II nucleotide sequences, 1992, *Immunogenetics,* 37, 79, 1992.
5. Ammer, H., Schwaiger, F.-W., Kammerbauer, C., Gomolka, M., Arriens, A., Lazary, S., and Epplen, J. T., Exonic polymorphism verses intronic simple repeat hypervariability in Mhc-DRB genes, *Immunogenetics,* 32, 330, 1992.
6. Schwaiger, F.-W., Weyers, E., Epplen, C., Brun, J., Ruff, G., Crawford, A. M., and Epplen, J. T. The paradox of Mhc-DRB exon/intron evolution: α-helix and β-sheet encoding regions diverge while hypervariable intronic simple repeats co-evolve with β-sheet codons, *J. Mol. Evol.,* 37, 260, 1993.
7. Sigurdardottir, S., Borsch, C., Gustafsson, K., and Anderson, L., Cloning and sequence analysis of 14 DRB alleles of the bovine major histocompatibility complex by using the polymerase chain reaction, *Anim. Genet.,* 22, 199, 1991.
8. O'hUigin, C., Bontrop, R., and Klein, J., Nonhuman primate Mhc-DRB sequences: a compilation, *Immunogenetics,* 38, 165, 1993.
9. Sarmiento, U. M., Sarmiento, J. I., and Storb, R., Allelic variation in the DR subregion of the canine major histocompatibility complex, *Immunogenetics,* 32, 13, 1990.
10. Hordvik, I., Grimholt, U., Fosse, V. M., Lie, O., and Endressen, C., Cloning and sequence analysis of cDNAs encoding the Mhc class II b chain in Atlantic salmon *(Salmo salar), Immunogenetics,* 37, 347, 1993.
11. Groenen, M. A. M., van der Poel, J. J., Dijkhof, R. J. M., and Giphart, M. J., The nucleotide sequence of bovine Mhc class II DQB and DRB genes, *Immunogenetics,* 31, 37, 1990.
12. Swarbrick, P. A., Schwaiger, F.-W., Buchan, G., Epplen, J. T., Griffen, J. F. T., and Crawford, A. M., Cloning and sequencing of expressed DRB genes of the red deer (Cervis elaphus) Mhc, *Immunogenetics,* 42, 1, 1995.
13. Burke, M. G., Stone, R. T., and Muggli-Cockett, N. E., Nucleotide sequence and northern analysis of a bovine major histocompatibility class II DRB like cDNA, *Anim. Genet.,* 22, 343, 1991.
14. Ballingall, K. T., Wright, H., Redmond, J., Dutia, B. M., Hopkins, J., Lang, J., Deverson, E. V., Howard, J. C., Puri, N., and Haig, D., Expression and characterisation of ovine major histocompatibility complex class II (OLA-DR) genes, Anim. Genet., 23, 347, 1992.

15. Hughes, A. and Nei, M., Nucleotide substitution at major histocompatibility complex, *Crit. Rev. Immunol.,* 6, 2953, 1989.
16. Kenter, M., Otting, N., Anholts, J., Jonker, M., Schipper, R., and Bontrop, R. E., Mhc-DRB diversity of the chimpanzee *(Pan troglodytes), Immunogenetics,* 37, 1, 1992.

CHAPTER 8

The Canine Major Histocompatibility Complex: Genetic Structure and Function of the Mhc of Dogs

Ulla M. Sarmiento

CONTENTS

0-8493-8032-4/96/$0.00+$.50

I. INTRODUCTION

The major histocompatibility complex (Mhc) in mammals contains two families of genes, which encode class I and class II molecules.[1] These cell surface glycoproteins regulate humoral and cellular immune responses by interacting with antigen-specific T cell receptors (TCR). The TCR ligand consists of a fragment from a processed antigenic protein bound to a self protein (an antigen-presenting molecule) encoded by the Mhc.[2] The Mhc molecules are of major importance in discrimination of self from non-self in that expression of fragments of self proteins maintains tolerance, while recognition of peptides derived from foreign antigens triggers antigen-specific T cell responses.[2,3] Because these molecules also determine the capacity to accept a graft from another individual (or tissue compatibility), they are termed histocompatibility molecules.[4] Although nonspecific immunosuppressants have been used to essentially overcome histocompatibility differences in solid organ transplantation, this strategy has been less successfully applied to bone marrow transplantation.[2]

The dog has been used as an animal model for bone marrow transplantation for over 2 decades.[5] The dog was the first random-bred species, in which the value of tissue typing was clearly demonstrated for the outcome of bone marrow grafts.[6,7] The work summarized in the first, second, and third international workshops on canine immunogenetics established that the DLA (dog leukocyte antigen) system is similar to the Mhc in other species.[8-11] The DLA can be divided into three serologically defined loci, designated DLA-A, DLA-B, and DLA-C, and a fourth region (termed DLA-D) defined by mixed leukocyte culture (MLC) reactivity.[10,11] Allelic polymorphism within the DLA complex has been studied by serology, mixed leukocyte culture, and more recently by restriction fragment length polymorphism (RFLP) and sequence analysis of DLA class I and class II genes.[8-19] Despite the widespread use of the dog model in transplantation biology, the DLA system has not been widely studied at the molecular level. This chapter reviews the available information on the structure, polymorphism, and haplotypic association of DLA class I and class II molecules. Finally, the role of DLA molecules in transplantation and as disease susceptibility markers will be discussed.

II. CHARACTERIZATION OF DLA CLASS I REGION

A. Serological Typing

The work summarized in the first, second, and third international workshops on canine immunogenetics has established that there are at least three genetic loci, termed DLA-A, DLA-B, and DLA-C, which encode for serologically defined antigens.[8-11] The DLA-A antigens are characterized as class I molecules by their association with beta-2-microglobulin and by the correlation of serologic DLA-A specificities with protein polymorphisms detected by two-dimensional electrophoresis, and by isoelectric focusing and immunoblotting.[20-22] Based on these studies, it has been suggested that the dog Mhc expresses a single major class I locus.[20-22] The DLA-B locus codes for class II antigens which are constitutively expressed on almost all circulating lymphocytes in the dog.[23] The DLA-C antigens are thought to be weakly expressed class I antigens.[21]

Serology is the most widely used method for studying Mhc class I polymorphism. Using a standardized microlymphocytotoxicity assay to evaluate anti-DLA sera in cell panels from unrelated mixed breed and Beagle dogs, the serological definitions for at least five DLA-A specificities (designated DLA-A2, A3, A9, Aw14, Aw15), four DLA-B specificities (DLA-B4, B5, B6, B13), and three DLA-C specificities (DLA-Cw1, Cw2, Cw3) were confirmed in the third workshop.[10] These studies also provided evidence of strong linkage disequilibrium between some alleles at the three loci. The strongest linkages were found between DLA-A2-B5-Cw1 and DLA-A9-B6-Cw2 alleles in both Beagle and mixed-breed populations.[10] Furthermore, it was determined that the antigen and gene frequencies of the DLA-A,B,C alleles differed between the Beagle and mixed-breed populations, although the frequencies of the "blank" (undetermined) specificities were similar in both groups.[10] This indicates that further efforts are required to seek serological reagents, which are equally sensitive in detecting alleles in major genetic groups of dogs. These findings are not entirely unexpected, since it is known that serologic HLA-A,B,C typing using antisera selected to define Caucasian HLA antigens does not distinguish many of the alleles found in non-Caucasian populations and that the inaccuracies and insensitivities in serotyping are greatest in populations with ethnic diversity.[24] Moreover, it has been found that patterns of serologic crossreactivity do not always reflect the underlying structural relationships between alleles, since serologically related antigens that share an epitope may be structurally distant based on nucleotide sequences.[24] These and other limitations to serologic typing have lead researchers to seek more accurate methods of tissue typing in dogs as in other species.

B. Restriction Fragment Length Polymorphism

The genetic organization and polymorphism of Mhc genes have been investigated in several species by genomic hybridizations (or Southern blot analysis) utilizing heterologous Mhc probes derived from human or mouse genes, which were the first Mhc genes to be cloned and fully characterized. Human class I cDNA probes crosshybridize strongly to genomic DNA from domestic animals, producing a complex and highly polymorphic pattern of restriction fragments, which can be used to genotype individual animals.

Human cDNA probes for classical class I genes HLA-A,B and a nonclassical HLA-E gene also have been used to analyze the restriction fragment length polymorphism (RFLP) of the class I region of the canine Mhc in 40 unrelated dogs, which had been serotyped for DLA-A,B,C antigens.[14] Depending on the enzyme used and the serotype studied, eight (TaqI) to fourteen (PstI, Pvu II) restriction fragments (range: 0.02 to 21.5 kb) hybridized with the HLA-B7 cDNA probe.[14] The DNA polymorphism detected by the HLA-B7 probe corresponded to the serologically defined DLA-A (but not DLA-B or DLA-C) allelic series. The most informative RFLP pattern was produced after PstI digestion of canine genomic DNA. Figure 1 illustrates the genomic polymorphism characterized by four restriction fragments: (1) a PstI 1.9-kb fragment correlated with the presence of the DLA-A2 serologic specificity; (2) a PstI 2.2-kb fragment correlated with the A7 specificity; (3) a PstI 2.5-kb fragment correlated with the A9 specificity; and (4) a PstI 3.1-kb fragment correlated with a unique A blank allele.[14] The assignment of DLA-A serologic specificities to particular fragment sizes was confirmed by two family studies in which the DLA-A specificities segregated with the PstI fragments. Interestingly, the DLA-A3, A8 and A "blank" serologic specificities did not correlate with any unique PstI restriction fragments but were genotypically associated with the DNA markers for the A2 allele (in the case of A3) or the A9 allele (in the case of A8 and A blanks). The additional fragments detected by the HLA-B7 probe were interpreted to represent crosshybridizing class I genes, most of which remain to be defined in the dog. However, the probe from the HLA-E gene (encoding a nonclassical class I antigen expressed in resting human T lymphocytes)[25] hybridized weakly to one to two nonpolymorphic restriction fragments with all enzymes except PstI, which gave six nonpolymorphic and two polymorphic bands at Pst 4.4 kb and 4.0 kb at low stringency conditions.[14] A simple biallelic polymorphism (at Pst 4.4/4.0 kb) similar to the human pattern (at Pst 4.2/3.8 kb) was detected by the HLA-E probe under high stringency conditions.[25]

In conclusion, the dog genome has been shown to contain class I DNA sequences, which clearly crosshybridize with the probes from human HLA-A,B genes encoding the classical transplantation antigens and the HLA-E gene, which encodes a nonclassical class I antigen.[25] Based on the smallest number of fragments detected by Southern blot analysis after digestion with nine restriction enzymes, the canine class I gene family has been estimated to have at least eight or more class I genes. Thus, the number of class I genes contained in the haploid genome of the dog may be smaller than that of human or mouse and resembles that of pig and rabbit.[26] Nevertheless, the dog genome contains more class I genes than serologically detectable transplantation antigens. The disparity may be explained by the presence of nonexpressed genes or by the inability of the available serologic reagents to distinguish between different class I gene products or to detect nonclassical class I antigens. Thus, RFLP analysis is particularly useful in genotyping dogs that are not clearly DLA-A typable by serology, since the genomic analysis is not constrained by the

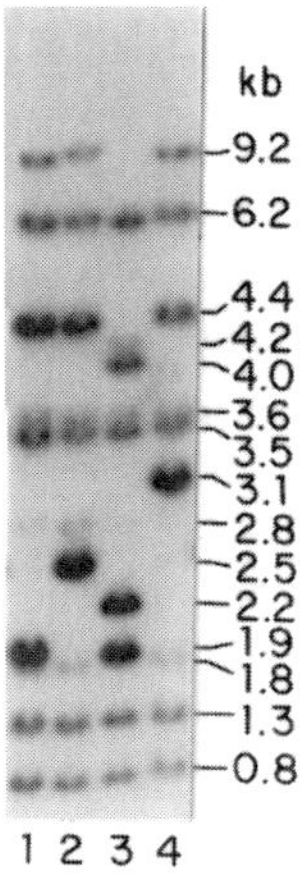

Figure 1 Autoradiogram of PstI-digested genomic DNA from four dogs hybridized with the human HLA-B7 class I probe. The DLA-A alleles illustrated include A2/A2 (lane 1), A9/A9 (lane 2), A2/A7 (lane 3), and a unique A blank (lane 4). The numbers refer to the sizes (in kilobases) of the restriction fragments. (From Sarmiento, U. M. and Storb, R., *Tissue Antigens*, 34, 158, 1989. With permission.)

level of cell surface expression or cell viability. This technique can also be used as a supplement to serotyping as a genotyping tool for antigenic specificities for which specific antisera are not available.

C. Molecular Characterization of Class I Genes

To define the molecular basis of polymorphism detected by serology and RFLP analysis, class I genes have been cloned and characterized in most domestic animals. In the dog, only one report of a class I cDNA sequence has been published to date.[16] The 1511-bp sequence was isolated from a cDNA library constructed from the spleen of a dog serotyped as DLA-A9, B6, Cw2.[16] The dog sequence had the characteristic organization of a class I gene and consisted of 8 exons encoding a signal (leader) peptide (24 residues), 3 extracytoplasmic domains (α1 domain: 90 residues, α2 domain: 93 residues, a3 domain: 92 residues), a transmembrane domain (35 residues), and an intracytoplasmic tail (28 residues). Surprisingly, the dog sequence had an insertion of 3 bp in exon 3 (at position designated as Leu-155) and a deletion of four codons from exon 5. Other conserved structural features included four cysteines (at positions 101, 165, 204, and 260) involved in disulfide bonds, a conserved site for N-linked glycosylation (Asn-Gln-Ser at positions 86 to 88), a serine phosphorylation site (SD/EXSL between residues 329 and 333), and the Arg-Phe-Asp-Ser sequence (at positions 35 to 38) found in other class I antigen alpha-1 domains and in Mhc class II antigen beta-1 domains.[16,27]

Comparison of the predicted amino acid sequences of class I genes from the dog, cat, ox, and human showed the DLA class I clone to be most similar to a cat class I gene, termed pFLA24.[28] The dog class I sequence also had high nucleotide similarity with the bovine BL3-7 clone and the human HLA-A11 allele.[29,30] The

DLA class I sequence contained 13 dog-specific residues at Asp-19, Thr-66, Ile-67, Glu-69, Arg-72, Leu-155 (3 bp insert), Thr-163, Asp-233, Val-284, Ala-289, Gly-298, Gln-308, and Arg-337. Based on the shared locus-specific nucleotides, percent nucleotide and amino acid similarity, and the length of the peptide encoded by the cDNA, the DLA class I gene was thought to represent the canine homolog of the HLA-A locus.[31] The dog gene has been termed the DLA-A gene and the allele has been designated as DLA-A9.[16] The DLA-A gene appeared to be functional (despite the insertion of a codon in exon 3) and may represent the only or major class I gene expressed in the dog, which is consistent with the results obtained by biochemical studies of DLA class I antigens.[21,22] The other class I genes detected by RFLP studies of dog genomic DNA may be pseudogenes or genes that are transcribed at very low levels in the adult spleen. Further sequence analysis of other DLA class I genes is clearly required to fully evaluate the degree of allelic variation in the class I region of the canine Mhc.

III. CHARACTERIZATION OF DLA CLASS II REGION

A. Serological Typing

The gene products of the Mhc are generally grouped into class I, class II, and class III antigens according to their expression, function, and structure. In most species, class II antigens have a restricted distribution on B lymphocytes, monocytes, and dendritic and endothelial cells. In the dog, however, class II antigens are widely expressed on both activated and resting T and B lymphocytes.[32-34] Since the serologically defined DLA-A, -B, and -C antigens are expressed on almost all peripheral blood lymphocytes, they were thought to represent class I antigens in the dog. However, recent studies have demonstrated that the DLA-B locus codes for a class II antigen with a polymorphic beta chain and an invariant alpha chain.[20,23] The constitutive expression of class II (DLA-B) antigens on peripheral blood lymphocytes is not unique to the dog but is also found in other carnivores, such as the cat and mink.[23] Although the functional relevance of the class II antigens remains to be determined, four DLA-B specificities (DLA-B4, B5, B6, B13) have been defined in the Beagle and mixed-breed populations in the third workshop.[10] The gene frequencies of "blanks" were quite high (0.555 to 0.660) but comparable in the two populations. This indicates that the currently available anti-DLA-B sera are limited in their ability to detect class II polymorphism.

B. Cellular Typing

Cellular techniques have been used to characterize polymorphism in the Mhc class II region in several species, including the dog. In the mixed leukocyte culture (MLC) reaction, T cells mount a proliferative response to non-self class II antigens. Work summarized in the three workshops has established that dogs have a fourth locus, termed DLA-D, which is responsible for MLC reactivity.[8-11] In the third workshop, ten different clusters of homozygous typing cells (HTC) were defined and assigned workshop specificities (termed DLA-Dw1 to -Dw10).[11] In addition, six provisional specificities (designated DLA-D1, -D9, -D10, -D16, R2B/MA3, and R5/50z) were well characterized within individual laboratories but represented by only two cells in the HTC panel.[11] Another cellular test used to define the DLA system is the primed lymphocyte test (PLT).[35,36] The PLT cells are generated by priming with appropriate HTC and tested by restimulation with a panel of HTC. This approach has allowed the recognition of supertypic determinants, which correlated with MLC-defined DLA-D types and DLA-B locus serologic specificities.[36] Although cellular typing is a useful tool for genetic analysis of dogs, its use is limited, since the HTC panels are not widely available. Moreover, further work is required to confirm many of the specificities in family studies, to determine the gene frequencies, and to analyze the antigens responsible for MLC and PLT reactivity at the molecular level.

C. Restriction Fragment Length Polymorphism

As in the class I region of the dog Mhc, Southern blot analysis has been used to define the genetic organization and polymorphism of DLA class II genes utilizing heterologous Mhc probes derived from human alpha and beta genes.[12,13]

1. Beta Genes

HLA class II cDNA probes have been used to analyze the restriction fragment length polymorphism (RFLP) of the DLA-D region in a study of 23 unrelated dogs.[12] The dogs studied represented nine different DLA-D types, which are thought to represent 47% of the alleles found in a random dog population.[37] Using nine restriction enzymes (BamHI, EcoRI, HindIII, PvuII, TaqI, RsaI, MspI, PstI, and BglII), genomic DNA was successively hybridized with radiolabeled HLA cDNA probes corresponding to DR, DQ, DP, and DO beta genes. The autoradiograms for all nine enzyme digests displayed multiple bands with the DRB, DQB, and DPB probes while the DOB probe hybridized with one to two bands. The RFLP patterns were highly polymorphic but consistent within each DLA-D type.

a. DR Beta

Depending on the enzyme used and the DLA-D type studied, four to ten restriction fragments (range: 0.3 to 22.7 kb) hybridized with the HLA-DRB probe, which detected unique restriction fragments (or allogenotopes) for DLA-Dw1, -D9, -D10, and -Dw3.[12] The DRB probe identified a genomic polymorphism characterized by a reciprocal and mutually exclusive relationship between two or three restriction fragments (representing two- or three-allele polymorphism) after BamHI, HindIII, PstI, and BglII digestion.[12] The sizes and distribution of the allelic variants are summarized in Table 1. Cluster analysis of the polymorphic restriction fragments detected by the DRB probe revealed four closely related supertypic groups or DLA-DR families: Dw3+Dw4+D1, Dw8+D10, D7+D16+D9, and Dw1 (Table 1). The DLA-DR families have been hypothesized to represent canine homologs of the

Table 1 Association of RFLP-Defined Class II Alpha and Beta Gene Alleles with Nine DLA-D Types

DLA-D type	DRb supertypic family[a]	DQb allele[b]	DPb allele	DOb allele	DRa allele[c]	DQa alleles[d]
Dw3	1	1	1	2	3	2,4
Dw4	1	1	1	2	3	2,4
D1	1	1	1	2	3	2,4
Dw8	2	2	?	2	2	2,3
D10	2	2	?	2	2	2,4
D9	3	3	?	1	3	2,4
D7	3	4	2	2	3	1,3
D16	3	4	2	2	1	1,3
Dw1	4	5	3	2	2	2,3

[a] Legend of DRb alleles (defining supertypic groups) and their characteristic restriction fragments detected by the HLA-DR beta gene probe (based on Reference 12):

DRb alleles (1) DRb1: Dw3+Dw4+D1 (HindIII 5.4 kb, RsaI 1.3 kb)
(2) DRb2: Dw8+D10 (PstI 16.5 kb)
(3) DRb3: D7+D9+D16 (HindIII 5.7 kb, TaqI 4.2 kb)
(4) DRb4: Dw1 (RsaI 1.1 kb)

[b] Legend of RFLP-defined DQb, DPb, DOb alleles detected by the HLA-DQ, DP, and DO beta gene probes:

DQb alleles (1) DQb1: Dw3+Dw4+D1 (Bg1II 0.7 kb, RsaI 1.3 kb)
(2) DQb2: Dw8+D10 (PstI 16.5 kb)
(3) DQb3: D9 (TaqI 3.2 kb)
(4) DQb4: D7+D16 (TaqI 8.0 kb)
(5) DQb5: Dw1 (RsaI 1.6, 1.2, 1.1, 0.9 kb, Bg1II 0.9 kb)
DPb alleles (1) DPb1: Dw3+Dw4+D1 (Bg1II 0.7 kb, RsaI 1.3 kb)
(2) DPb2: D7+D16 (Bg1II 3.4 kb)
(3) DPb3: Dw1 (RsaI 3.6, 0.8 kb, Bg1II 0.9 kb)
DOb alleles (1) DOb1: D9 (EcoRI 3.7 kb, PvuII 13.1 kb)
(2) DOb2: Dw1+Dw3+Dw4+Dw8+D1+D7+D10+D16 (EcoRI 3.6 kb, PvuII 3.6 kb)

[c] Legend of restriction fragments defining DRa alleles detected by the HLA-DR alpha gene probe (based on Reference 13):

DRa alleles: (1) DRa1: D16 (EcoRI 6.7+3.0 kb)
(2) DRa2: Dw1+Dw8+D10 (EcoRI 11.9 kb)
(3) DRa3: Dw3+Dw4+D1+D7+D9 (EcoRI 14.4 kb)

[d] Legend of restriction fragments defining DQa alleles detected by the HLA-DQ alpha gene probe:

DQa alleles (1) DQa1: D7+D16 (BamHI 11.9 kb, PvuII 3.5 kb, TaqI 12.7 kb)
(2) DQa2: Dw1+Dw3+Dw4+Dw8+D1+D9+D10 (BamHI 3.9 kb, PvuII 9.3 kb, TaqI 11.5 kb)
(3) DQa3: Dw1+D7+D16 (RsaI 2.1+1.0 kb)
(4) DQa4: Dw3+Dw4+D1+D9+D10 (RsaI 0.4 kb)

supertypic groups DRw52 and DRw53 described in humans, and appeared to correlate with B locus serologic specificities.[12,38] Recently, a canine DRB cDNA probe has also been used for RFLP testing of dog families to confirm the cosegregation of RFLP alleles with DLA-D specificities.[39]

b. DQ Beta

Six to eleven restriction fragments (range: 0.4 to 16.5 kb) hybridized with the HLA-DQB probe. Generally a strongly hybridizing nonpolymorphic band (interpreted to represent the canine homolog of the DQB2, previously DXB gene) and one or more polymorphic bands (corresponding to the DQB1-like gene with two- or three-allele polymorphism) were detected after TaqI, BglII, PstI, and BamHI digestion.[12] The DQB probe could be used to identify the following DLA-D types by specific restriction fragments: Dw1, D10, and D9 (Table 1).

c. DP Beta

Depending on the enzyme used, 6 to 14 light to strong intensity bands (range: 0.3 to 22.7 kb) hybridized with the HLA-DPB probe. There was considerable crosshybridization between the major bands detected by the DRB, DQB, and DPB probes.[12] Two- to four-allele polymorphism was detected by the DPB probe after RsaI, BglII, and MspI digestion. The DPB probe could be used to identify the following DLA-D types: Dw1, Dw3, and D16 (Table 1). The most informative probe-enzyme combination for discriminating all nine different DLA-D types was the DPB probe hybridized with RsaI and PstI digests of genomic DNA (Figure 2).[12]

d. DO Beta

After EcoRI, PvuII, and TaqI digestion, two-allele polymorphism of a single gene was detected by the HLA-DOB probe (Table 1). DNA digests with the other restriction enzymes were characterized by a single nonpolymorphic band in all D types studied.[12]

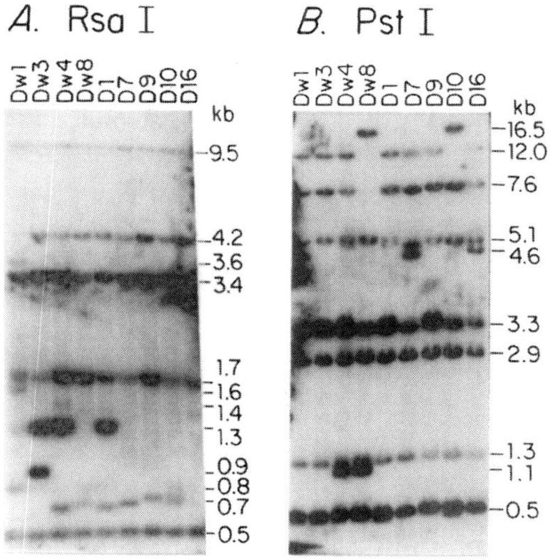

Figure 2 Autoradiogram of (A) RsaI- and (B) PstI- digested genomic DNA from nine DLA-D homozygous dogs hybridized with HLA-B7 class II DP beta probe. The numbers refer to the sizes (in kilobases) of the restriction fragments. (From Sarmiento, U. M. and Storb, R. F., *Immunogenetics*, 28, 117, 1988. With permission.)

In summary, standard RFLP patterns have been established for nine DLA-D types, which can be discriminated by using two enzymes (RsaI and PstI) and the HLA-DPB probe. This provides the basis for DLA-D genotyping at a population level. It remains to be determined whether the DLA-D specificities reflect DRB gene allelic polymorphisms or the composite contributions of DR, DQ, and DP beta gene products.

2. Alpha Genes

Human HLA cDNA probes have also been used to analyze RFLP of the alpha genes of the DLA-D region in dogs in a study.[13] Genomic DNA from 23 unrelated DLA-D homozygous dogs representing 9 DLA-D types was digested with nine restriction enzymes (BamHI, EcoRI, HindIII, PvuII, TaqI, RsaI, MspI, PstI, and BglII). Southern blots were successively hybridized with radiolabeled HLA cDNA probes corresponding to DQ, DP, DZ, and DR alpha genes.[13]

a. DR Alpha

Depending on the enzyme used and the DLA-D type studied, one to four restriction fragments (range: 0.02 to 16.6 kb) hybridized with the HLA-DRA cDNA probe.[13] The RFLP pattern was characterized by three alleles tentatively designated as DRα1 found in the D16 type, DRα2 found in Dw8+D10+Dw1 types, and DRα3 present in Dw3+Dw4+D1+D9+D7 types (Table 1). This polymorphism is consistent with a model in which the fragments are reciprocal allelic variants of a single gene, since DNA from each of the DLA-D homozygous dogs displayed only one of the three fragments. DNA digests with the other restriction enzymes were characterized by a single, nonpolymorphic band present in all D types.[13]

b. DQ Alpha

Two to seven fragments (range: 0.02 to 24.4 kb) hybridized with the HLA-DQA probe. Generally, a strongly hybridizing polymorphic band and a moderately intense nonpolymorphic band were present.[13] Two distinct RFLP patterns were evident (Table 1). With the enzymes BamHI, PvuII, or TaqI, a biallelic pattern was present with one allele (designated DQα1) found in D7+D16 types and a second allele (DQα2) found in the D types Dw3+Dw4+D1+Dw8+D10+D9+Dw1 (Table 1). A second pattern detected in RsaI-digested DNA was characterized by an allele (DQα3) found in D types Dw8+Dw1+D7+D16 while a reciprocal allele (DQα4) was present in Dw3+Dw4+D1+D10+D9 types (Table 1).[13]

c. DP Alpha

Depending on the enzyme used, three to eight bands of light to moderate intensity (range: 0.02 to 24.2 kb) were detected by the HLA-DPA probe.[13] Most of the bands were crosshybridizing fragments detected with greater intensity by the HLA-DRA or DQA probes. However, the detection of a unique band (of light intensity in PstI digests) or a fragment (in BamHI, EcoRI, PvuII, TaqI, RsaI, PstI and BglII digests), which hybridized more strongly with the DPA probe than with the DQA probe, provided presumptive evidence for the presence of at least one DPA-like gene, which appeared to be nonpolymorphic with all nine enzymes tested. The existence of a second nonpolymorphic DPA gene, which may not have been detected under stringent conditions, could not be ruled out on the basis of this study. [13]

d. DN Alpha (DZA)

Depending on the enzyme used, the HLA-DNA (previously DZA) probe hybridized weakly or not at all to canine DNA.[13] The fragments detected were interpreted to represent crosshybridizing nonpolymorphic DQA- or DPA-like bands. Thus, the existence of a DNA-like gene in the canine genome has not be confirmed.[13]

3. Alpha and Beta Gene Allelic Associations with DLA-D Types

Table 1 summarizes the associations of RFLP-defined allelic forms of class II alpha and beta genes (or preferential allelic associations) with the nine DLA-D types studied thus far.[13] The strong linkage disequilibria between the DLA-D types and the RFLP-defined allelic forms allowed the construction of preferential allelic associations extending across the DLA-D region. The RFLP analysis of the four supertypic groups or DLA-DR families revealed that the Dw3+Dw4+D1 group had identical alpha and beta gene alleles (Table 1). Similarly, the Dw8+D10 supertypic family shared all but the DQa alleles. In contrast, the D9+D7+D16 family had more variation in the alpha and beta gene alleles with the D9 type displaying DQa, DQb, DPb, and DOb alleles distinct from those of the D7+D16 subgroup which differed from each other in the DRa alleles. The Dw1 type had unique DRb, DQb, and DPb alleles although the alpha gene profile was identical to that of the Dw8 type (Table 1).[13] Based on these studies, it has been concluded that most of the DLA-D region polymorphism appears to be derived from DR, DQ, and DP beta gene allelic variation, although limited polymorphism may also be contributed by DR and DQ alpha genes. It is possible that the combination of cell surface antigens encoded by DR, DQ and DP loci may form the basis for mixed leukocyte culture stimulating determinants in the dog as in humans.[40]

These studies also have provided information about the genetic organization of the DLA-D region, which appears to closely resemble that of the HLA complex. Based on the RFLP analysis of DLA class I and class II genes, a hypothetical schematic map of the canine Mhc has been constructed (Figure 3). The class I region is thought to consist of at least eight class I genes, two of which may correspond to serologically defined DLA-A and DLA-C loci, and at least one gene may represent a canine homolog of a HLA-E-like gene (Figure 3). In the class II region, three major subregions corresponding to DR, DQ, and DP have been identified at the molecular level. The DR subregion has been estimated to have at least two DRB-like genes and one DRA-like gene. The DLA-DQ subregion has been estimated to have two beta genes: a highly polymorphic DQB1-like gene and a nonpolymorphic DQB2-like (previously DXB) gene.[12] The DQ locus also contains two alpha genes,

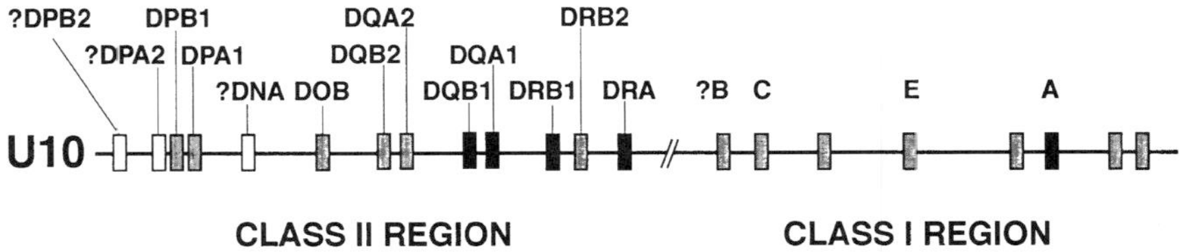

Figure 3 A hypothetical schematic map of the canine Mhc on chromosome U10. The class I and class II gene locations and numbers represent estimates from genomic hybridization studies. The genes shown include known expressed loci (black boxes), genes detected by RFLP analysis (gray boxes), and genes of unknown status (white boxes).

presumably corresponding to an DQA1-like gene and a DQA2-like (previously DXA) gene. The DP subregion has been estimated to have at least one DPB1-like gene and one nonpolymorphic DPA-like gene.[12] While the DO locus has a DOB-like gene, the DNA (previously DZA) gene has not been clearly demonstrated in canine genomic DNA. The canine DNA gene may have been lost during evolution or have diverged sufficiently from the human DNA gene to have escaped detection with the human probe. Nevertheless, comparison of the number of genes in the canine Mhc system with that of humans demonstrates remarkable structural conservation of class I and class II regions.

D. Molecular Characterization of Class II Genes

Recent DNA sequencing studies of class II genes in humans have provided considerable insight into the nature and extent of allelic polymorphism in the HLA-D region. Mixed leukocyte culture typing differences have been defined and localized to the hypervariable regions of the HLA-DRB molecule.[38,39] Despite the widespread use of the dog model in biomedical research, the DLA system has not been widely studied at a molecular level. The following section will summarize the available data on molecular studies used to clone and sequence the first DRB, DQB, and DQA genes and their alleles in the dog.[15,17-19]

1. DR Subregion

a. DLA-DRB1 Gene

To isolate DLA-DRB cDNA clones, a lambda ZAP phage cDNA library (constructed from the spleen of a dog cellularly typed as a DLA-Dw3/Dw3 homozygote by MLC) has been screened by plaque hybridization using a human DRB cDNA clone as a probe.[15] The nucleotide and predicted amino acid sequence of the 1184-bp DLA-DRB cDNA clone were typical of a functional Mhc class II beta chain, which consisted of a signal peptide (29 residues), the extracellular portion of the beta chain (b1 domain: 94 residues, b2 domain: 104 residues), a transmembrane region (21 residues), and a cytoplasmic tail (18 residues).[15] Other conserved structural features included four cysteines in the extracellular portion (at positions 15, 79, 117, and 173), a potential N-glycosylation site (Asn-Gly-Thr at positions 19 to 21), and the

Arg-Phe-Asp-Ser sequence (at positions 39 to 42), which is found in other Mhc class II antigen beta-1 domains and in class I antigen alpha-1 domains.[15,27]

Comparison of predicted amino acid sequences of class II genes from dog, human, mouse, and ox has shown the dog sequence to be most similar to the HLA-DRB1 gene sequence with 87% nucleotide and 81% amino acid sequence identity with the DR5 (DRB1*05) allele.[15] The dog gene has been termed the DLA-DRB1 gene and the allele has been designated as DLA-DRB1*Dw3.[15] The gene appeared to be functional and probably represents the only DRB gene expressed in the dog. This suggests that the second DRB-like gene detected by RFLP studies of dog genomic DNA may represent a pseudogene or a gene that is transcribed at very low levels in the spleen.[15]

b. DLA-DRB1 Gene Polymorphism

Allelic variation in the DR subregion of the canine major histocompatibility complex (DLA) has been analyzed by nucleic acid sequencing of cDNA clones of DRB genes amplified *in vitro* by the reverse transcriptase-polymerase chain reaction (RT-PCR).[17] Sequence analysis of a panel of 19 DLA-D homozygous dogs (representing 12 different DLA-D types) has demonstrated the presence of one expressed DRB locus with at least nine distinct alleles (Figure 4). Unique DLA-DRB alleles were found in the DLA-D types Dw1, Dw3, Dw4, Dw8 (workshop assignments), and D4, D6, D7, D8, and D9 (Seattle assignments). Interestingly, the DRB genes of three DLA-D types (D1, D10, and D16) were identical to those of Dw3/Dw4 (for D1), Dw8 (for D10), and D6 (for D16).[17] The DLA-DRB alleles had 96 to 99% overall nucleotide sequence similarity and 93 to 99% amino acid sequence similarity with each other. The nucleotide sequences of all nine DLA-DRB alleles were typical of functional Mhc class II beta chains and contained three allelic hypervariable regions in the beta-1 domain at positions 8 to 16, 26 to 39, and 57 to 74. The predicted amino acid sequences of the DLA-DRB clones are shown in Figure 4.[17] At each variable residue, two to five amino acid substitutions were found. The most polymorphic residues were located at positions 37 (with 5 amino acid substitutions), 11, 13, 28, and 71 (each with 4 substitutions) (Figure 5).

Cluster analysis of the nucleotide and predicted amino acid sequences has been used to subdivide the DLA-DRB alleles into three major allelic groups (Figure 6), which may represent the canine counterparts of the supertypic groups described in humans.[17] Group 1 contained the alleles *Dw4, *D9, and *D4; group 2 was composed of the *D6 and *D7 alleles; and group 3 consisted of the *Dw1, *Dw8, and *D8 alleles (Figure 6). Further analysis of the groups illustrated that the hypervariable regions (HVRs) were frequently shared between members of the same major group (Figure 4). For example, Group 1 alleles *Dw3 and *Dw4 shared all three HVRs and differed only at position 47 (Phe in *Dw3; Tyr in *Dw4). The *D9 allele shared the first HVR with *Dw3 but had one substitution in the second HVR and a unique third HVR. The *D4 allele had a unique first HVR and only one to two substitutions in the second and third HVRs. Group 2 alleles *D6 and *D7 were identical to each other except at position 37 (Phe in *D7; Tyr in *D6). In Group 3, the *Dw1 allele shared the first HVR with the *Dw8 allele; the second HVR

DRB

```
                                                                                                              DLA-D types^a
           1        10        20        30        40        50        60        70        80        90
DRB1*Dw3   RDTPPHFLEVAKSECYFTNGTERVRFVERYIHNREEFVRFDSDVGEFRAVTELGRPVAESWNGQKEILEQERATVDTYCRHNYGVIESFTVQRR   Dw3, D1
DRB1*Dw4   ----------------------------------------------Y-----------------------------------------------   Dw4, D1
DRB1*D9    ------------------------------------N---------Y---------D-----R---F--RK--A--------------------   D9
DRB1*D4    -------EYQF-P--H-------------H----------------Y---------D---------L---------------------------   D4
DRB1*D7    ------------A--H---------------Y--------------Y---------D--Y--P---L--RG--A-----------G--------   D7, D6
DRB1*D6    ------------A--H---------------Y----Y---------Y---------D--Y--P---L--RG--A-----------G--------   D6, D16
DRB1*D8    -------VKMY-A--H---------YLM-D-Y------------------------D--Y---------RK--A-----------G--------   D8
DRB1*Dw8   ---------MV-F--H---------LLV-D-Y----H-P-------Y---------D--Y------L---R--E---V----------------   Dw8, D10
DRB1*Dw1   ---------MV-F--H---------YLA-D-Y----IL--------Y---------I-----R-------R--A---------------A----   Dw1

                 |  βHV1  |         |   βHV2   |                |   βHV3   |
```

DQB

```
                                                                                        DLA-D types^b
             14    20        30        40        50        60        70        80        90
DQB1*0101    ECYFTNGTERVRLLTRDIYNREEHVRFDSDVGEYRAVTELGRPDAEYWNGQKELLERRRAEVDTVCRHNYGREELTTLQRR   Dw1, Dw3, Dw8, D9
DQB1*0201    ---------------KY------F---------F---------------R--DEMD-V---L-------------------   Dw1, Dw3, Dw4, Dw8, D1
DQB1*0301    ---------------KY------Y-------------------S-----P--DEMD-V---L---------L---------   D10
DQB1*0401    ---------------KY------Y-------------------W-----P--DEMD-V---L---------L---Y-----   D1, D8

       |  βHV1  |      |    βHV2    |       |  βHV3  |  |   βHV4   |
```

DQA

```
                                                                                        DLA-D types^c
                     20        30        40        50        60        70        80
DQA1*0101    GINVYQSYGPSGQYTHEFDGDEEFYVDLEKKETVWRLPVFSTFRSFDPQGALRNLAIIKQNLNIMTKRSNQTAATN         Dw3, D4, D6, D10, D16
DQA1*0102    -------------F--------------------------------------------------------------         Dw8
DQA1*0201    -------------------------------------------T-------------T------------K-----         D4
DQA1*0202    -------------F-----------------------------T-------------T------------K-----         D7

           |     αHV1     |      |   αHV2   |     |   αHV3   |
```

Figure 4 Alignment of the predicted amino acid sequences of the first domains of DRB, DQB, and DQA sequences obtained from nine DLA-D homozygous dogs. The standard one-letter amino acid code is used. A dash indicates identity with the top sequence. The first (HV1), second (HV2), third (HV3), and fourth (HV4) hypervariable regions are indicated below the sequences for alpha and beta chains. The allele designations are shown on the left, and the DLA-D types in which the alleles have been found are indicated to the right of the sequences. (Modified from References 17, 18, and 19. With permission.)

DLA-DR molecule

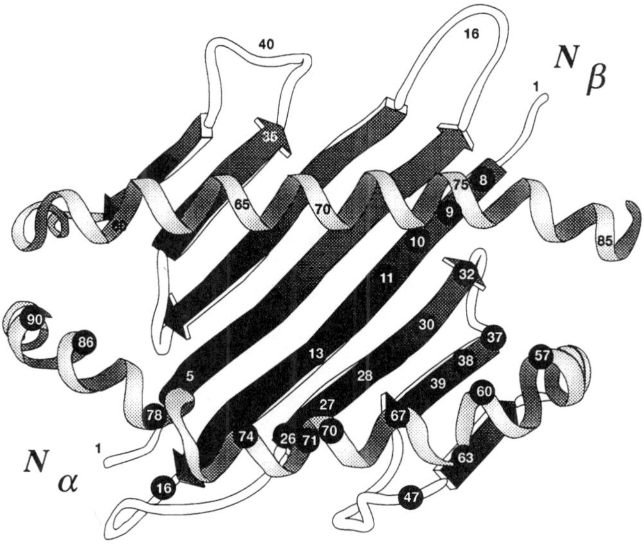

DLA-DQ molecule

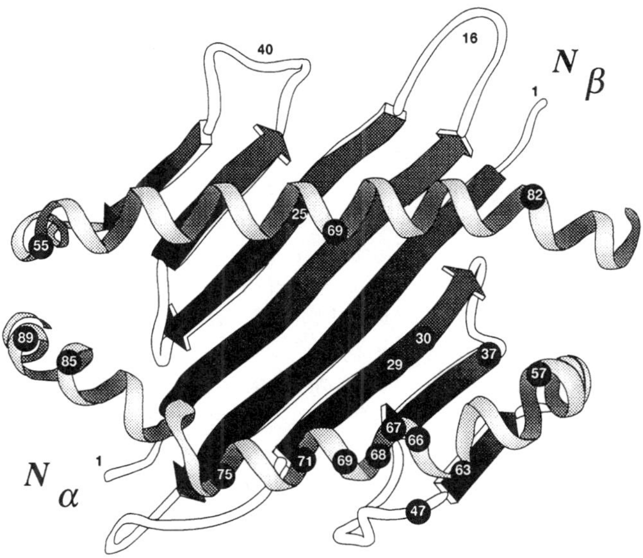

Figure 5 A model of the antigen-binding sites of DLA-DR and DLA-DQ molecules . The residues on the alpha (α) and beta (β) chains are numbered. The positions of polymorphic residues are indicated on the alpha and beta chains (black dots). Structure based on the hypothetical model of class II antigen-binding site proposed by Brown et al.[51]

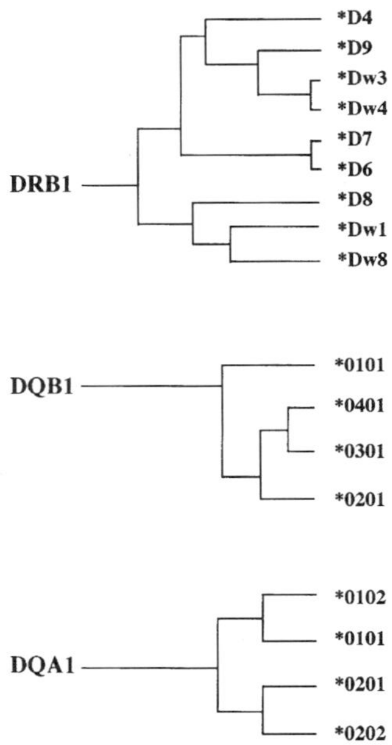

Figure 6 Complete linkage dendrograms (or farthest neighbor tree diagrams) for DLA-DRB1, -DQB1 and -DQA1 gene alleles obtained by cluster analysis of first domain amino acid sequences (shown in Figure 4). (Modified from References 18 and 19. With permission.)

resembled that of the *D8 allele while the third HVR was unique. Moreover, the second and third HVRs of both *D8 and *Dw8 alleles were distinct from other alleles (Figure 4).

Comparison of the predicted amino acid sequences of DLA-DRB alleles with known HLA-DRB molecules has demonstrated that the types of amino acid substitutions found at each variable residue are highly conserved between the dog and human alleles even at the most polymorphic locations, such as position 37, and at positions 67, 70, 71, and 74, which are thought to be of primary importance in T-cell recognition of HLA-DRB-associated alloepitopes.[40-43] Moreover, comparison of residues shared in the first, second, and third HVRs (bHV3) of the beta-1 domains of human and canine DRB molecules has revealed that some of the major groups of DLA-DRB alleles appear to correspond to distinct HLA supertypic groups.[17] Group 1 DLA-DRB alleles most resemble the HLA-DRB alleles DR3, DR5, and DR6 (of the HLA-DRw52 supertypic group), and group 2 alleles have segmental sequence similarity with the HLA-DRB4 gene and -DRB1 gene alleles, DR4 and DR10, while group 3 alleles are most closely related to the HLA-DR1 allele.[17] Furthermore, comparison of the HVRs of dog DRB alleles to each other has also revealed that some of them are more closely related to certain human DRB alleles than to dog DRB alleles belonging to distinct major groups.[17]

The presence of identical amino acid substitutions among alleles of two divergent species, such as human and dog, might be argued to reflect convergent evolution of unrelated alleles. Since the DRB first domain polymorphism is likely driven by positive selection at the protein level, a large number of synonymous nucleotide substitutions would be expected between human and dog alleles in convergent evolution of the amino acid sequences.[44] It has been found, however, that the majority of the shared polymorphic residues are encoded by completely identical codons despite presumably unequal codon usage in humans and dogs. This finding argues strongly for the existence of common ancestral alleles before the divergence of the carnivore and primate lineages, as hypothesized by the trans-species mode of evolution, in which a group of major alleles is thought to be passed on in phylogeny from one species to another.[45] Thus, the evolution of DRB polymorphism has been postulated to be similar to the DQB locus, in which many of the major allelic types shared by mammals are thought to be ancient and maintained by some form of balancing selection while species-specific subtypic diversity has been generated more recently after speciation.[17,46]

2. DQ Subregion

a. DLA-DQA1 Gene Polymorphism

Based on the RFLP studies, the DLA-D region has been documented to have at least two DQ alpha chain genes.[13] The molecular studies have also been extended to the DLA-DQ subregion. Recently, the allelic diversity of DQA cDNA sequences has been determined in eight unrelated DLA-D homozygous dogs by the reverse transcriptase-polymerase chain reaction (RT-PCR) as described.[18]

DNA sequence analysis of the cDNA clones from eight dogs (representing seven different DLA-D types) identified four class II alpha chains, which were designated as DLA-DQA molecules based on homology to HLA-DQA genes.[47] The 564-bp PCR product included partial sequence for the α1 domain (residues 12 to 87) and α2 domain (residues 88 to 180).[18] The signal and connecting peptides, transmembrane, and cytoplasmic domains were not amplified. The dog sequences had structural features typical of functional class II alpha chains, including two cysteines involved in disulfide bonds (at positions 110 and 166) and two potential N-linked glycosylation sites (Asn-X-Thr at positions 81 to 83 and 121 to 123).[18] The DLA-DQA sequences had limited polymorphism with 98.8 to 99.8% overall nucleotide and 94.7 to 98.7% amino acid sequence similarity. The structural relatedness of the sequences is illustrated by cluster analysis of the amino acid sequences to construct a farthest neighbor tree diagram shown in Figure 6. These data suggest that the four sequences represent closely related alleles of a single locus rather than products of two different loci. The sequences have been designated as DLA-DQA1*0101 (found in DLA-Dw3, -D4, -D6, -D10, and -D16), DQA1*0102 (in DLA-Dw8), DQA1*0201 (in DLA-D4), and DLA-DQA1*0202 (in DLA-D7). Figure 4 summarizes the haplotypic associations and sequence differences between the DLA-DQA1 alleles, which occurred at codons 25 (Tyr/Phe), 55 (Arg/Thr), 69 (Ile/Thr), 81 (Gln/Lys), and 141 (Phe/Leu). It is of interest to note that four out of five polymorphic residues in the canine DQA sequences are localized to the α1 domain HVRs, which have been termed aHV1 (codons 11 to 35), aHV2 (codons 44 to 59), and aHV3 (codons 66 to 83) (Figures 4 and 5).[48] Moreover, as in primates, the allelic diversity in the α1 domain of the dog (at positions 25, 55, 69, and 82) can be attributed to point mutations, all of which affect either the first or second codon positions and lead to amino acid replacements rather than silent substitutions.[49,50] These data have provided further evidence for the hypothesis that there is positive selection for replacement substitutions, which promote charge profile diversity in the antigen-binding site of Mhc class II genes (Figure 5) in the dog as in other species.[48-51] The lack of silent substitutions and the limited diversity of the DLA-DQA locus suggest that the alleles may have diverged recently. Presumably selection for a/b chain interactions may constrain the rate of protein evolution in the dog as in other mammals.[48-50]

Comparison of predicted amino acid sequences of the α1 domains of DQA or equivalent genes from the dog, pig, human, rabbit, ox, rat, and mouse has revealed that the DLA-DQA alleles are most similar to the pig SLA-DQAc allele with 80 to 86% amino acid sequence identity in the α1 domain.[18] Interestingly, the DLA-DQA sequences contained five residues unique to the dog at positions 53 (Thr), 66 (Leu), 71 (Gln), and 82 (Gln/Lys). Based on the sequences of the DQA clones amplified from the eight HTC dogs studied, it has been presumed that dogs express a single functional DQA gene with at least four distinct alleles.[18] The second DQA gene detected by previous RFLP studies of dog genomic DNA is thought to represent a pseudogene or a gene that is not transcribed (such as the HLA-DQA2 or DXA gene) (Figure 3). Thus, the dog DQA gene reported has been termed the DLA-DQA1 gene and the four sequences have been designated as the DLA-DQA1*0101, *0102, *0201, and *0202 alleles.[18]

b. DLA-DQB1 Gene Polymorphism

Based on the genomic hybridization studies, the DLA-D region has been documented to have at least two DQ beta chain genes.[12] In the DQ subregion of the dog Mhc, dogs have been found to have a single functional DQA gene with at least four distinct alleles.[17] Recently, molecular studies also have been extended to the DLA-DQB genes to analyze the allelic diversity of DQB cDNA sequences amplified from 12 DLA-D homozygous dogs using the reverse transcriptase-polymerase chain reaction (RT-PCR).[19] Interestingly, the amplification of DQB gene segments failed in 4 of 12 dogs presumably due to sequence divergence at the PCR primer sites of DLA-D types, D4, D6, D7, and D16, even though the DQA1 genes of the same dogs had been successfully amplified in previous experiments.[18]

DNA sequence analysis of the cDNA clones from eight dogs identified four class II beta chains, which were designated as DLA-DQB molecules based on homology to HLA-DQB genes.[44] The 651-bp PCR product included partial sequence for the b1 domain (residues 14 to 94), b2 domain (residues 95 to 188), and transmembrane domain (189 to 215) (Figure 4).[19] The signal peptide and cytoplasmic domain were not amplified. The dog sequences had features typical of functional class II beta chains, including four cysteines involved in disulfide bonds (at positions 15, 79, 117, and 173) and a potential N-linked glycosylation site (Asn-Gly-Thr at positions 19 to 21). The DLA-DQB molecules had 89.3 to 96.7% overall nucleotide and 85 to 97.5% amino acid sequence similarity. Cluster analysis of the first domain amino acid sequences demonstrated the structural relatedness of the alleles, which is shown in the tree diagram in Figure 6. These data suggest that the four sequences represent closely related alleles of a single locus rather than products of two different loci. The sequences have been designated as DLA-DQB1*0101 (found in DLA-Dw1, -Dw3, -Dw8, -D9), DQB1*0201 (in DLA-Dw1, -Dw3, -Dw4, -Dw8, -D1), DQB1*0301 (in DLA-D10), and DQB1*0401 (in DLA-D1, -D8).

Figure 4 summarizes the haplotypic associations and predicted amino acid sequence differences between the DLA-DQB alleles.[19] The polymorphism in the DLA-DQB sequences was restricted mostly to the first (b1) domain. The most polymorphic residues were located at positions 37, 57, and 63 (each with three substitutions), 29, 30, 47, 66, 67, 68, 69, 71, 75, 85, 89, 102, 105, and 141 (each with two substitutions) (Figure 4). The DLA-DQB first domain sequences had 14 replacement substitutions and 3 silent substitutions. Moreover, five out of seven radical substitutions (which involve changes in charge) were located at the putative antigen-binding site of the class II molecule (Figures 4 and 5).[51] These data have provided further evidence for the hypothesis that there is positive selection for variability and charge profile diversity in the antigen recognition site of class II molecules.[52,53] Although many of the residues have been conserved between the DQB sequences of dogs and other mammals, ten substitutions apparently unique to the dog are found at nonpolymorphic positions 84 (G), 88 (L), and at polymorphic positions 30 (D), 63 (R/P), 67 (E), 69 (D), 71 (V), 85 (R), and 89 (Y). Moreover, positions 29 (R/K) and 69 (E/D) are unique in that they are polymorphic only in the dog. It is also interesting to note that codon 63 is highly polymorphic (G/R/P) in both DQB and DRB chains of the dog, as in the DQB molecules of the mouse, rat, miniature swine, and macaque, although this site is nonpolymorphic in other species, such as human, ox, sheep, and horse.[17,19,43,46,53-55]

In conclusion, based on the sequences of the DQB clones amplified from the eight HTC dogs, it has been presumed that dogs express a single functional DQB gene with at least four distinct alleles.[19] The second DQB gene detected by RFLP studies of dog genomic DNA may be a pseudogene or a gene that is not transcribed (such as the HLA-DQB2 or DXB gene). The dog DQB gene has been termed the DLA-DQB1 gene and the four sequences have been designated as the DLA-DQB1*0101, *0201, *0301, and *0401 alleles.[19]

E. Haplotypic Diversity in DLA Class II Region

Recent gene cloning and sequence analysis of the DLA class II region has revealed that the DLA-DR subregion has a single expressed DRB1 gene with at least nine alleles (termed DLA-DRB1*Dw1, *Dw3, *Dw4, *Dw8, *D4, *D6, *D7, *D8, and *D9).[17] Similar studies of the DLA-DQ subregion have demonstrated that the dog has a single functional DLA-DQA1 gene (with at least four distinct alleles, designated DLA-DQA1*0101, *0102, *0201, and *0202) and one functional DLA-DQB1 gene (with at least four alleles, termed DLA-DQB1*0101, *0201, *0301, and *0401).[18,19] Since the haplotypic associations of the known DLA class II alpha and beta gene alleles and their correlation to the cellular and serologic typing results have not been previously reported, they will be discussed briefly in this review.

The haplotype analysis presented here is based on previously published DLA-DRB1, -DQB1, and -DQA1 sequences obtained from 12 unrelated DLA-D homozygous dogs representing 12 different DLA-D types.[17-19] Table 2 lists the dogs by their DLA-B serotypes, DLA-D types, and class II alleles determined by nucleotide sequence analysis of DLA-DRB1, -DQA1, and -DQB1 genes or gene fragments. Table 2 also demonstrates that phenotypic homozygosity, defined as the DLA-D type (by the mixed leukocyte culture reaction), tends to correlate with homozygosity at the DRB1 locus. This finding presumably reflects that most of the alloreactive T-cell clones in mixed leukocyte cultures are directed against the highly polymorphic and immunodominant DR molecule, which is likely expressed at higher levels than the DQ molecule in dogs as in humans.[42]

Table 2 Comparison of 12 Homozygous Typing Cell Dogs by their DLA-D Specificities and by their DLA Class II Haplotypes (Combination of Alleles at DLA-DRB1-DQA1-DQB1 loci)

Dog no.	DLA-B serology[a]	DLA-D type[b]	DLA class II haplotype[c]		
C624	B6	Dw3	DRB1*Dw3	DQA1*0101	DQB1*0101
			DRB1*Dw3	DQA1*0101	DQB1*0201
B334	B6	Dw4	DRB1*Dw4	nd	DQB1*0201
			DRB1*Dw4	nd	DQB1*0201
B377	B6	D1	DRB1*Dw4	nd	DQB1*0201
			DRB1*Dw4	nd	DQB1*0401
C843	B6	Dw8	DRB1*Dw8	DQA1*0102	DQB1*0101
			DRB1*Dw8	DQA1*0102	DQB1*0201
C334	B4,5,6	D10	DRB1*Dw8	DQA1*0101	DQB1*0301
			DRB1*Dw8	DQA1*0101	DQB1*0301
C535	B6	D8	DRB1*D8	nd	DQB1*0401
			DRB1*D8	nd	DQB1*0401
C648	B5	Dw1	DRB1*Dw1	nd	DQB1*0101
			DRB1*Dw1	nd	DQB1*0201
PK28	B5	D4	DRB1*D4	DQA1*0101	nd
			DRB1*D4	DQA1*0201	nd
C294	B5	D9	DRB1*D9	nd	DQB1*0101
			DRB1*D9	nd	DQB1*0101
C170	B4,14	D6	DRB1*D6	DQA1*0101	nd
			DRB1*D6	DQA1*0101	nd
B906	B4	D16	DRB1*D6	DQA1*0101	nd
			DRB1*D6	DQA1*0101	nd
C461	B4,6	D7	DRB1*D7	DQA1*0202	nd
			DRB1*D7	DQA1*0202	nd

[a] Nomenclature for DLA-B serologic specificities is based on the third international workshop on canine immunogenetics.[10]

[b] Nomenclature for DLA-D specificities (determined by one-way mixed leukocyte culture) is based on the third international canine immunogenetics workshop.[11]

[c] Based on previously published nucleotide sequences for DLA-DRB1,[17] -DQA1,[18] and -DQB1 alleles[19] nd indicates sequence not determined.

Since most MLC typing differences in humans have been localized to the third hypervariable region of the HLA-DR beta chain, the polymorphic amino acid positions of the DLA-DRB1 sequences (shown in Figure 4) also have been compared in these dogs.[41] Table 2 shows that some DLA-DRB1 alleles (such as *Dw1, *D4, *D8, and *D9) corresponded to a single DLA-D type, while other alleles (including *Dw8, *Dw3, *Dw4, *D6, and *D7) were found in association with more than one DLA-D type. For example, the DRB1*Dw8 allele was found in DLA-Dw8 and -D10. Similarly, the DRB1*Dw4 allele was present in DLA-Dw4 and -D1 (Table 2). Since the MLC typing differences for the latter DLA-D types could not be explained on the basis of their DRB1 gene polymorphism, other class II genes had to be investigated for their contributions to T-cell allorecognition patterns in mixed leukocyte cultures.

Comparison of the class II haplotypes of the dogs shown in Table 2 revealed additional polymorphism in the DQ subregion. For example, dog C843 (cellularly typed as a DLA-Dw8 homozygote) was by nucleotide sequence analysis a DQB heterozygote with two different beta gene alleles, which could combine with a single alpha gene allele to form at least two different DQ molecules (Table 2). Similarly, dog C334 (typed as DLA-D10) was a DQB1 homozygote with unique DQ alpha and beta gene alleles, which differed from those of dog C843. Although the latter two dogs carried an identical DR product (DRB1*Dw8), they expressed strikingly different DQ molecules on their cell surface, which were apparently recognized in the mixed leukocyte culture reaction. Similar comparison of dogs representing DLA-D types Dw4 and D1 showed that they had identical DRB1 alleles, and that one dog was a DQB1 homozygote, while the other dog was a DQB1 heterozygote (Table 2). Although sequence data could not be obtained for all the dogs studied, these data strongly suggest that the molecular basis for the MLC typing differences may lie in the DQ as well as DR molecules.

The identification of putative antibody-binding epitopes corresponding to the DLA-B serologic specificities has been difficult in the dog for two reasons. First, there are no dog-specific monoclonal antibodies available for serologic typing in the dog. Second, the currently used polyclonal reagents tend to be multispecific and may not represent the major genetic groups of dogs equally due to differences in their ability to detect identical alleles in mixed-breed vs. Beagle dogs.[10] Since previous RFLP studies have suggested that the DLA-B serologic specificities may correlate with supertypic DLA-DR families, the dogs (shown in Table 2) were also analyzed with respect to their DLA-B serologic specificities and DLA-DRB chain first domain polymorphism (shown in Figure 4) in search for shared sequence motifs, which might function as antibody-binding sites. From these data, it may be hypothesized that position 63 (in the third HVR) of the DLA-DRB molecule may constitute or at least participate in forming a binding site for DLA-B antibodies, since examination of the sequences revealed that dogs carrying the DLA-B6 serologic specificity generally have a glycine (G) residue at that site, while those that carry the DLA-B5 specificity have an arginine (R) residue, and those with the DLA-B4 specificity have a proline (P) residue at that site (Figure 4, Table 2). Moreover, position 63 is highly polymorphic (with three substitutions) in both DRB and DQB chains of the dog and is located on the kink between helices and predicted to be solvent-accessible based on the hypothetical class II model (Figure 5).[51] Future site-directed mutagenesis studies will be necessary to identify the relationship of DLA-B locus serologic specificities to structural differences in the DLA class II molecules.

In conclusion, these data confirm that the DLA-D specificity determined by the mixed leukocyte culture reaction is defined by the composite response of T lymphocytes to epitopes present on all class II products, including DR, DQ (and presumably DP) molecules and thus represents a "haplotype designation" in dogs as in humans.[56] Finally, it may be concluded that, while phenotypic DLA-D homozygosity (defined by mixed leukocyte culture) tends to correlate with homozygosity at the DRB1 locus, it may not always be accompanied by homozygosity at the DQB1 or DQA1 loci in the dog. From these considerations, it is clear that certain applications, such as transplantation studies in unrelated dogs, may require DLA typing based on DNA sequences to accurately assess the role of histocompatibility matching. Future studies of larger populations of outbred dogs and novel class II genes will be required to fully characterize the degree of haplotypic diversity at the DLA class II region of the dog.

IV. Mhc-DISEASE ASSOCIATIONS

The molecular basis of Mhc-disease associations has been the focus of major research efforts ever since the observation that certain HLA class I and class II alleles were strongly associated with susceptibility to over 40 human diseases, most of which were autoimmune in nature. [41,57] Although Mhc molecules are thought to play a permissive rather than a causative role in these polygenic diseases, discrete hypervariable domains within class II molecules appear to represent the "susceptibility determinant" for many autoimmune diseases.[58] However, the contribution of the Mhc to the genetic basis of these diseases may vary substantially and involve different Mhc molecules.[57,58]

Very few DLA-disease association studies have been described in the dog in the past 2 decades.[59] The nomenclature used by the early workers reflects the local laboratory or first workshop designations for DLA types. For example, dogs with leukosis (uncharacterized leukocyte malignancy) were found to carry the DL A7 (long) serotype, while diabetes mellitus appeared to be associated with dogs typed as DL A3 and A7.[60] Canine atopy has been reported in dogs typed as DL3/R15.[61] A subsequent study of diabetes mellitus has revealed a greater incidence of DLA A10 and B4 in affected Keeshond dogs.[62] There have been two reports of DLA association with canine transmissible venereal tumors, which were found to have DL3, 10, and 8 antigens on their cell surface.[63,64] More recent analyses of narcolepsy have failed to demonstrate DLA-D association (determined by mixed leukocyte culture and RFLP analysis) with the disease in several breeds of dogs.[65,66] However, systemic lupus erythematosus has been found to be associated with the class I antigen DLA-A7 in German shepherd dogs.[67]

The general lack of data on DLA-disease associations indicates the need for more extensive studies of canine diseases using currently available DLA typing methods. Given the variety of diseases (such as rheumatoid arthritis, dermatomyositis, pemphigus vulgaris, myasthenia gravis, Addison's disease, chronic active hepatitis, etc.) found in both humans and dogs (often with breed-related predispositions), increasing our understanding of the pathogenetic mechanisms of these conditions would undoubtedly benefit both species in terms of potential immunotherapeutic interventions.

REFERENCES

1. Klein, J., *Natural History of the Major Histocompatibility Complex,* John Wiley & Sons, New York, 1986.
2. Barber, L. D. and Parham, P., Peptide binding to major histocompatibility complex molecules, *Annu. Rev. Cell Biol.,* 9, 163, 1993.
3. Schwartz, R. H., T-lymphocyte recognition of antigen in association with gene products of the major histocompatibility complex, *Annu. Rev. Immunol.,* 3, 237, 1985.
4. Klein, J., Takahata, N., and Ayala, F. J., Mhc polymorphism and human origins, *Sci. Am.,* December 1993.
5. Storb, R. and Deeg, H. J., Contributions of the dog model in marrow transplantation, *Plasma Ther. Transfus. Technol.,* 6, 303, 1985.
6. Storb, R., Epstein, R. B., Bryant, J., Ragde, H., and Thomas, E. D., Marrow grafts by combined marrow and leukocyte infusions in unrelated dogs selected by histocompatibility typing, *Transplantation,* 6, 587, 1968.
7. Epstein, R. B., Storb, R., Ragde, H., and Thomas, E. D., Cytotoxic typing antisera for marrow grafting in littermate dogs, *Transplantation,* 6, 45, 1968.
8. Vriesendorp, H. M., Westbroek, D. L., D'Amaro, J., van der Does, J. A., van der Steen, G. J., van Rood, J. J., Albert, E., Bernini, L., Bull, R. W., Cabasson, J., Epstein, R. B., Erikson, V., Feltkamp, T. E. W., Flad, H. D., Hammer, C., Lang, R., Largiader, F., von Loringhoven, K., Los, W., Meera Khan, P., Saison, R., Serrou, B., Schnappauf, H., Swisher, S. N., Templeton, J. W., Uhlschmidt, G., and Zweibaum, A., Joint report of the first international workshop on canine immunogenetics, *Tissue Antigens,* 3, 145, 1973.
9. Vriesendorp, H. M., Albert, E. D., Templeton, J. W., Belotsky, S., Taylor, B., Blumenstock, D. A., Bull, R. W., Cannon, F. D., Epstein, R. B., Ferrebee, J. W., Grosse-Wilde, H., Hammer, C., Krumbacher, K., Leon, S., Meera Khan, P., Mickey, M. R., Matola, M., Rapaport, F. T., Saison, R., Schnappauf, H., Scholz, S., Schroeder, M. L., Storb, R., Wank, R., Westbroek, D. L., and Zweibaum, A., Joint report of the second international workshop on canine immunogenetics, *Transplant Proc.,* 8, 289, 1976.
10. Bull, R. W., Vriesendorp, H. M., Cech, R., Grosse-Wilde, H., Bijma, A. M., Ladiges, W. L., Krumbacher, K., Doxiadis, I., Ejima, H., Templeton, J., Albert, E. D., Storb, R., and Deeg, H. J., Joint report of the third international workshop on canine immunogenetics. II. Analysis of the serological typing of cells, *Transplantation,* 43, 154, 1987.
11. Deeg, H. J., Raff, R. F., Grosse-Wilde, H. I., Bijma, A. M., Buurman, W. A., Doxiadis, I., Kolb, H. J., Krumbacher, K., Ladiges, W., Losslein, K. L., Schoch, G., Westbroek, D. L., Bull, R. W., and Storb, R., Joint report of the third international workshop on canine immunogenetics. I. Analysis of homozygous typing cells, *Transplantation,* 41, 111, 1986.
12. Sarmiento, U. M. and Storb, R., Restriction fragment length polymorphism of the major histocompatibility complex of the dog, *Immunogenetics,* 28, 117, 1988a.

13. Sarmiento, U. M. and Storb, R., Characterization of class II alpha genes and DLA-D region allelic associations in the dog, *Tissue Antigens,* 32, 224, 1988b.
14. Sarmiento, U. M. and Storb, R., RFLP analysis of DLA class I genes in the dog, *Tissue Antigens,* 34, 158, 1989.
15. Sarmiento, U. M. and Storb, R., Nucleotide sequence of a dog DRB cDNA clone, *Immunogenetics,* 31, 396, 1990a.
16. Sarmiento, U. M. and Storb, R., Nucleotide sequence of a dog class I cDNA clone, *Immunogenetics,* 31, 400, 1990b.
17. Sarmiento, U. M., Sarmiento, J. I., and Storb, R., Allelic variation in the DR subregion of the canine major histocompatibility complex, *Immunogenetics,* 32, 13, 1990.
18. Sarmiento, U. M., DeRose, S., Sarmiento, J. I., and Storb, R., Allelic variation in the DQ subregion of the canine major histocompatibility complex. I. DQA, *Immunogenetics,* 35, 416, 1992.
19. Sarmiento, U. M., DeRose, S., Sarmiento, J. I., and Storb, R., Allelic variation in the DQ subregion of the canine major histocompatibility complex. II. DQB, *Immunogenetics,* 37, 148, 1993.
20. Krumbacher, K., van der Feltz, M. J. M., Happel, M., Gerlach, C., Losslein, L. K., and Grosse-Wilde, H., Revised classification of the DLA loci by serological studies, *Tissue Antigens,* 27, 262, 1986.
21. van der Feltz, M. J. M. and Ploegh, H. L., Immunochemical analysis of glycosylated and non-glycosylated DLA class I antigens, *Immunogenetics,* 19, 95, 1984.
22. Neefjes, J. J., Doxiadis, I., Stam, N. J., Beckers, C. J., and Ploegh, H. L., An analysis of class I antigens of man and other species by one-dimensional IEF and immunoblotting, *Immunogenetics,* 23, 164, 1986.
23. Doxiadis, I., Krumbacher, K., Neefjes, J. J., Ploegh, H. L., and Grosse-Wilde, H., Biochemical evidence that the DLA-B locus codes for a class II determinant expressed on all canine peripheral blood lymphocytes, *Exp. Clin. Immunogenet.,* 6, 219, 1989.
24. Parham, P., HLA, anthropology, and transplantation, *Transplant. Proc.,* 25, 159, 1993.
25. Koller, B. H., Geraghty, D. E., Shimizu, Y., DeMars, R., and Orr, H. T., HLA-E, a novel HLA class I gene expressed in resting T lymphocytes, *J. Immunol.,* 141, 897, 1988.
26. Kindt, T. J. and Singer, D. S., Class I major histocompatibility complex genes in vertebrate species: what is the common denominator?, *Immunol. Res.,* 6, 57, 1987.
27. Young, J. A. T., Wilkinson, D., Bodmer, W. F., and Trowsdale, J., Sequence and evolution of HLA-DR7- and -DRw53-associated beta-chain genes, *Proc. Natl. Acad. Sci. U.S.A.,* 84, 4929, 1987.
28. Yuhki, N., Heidecker, G. F., and O'Brien, S. J., Characterization of Mhc cDNA clones in the domestic cat, *J. Immunol.,* 142, 3676, 1989.
29. Ennis, P. D., Jackson, A. P., and Parham, P., Molecular cloning of bovine class I Mhc cDNA, *J. Immunol.,* 141, 642, 1988.
30. Cowan, E. P., Jelachich, M. L., Biddison, W. E., and Coligan, J. E., DNA sequence of HLA-A11: remarkable homology with HLA-A3 allows identification of residues involved in epitopes recognized by antibodies and T cells, *Immunogenetics,* 25, 241, 1987.
31. Parham, P., Lawlor, D. A., Lomen, C. E., and Ennis, P. D., Diversity and diversification of HLA-A,B,C alleles, *J. Immunol.,* 142, 3937, 1989.
32. Deeg, H. J., Wulff, J. C., and DeRose, S., Unusual distribution of Ia-like antigens on canine lymphocytes, Immunogenetics, 16, 445, 1982.
33. Doveren, R. F. C., Buurman, W. A., Schutte, B., Groenewegen, G., and van der Linden, C. J., Class II antigens on canine T lymphocytes, *Tissue Antigens,* 25, 255, 1985.

34. Doveren, R. F. C., van der Linden, C. J., Spronken, E. E. M., Groenewegen, G., and Buurman, W. A., Canine Mhc-class II antigens on B and T lymphocytes, *Tissue Antigens,* 27, 87, 1986.
35. Weiden, P. L., Gochoel, K., Deeg, J., and Storb, R., Primed lymphocyte test in dogs: restimulation by non-DLA-D determinants and failure to detect lymphoma associated antigens, *Immunogenetics,* 8, 529, 1979.
36. Ladiges, W. C., Raff, R. F., Brown, S., Deeg, H. J., and Storb, R., The canine major histocompatibility complex, supertypic specificities defined by the primed lymphocyte test (PLT), *Immunogenetics,* 19, 359, 1984.
37. Raff, R. F., Deeg, H. J., Farewell, V. T., DeRose, S., and Storb, R., The canine major histocompatibility complex: population study of DLA-D alleles using a panel of homozygous typing cells, *Tissue Antigens,* 21, 360, 1983.
38. Bell, J. I., Denney, D., Foster, L., Belt, T., Todd, J. A., and McDevitt, H. O., Allelic variation in the DR subregion of the human major histocompatibility complex, *Proc. Natl. Acad. Sci. U.S.A.,* 84, 6234, 1987.
39. Burnett, R. C., DeRose, S. A., and Storb, R., A simple restriction fragment-length polymorphism assay for Mhc class II gene testing of dog families, *Transplantation,* 57, 280, 1994.
40. Moller, E., Carlsson, B., and Wallin, J., Implication of structural class II gene polymorphism for the concept of serologic specificities, *Immunol. Rev.,* 85, 107, 1985.
41. Gregersen, P. K., HLA class II polymorphism: implications for genetic susceptibility to autoimmune disease, *Lab. Invest.,* 61, 5, 1989.
42. Reinsmoen, N. L. and Bach, F. H., Structural model for T-cell recognition of HLA class II-associated alloepitopes, *Hum. Immunol.,* 27, 51, 1990.
43. Marsh, S. G. E. and Bodmer, J. G., HLA-DR and -DQ epitopes and monoclonal antibody specificity, *Immunol. Today,* 10, 305, 1989.
44. Fan, W., Kasahara, M., Gutknecht, J., Klein, D., Mayer, W. E., Jonker, M., and Klein, J., Shared class II Mhc polymorphisms between humans and chimpanzees, *Hum. Immunol.,* 26, 107, 1989.
45. Klein, J., Origin of the major histocompatibility complex polymorphism: The trans-species hypothesis, *Hum. Immunol.,* 19, 155, 1987.
46. Gyllensten, U. B., Lashkari, D., and Erlich, H. A., Allelic diversification at the class II DQB locus of the mammalian major histocompatibility complex, *Proc. Natl. Acad. Sci. U.S.A.,* 87, 1835, 1990.
47. Horn, G. T., Bugawan, T. L., Long, C. M., Manos, M. M., and Erlich, H. A., Sequence analysis of HLA class II genes from insulin-dependent diabetic individuals, *Hum. Immunol.,* 21, 249, 1988.
48. Braunstein, N. S., Germain, R. N., Loney, K., and Berkowitz, N., Structurally interdependent and independent regions of allelic polymorphism in class II Mhc molecules, *J. Immunol.,* 145, 1635, 1990.
49. Gyllensten, U. B. and Erlich, H. A., Ancient roots for polymorphism at the HLA-DQa locus in primates, *Proc. Natl. Acad. Sci. U.S.A.,* 86, 9986, 1989.
50. Erlich, H. A. and Gyllensten, U. B., The evolution of allelic diversity at the primate major histocompatibility complex class II loci, *Hum. Immunol.,* 30, 110, 1991.
51. Brown, J. H., Jardetzky, T., Saper, M. A., Samraoui, B., Bjorkman, P. J., and Wiley, D. C., A hypothetical model of the foreign antigen-binding site of class II histocompatibility molecules, *Nature,* 332, 845, 1988.
52. Hughes, A. L. and Nei, M., Nucleotide substitution at major histocompatibility complex class II loci: evidence for overdominant selection, *Proc. Natl. Acad. Sci. U.S.A.,* 86, 958, 1989.

53. Hughes, A. L. and Nei, M., Evolutionary relationships of class II major-histocompatibility-complex genes in mammals, *Mol. Biol. Evol.*, 7, 491, 1990.
54. Fujii, H., Kakinuma, M., Yoshiki, T., and Natori, T., Polymorphism of the class II gene of rat major histocompatibility complex, *RT1*: partial sequence comparison of the first domain of the *RT1.Bb 1* alleles, *Immunogenetics*, 33, 399, 1991.
55. Gustafsson, K., LeGuern, C., Hirsch, F., Germana, S., Pratt, K., and Sachs, D. H., Class II genes of miniature swine. IV. Characterization and expression of two allelic class II *DQB* cDNA clones, *J. Immunol.*, 145, 1946, 1990.
56. Bach, F. H., The HLA class II genes and products: the HLA-D region, *Immunol. Today*, 6, 89, 1985.
57. Bell, J. I., Todd, J. A., and McDevitt, H. O., The molecular basis of HLA-disease association, *Adv. Hum. Genet.*, 18, 1, 1989.
58. Dalton, T. A. and Bennett, J. C., Autoimmune disease and the major histocompatibility complex: therapeutic implications, *Am. J. Med.*, 92, 183, 1992.
59. Day, M. J. and Penhale, W. J., A review of major histocompatibility complex-disease associations in man and dog, *Vet. Res. Commun.*, 11, 119, 1987.
60. Vriesendorp, H. M., D'Amaro, J., van der Does, J. A., Westbroek, D. L., and Epstein, R. B., Analysis of the DL-A system in families and populations of healthy and diseased individuals, *Transplant. Proc.*, 5, 311, 1973.
61. Vriesendorp, H. M., Smid-Mercx, B. M. J., Visser, T. P., Haliwell, R. E. W., and Schwartzman, R. M., Serological DL-A typing of normal and atopic dogs, *Transplant. Proc.*, 7, 375, 1975.
62. Kramer, J. W., Nottingham, S., Robinette, J., Lenz, G., Sylvester, S., and Dessouky, M. I., Inherited, early onset, insulin requiring diabetes mellitus in Keeshond dogs, Diabetes, 29, 558, 1980.
63. Epstein, R. B. and Taylor Bennett, B., Histocompatibility typing and course of canine venereal tumors transplanted into unmodified random dogs, *Cancer Res.*, 34, 788, 1974.
64. Taylor Bennett, B., Taylor, Y., and Epstein, R., Segregation of the clinical course of transmissible venereal tumor with DL-A haplotypes in canine families, *Transplant. Proc.*, 7, 503, 1975.
65. Dean, R. R., Kilduff, T. S., Dement, W. C., and Grumet, F. C., Narcolepsy without unique Mhc class II antigen association: studies in the canine model, *Hum. Immunol.*, 25, 27, 1989.
66. Motoyama, M., Kilduff, T. S., Lee, B. S. M., Dement, W. C., and McDevitt, H. O., Restriction fragment length polymorphism in canine narcolepsy, *Immunogenetics*, 29, 124, 1989.
67. Teichner, M., Krumbacher, K., Doxiadis, I., Doxiadis, G., Fournel, C., Rigal, D., Monier, J. C., and Grosse-Wilde, H., Systemic lupus erythematosus in dogs: association to the major histocompatibility complex class I antigen DLA-A7, *Clin. Immunol. Immunopathol.*, 55, 255, 1990.

CHAPTER 9

The Swine Major Histocompatibility Complex

Lawrence B. Schook, Mark S. Rutherford, Jong-Keuk Lee, Yun-Chao Shia, Marite Bradshaw, and Joan K. Lunney

CONTENTS

0-8493-8032-4/96/$0.00+$.50

I. INTRODUCTION

This review is intended as an update on current knowledge of the swine major histocompatibility complex (Mhc) or swine leukocyte antigen (SLA) complex, including its genomic structure and the molecular and monoclonal antibody (mAb) probes available for analysis of SLA gene expression. The importance of Mhc genes in regulating immune responses has been well documented for many species. This review will present current data on the role of SLA genes in regulating disease and immune responses, and the association of specific SLA haplotypes and alleles with economically important traits.

It has been 25 years since the existence of the swine Mhc or SLA complex was clearly established by Vaiman and Binns and their colleagues (Vaiman et al., 1970; Viza et al., 1970). SLA encoded control of swine antibody responses to defined antigens were first noted by Vaiman et al., (1978) and confirmed by Lunney et al. (1986) and Mallard et al. (1989c). Because of the importance of the SLA complex in controlling immune responses, several laboratories developed herds of swine bred specifically for individual SLA haplotypes. SLA defined herds were subsequently developed in England by Binns and colleagues (Viza et al., 1970), in the U.S. by Sachs et al. (1976), in the Czech Republic by Hruban and colleagues (Hradecky et al., 1985), and in the U.S. in outbred pigs by Smith (University of Oklahoma, unpublished). By international agreement (White and Lunney, 1979) unique letters were assigned for new SLA haplotypes; the NIH minipigs were originally assigned haplotypes *a, c,* and *d* and the British pigs haplotypes b and *e*. As SLA recombinant chromosomes have been identified in the NIH minipigs other haplotypes have been added, some of which are shown in Figure 1. These herds of inbred pigs have enabled scientists to assess in detail immune responses that regulate organ transplantation (reviewed in Sachs, 1994; Kenmochi et al., 1994).

II. THE GENOMIC STRUCTURE OF THE SLA COMPLEX

The SLA complex is a set of linked genes (or haplotype) that was first mapped to swine chromosome 7 by Rabin et al. (1985). More recently extensive research has shown that the SLA complex encodes an increasingly larger number of genes including the class I and class II SLA glycoproteins, complement components, tumor necrosis factor, and heat shock proteins (Lunney, 1994). Figure 2 provides a map of the SLA complex in comparison to the human leukocyte complex (HLA) as published by Trowsdale et al. (1991) and Campbell and Trowsdale (1993). The SLA map is based on data from colleagues worldwide and was recently updated at the SLA workshop of the International Society of Animal Genetics in August, 1994. Early mapping of the SLA complex, based on the SLA serological typing of large numbers of individuals, indicated a short distance (0.3 to 0.4 cm between SLA class I and class II loci) (Vaiman et al., 1978; Lunney et al., 1986). Edfors-Lilja et al. (1993) recently published linkage studies on a smaller number of pigs that suggested longer distances, i.e., that SLA I was 0.8 cm from TNF which was 0.7 cm from DQB on the short arm of swine chromosome 7.

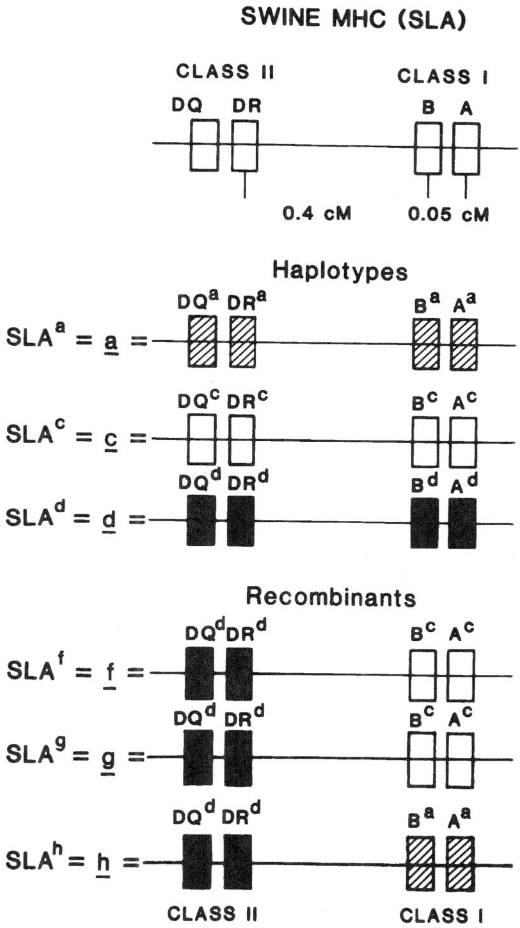

Figure 1 SLA haplotypes of NIH minipigs.

The introduction of pulse field gel electrophoretic (PFGE) techniques has enabled scientists to more precisely measure molecular distances within the SLA complex. The data from Xu et al. (1992) set the framework for the pulse field results summarized in Figure 1; this was extended by unpublished results from several laboratories (Wu et al., submitted; Singer, LeGuern, Chardon, personal communications). Figure 2 compares published data on known genes in the SLA complex map to those in the human HLA complex map prepared by Campbell and Trowsdale (1993). Maximal distances (from pulse field studies) are noted by **x** on the SLA map; unknown distances are assumed to be homologous to the human distance. From the PFGE data, it is clear that the distance from the SLA class I loci to the SLA class III loci is at least 100 kb shorter than for the homologous human genes; similarly, the distance between SLA class III and class II loci is 100 kb shorter. More detailed data within each region of the SLA complex are presented within the appropriate section below.

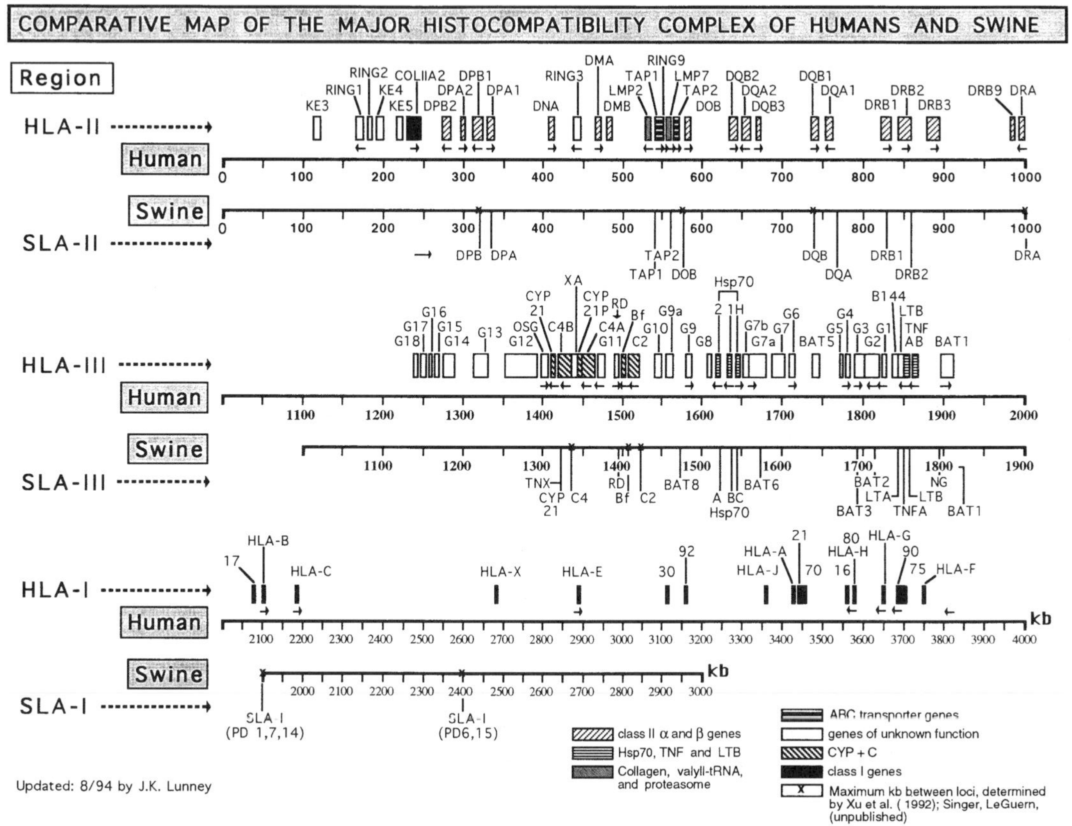

Figure 2 Comparative map of the major histocompatibility complex of humans and swine. The HLA map (Trowsdale et al., 1991; Campbell and Trowsdale, 1993) has been compared to the SLA map generated by Xu et al. (1992) and expanded with earlier published data and more recent unpublished data from Drs. Chardon, LeGuern, Peelman, Rothschild, Sachs, Singer, Vaiman, and Warner (adapted from Lunney, 1994). The three regions of the SLA complex, SLA-I, II, and III are noted; distances are given in kilobasepairs (kb). Homologous loci are aligned unless data in the swine has proven that there is an alternate location. The x indicates the maximal distance as determined by pulse field gel electrophoresis or by overlapping cosmid mapping.

A. SLA Class I Region

1. Genes

The traditional, expressed class I locus products are designated SLA-A, B, C, encode for 45,000 Mr glycoproteins that noncovalently associate with the 12,000 Mr beta$_2$-microglobulin, and are expressed on the surface of most cells (Vaiman et al., 1979, Lunney and Sachs, 1978). Most cells express class I antigens, and most mature swine lymphocytes express high levels of class I antigens. Serologic evidence clearly proves the existence of two expressed SLA class I gene products and indicates the expression of a third locus in some SLA haplotypes (Vaiman et al., 1979). Singer and colleagues were the first to clone and express a swine class I gene (Singer et al., 1982). Using specific restriction fragment length polymorphisms (RFLP) of genomic DNA, her group then proceeded to demonstrate that the SLA^d haplotype encodes six to eight class I genes, although only two genes (PD1, PD14) were demonstrated to be expressed at the protein level while a third (PD6) was only detected at the mRNA level (Singer et al., 1988; Ehrlich et al., 1987). Through DNA sequence analyses, PD1, PD7, and PD14 are 80 to 85% similar with each other but only 49 and 55% similar to PD6 and PD15. Moreover, PD6 shares only 45% similarity with PD6. These comparative sequencing results indicate that the SLA class I genes can be divided into at least three subfamilies. Most recently Gaycken et al. (1994) identified 9 class I genes in their screening of a porcine cosmid library with human cDNA probes and noted that the class I positive fragments ranged from 3.6 to 5.8 Kb and were separated by 11.9 to 22.2 Kb.

Chardon, Vaiman, and their colleagues also confirmed that swine have a limited number of class I genes when they analyzed DNA from a wide range of SLA haplotyped swine. Their findings were consistent with seven to ten SLA class I genes (Chardon et al., 1985 a,b). As expected, considerable polymorphisms were observed between distinct haplotypes; polymorphic RFLP fragments segregated in accordance with serological typing results. Other scientists have utilized RFLP patterns to characterize SLA haplotypes (Flanagan et al., 1988; Jung et al., 1989b; Shia et al., 1991). Seven serologically defined SLA class I alleles were observed in four Swiss Large White and American Hampshire families characterized by Southern blot-RFLP hybridization to distinguish pattern types (Shia et al., 1991). A porcine class I probe, PD1A, was hybridized to genomic DNA digested with one of four restriction enzymes, and seven allelic patterns were identified. Depending on the restriction enzyme, 19 to 24 bands were observed, of which 10 to 16 were polymorphic. Based on the segregation of RFLP pattern types, several serotypes were shown to correlate to specific bands generated by Southern blotting. For example, the H8 serotype segregated with a 3.4-Kb *Taq*I fragment, and the H16 serotype segregated with a 1.3-Kb *Taq*I band. Additionally, several serotypes shared restriction fragments. Unfortunately, the PD1A probe utilized was not locus specific and hybridized to several class I loci. Thus, it was not possible to distinguish which class I locus polymorphism was detected using this probe. Locus-specific RFLP genotyping for SLA class I alleles has not been reported.

2. Region Map

Singer and colleagues (unpublished data) have used PFGE to show that the three class I SLA genes, PD1, PD7, and PD14, all reside on the same DNA fragment (<320 Kb) and that they are separated by no more than 500 Kb from the fragment that contains PD6 and PD15. This spacing is noted in the comparative map (Figure 2). As mentioned above, the SLA class I genes are fewer in number and appear to be more compressed in swine than in humans or mice, even following efforts to demonstrate nonclassical class I Mhc loci (Singer et al., personal communication).

3. Serological Reagents

SLA typing alloantisera have been developed in laboratories worldwide; the most recent international comparative test was reported by Renard et al. (1988). More recently monoclonal antibodies (mAb) have been produced and tested for SLA class I antigen reactivity as has been discussed in detail in Lunney et al. (1988). Table 1 lists the anti-class I SLA mAbs and their reactivity on cells from NIH minipigs and from outbred swine and on transfected mouse L cell lines that express specific SLA^d class I gene products. Most of these mAb when tested on outbred swine populations appear to detect determinants common to a large number of haplotypes and, thus, were designated as being broadly polymorphic or as reacting with "public" SLA class I specificities (Kristensen et al., 1993).

Most mAbs did not demonstrate allelic or "private" SLA class I specificities, but several mAbs should prove useful for improving the definition of specific SLA haplotypes or as specific SLA typing reagents in restricted (SLA inbred or defined) swine populations, e.g., mAbs 2.12.3 and 16.7E4. Indeed there are now a few mAbs that exhibit more restricted reactivity and may be recognizing unique, "private" SLA class I specificities, e.g., 2.32.1 with SLA-I specificity W9 as defined by its reactivity with cells from swine with SLA haplotypes with specificity W9 (H04 and H30 in tested breeds) (Table 1) (Kristensen et al., 1993). The mAb specifically reactive with cells from SLA^c swine may also prove to be reactive with a private specificity, as evidenced by their restricted reactivity with Gottingen minipigs. mAbs that react with monomorphic determinants of SLA class I genes can be divided into several groups: (1) those that probably recognize a monomorphic determinant of all classical SLA class I genes, e.g., PT85; (2) those that recognize a monomorphic determinant of possibly a single genetic locus; the early example of this class, 2.28.1, turned out to be polymorphic, not monomorphic, when tested on the broader outbred swine populations; and (3) those that recognize monomorphic determinants of nonclassical SLA class I genes, e.g., 76-7-4 detecting swine CD1 (Lunney, 1993).

B. SLA Class II Region

1. Genes

The SLA class II region encodes a variety of class II genes (Figure 2), of which only the SLA-DR and SLA-DQ appear to be expressed at the protein level. The expressed proteins consist of 33,000 to 35,000 Mr alpha (A) chains associated with 27,000 to 28,000 Mr beta (B) chains (Lunney et al., 1978, 1983; Chardon et al., 1981). Whereas, all mature swine lymphocytes express high levels of class I antigens, there is a differential expression of class II antigens. As expected, B cells and macrophages both express SLA-DR and -DQ antigens but, unexpectedly, T cells express higher levels of SLA-DR than -DQ antigens; moreover, there is preferential expression of class II antigens on the $CD8^+$ T cell subset (Pescovitz et al., 1985; Lunney and Pescovitz, 1987). The importance/relevance of this unusual class II expression on T cells has yet to be fully appreciated. cDNA clones have been isolated for several alleles of DRB, DBA, DQB, and DQA; sequence analyses of four class II cDNAs, designated pdDQA, pdDQB, pdDRA, and pdDRB, showed 75 to 87% identity to the corresponding HLA genes (Gustafsson et al., 1990 a,b; Pratt et al., 1990).

2. Regional Map

RFLP analysis of SLA class II encoded genes has revealed a high degree of polymorphism. Analyses using class II porcine probes revealed a greater number of restriction fragments, as compared to human class II probes, and indicated that up to three alpha chain genes and five to eight beta chain genes may be encoded by each SLA haplotype (Chardon et al., 1985 a,b; Vaiman et al., 1986; Warner, 1986; Sachs et al., 1988; Shia et al., 1991). Extensive evidence from LeGuern, Sachs, and colleagues on genomic DNA from NIH minipigs indicates that, for each SLA haplotype, there is one monomorphic DRA and two polymorphic DRB genes; one (or more) polymorphic DQA and DQB gene(s); no more than one DPA and one DPB gene, but no evidence for an expressed SLA-DP protein; evidence only for a DOB and DZA gene, and none for DX genes (Sachs et al., 1988; Hirsch et al., 1990, 1992; LeGuern, personal communication).

In contrast to NIH minipigs, studies in Swiss Large White and American Hampshire breeds using a porcine DRA probe identified two to five polymorphic fragments. The porcine DQA cDNA probe detected seven different restriction fragments using *Taq*I digests of genomic DNA (Shia et al., 1991). However, in each haplotype only one or two restriction fragments were observed, further indicating that the porcine DQA region most likely contains a single gene. No cross-hybridization between DRA and DQA was observed. In contrast, porcine probes against DRB and DQB cross-hybridized. Whereas the RFLP patterns for DQA, DRA, and DQB were not complex, suggesting the existence of only a few loci for these genes, the complexity observed for DRB suggested that several genes exist for this locus. Overall, Southern blot analysis of SLA class II genes using porcine cDNA probes revealed the presence of six allelic RFLP patterns for DQA and DQB, five allelic RFLP patterns for DRA, and seven allelic RFLP patterns for DRB (Shia et al., 1991).

Mapping of the outer limits of the SLA class II region indicates that the SLA-DP gene loci are ~300 kb from SLA-DO and another 150 kb from the SLA-DQ loci (LeGuern, personal communication) (Figure 2). The class II region of the Mhc encodes an additional series of non-Mhc genes which are involved in antigen processing and the class I presentation pathway. Recent investigations have identified polymorphisms at the transporter associated with antigen processing (TAP)-1 and TAP-2

Table I Monoclonal Antibodies Reactive with SLA Class I Antigens

MAb	Ig Class	NIH minipig dd	NIH minipig cc	NIH minipig gg	NIH minipig aa	Cellular Reactivity[a]: Outbred[b] (SLA Haplotype)	Cellular Reactivity[a]: Transfectants[c] PD1	Cellular Reactivity[a]: Transfectants[c] Pd14	Ref.
Polymorphic reactivity									
74-11-10	IgG2b	+	+	+	+	P (FJ1,B1,B3,W9) (FJ13,W16,a;Yuc-x)	+	+	Pescovitz et al., 1984
2.2.13		+	+	+	+	P (Yuc-x,y,z)	–	–	Ivanoska et al., 1991
2.12.3	IgM	+	–	–	–	P (W9,H14;Yuc-x)	–	+	Ivanoska et al., 1991
2.27.3	IgG	+	+	NT[d]	+	P (B3,W9;Yuc-x,y,z)	+	+	Ivanoska et al., 1991
2.28.1	IgM	+	+	+	+	P (broad)	+	–	Ivanoska et al., 1991
2.32.1	IgG	+	–	–	–	P (W9;Yuc-x)	–	+	Ivanoska et al., 1991
16.7D4	IgG1	–	+	+	–	NT	–	–	Lunney et al., 1988
16.7E4	IgG1	–	+	+	–	P (B1,W15)	–	–	Lunney et al., 1988
16.7F6	IgG1	–	+	+	–	NT	–	–	Lunney et al., 1988
H21A	IgM	–	–	–	–	P		NT	Davis et al., 1984, 1987
CA48P-A	IgM	+	+	+	+	P		NT	Davis et al., 1984, 1987
CA4C-A	IgM	+	+	+	+	P		NT	Davis et al., 1984, 1987
H11A	IgG2a	+	+	+	+	P		NT	Davis et al., 1984, 1987
H58A	IgG2a	+	+	+	+	P		NT	Davis et al., 1984, 1987
H17A	IgG2	–	–	–	–	P		NT	Davis et al., 1984, 1987
H22A	IgG2a	NT	NT	NT	NT	P		NT	Davis et al., 1984, 1987
5959[e]	IgM	+	–	NT	–	NT (Yuc-x)		NT	Smith (unpublished)
2380[e]	IgM	+	–	NT	–	NT (Yuc-x)		NT	Smith (unpublished)

Monomorphic reactivity

PT85	IgG2a	+	+	+	+	M	+	+	Davis et al., 1984, 1987
2.13.3		+	+	NT	+	NT		NT	Ivanoska et al., 1991
FMD2	IgG2b	NT	NT	NT	NT	M		NT	Letesson et al., 1986
IIB2	IgG2a	NT	NT	NT	NT	M		NT	Letesson et al., 1986
7-34-1	IgG2a	+	+	+	+	M	–	–	Lie et al., 1988
76-7-4	IgG	+	+	+	+	M		NT	Pescovitz et al., 1984

[a] mAb reactivity was assessed by complement mediated mAb cytotoxicity assay, cell ELISA assay, or flow cytometric analysis on cells from NIH minipigs.

[b] mAb reactivity was assessed by complement mediated mAb cytotoxicity assay on lymphocytes from SLA class I typed animals of 18 different breeds in Denmark and France (Kristensen et al., 1993). Reactivity with outbred swine is designated as polymorphic (P) or monomorphic (M); SLA haplotype or specificity reactions are noted as well as additional (a) uncharacterized specificities. Additional testing was performed by D. Smith (University of Oklahoma Medical School) on three lines of SLA-defined Yucatan swine (Yuc-x,y,z).

[c] Reactivity of mAb on mouse L cell fibroblasts that express the PD1 or PD14 class I SLA gene products (D. Singer et al., 1982, 1988) were assayed as described above for mAb binding.

[d] NT = not tested.

[e] MAb produced by One Lambda, Inc., Canoga Park, CA, found to be reactive with HLA-A34.

Table 2 Monoclonal Antibodies Reactive with SLA class II Antigens

MAb	Ig Class	SLA Specificity[a] *dd*	*cc*	*gg*	*aa*	Outbred	Ref.
SLA-DR reactivity							
MSA3	IgG2b, k	+	+	+	+	+	Hammerburg and Schurig, 1986
ISCR3	IgG2b, k	+	+	+	+	+	Watanabe et al., 1983
40B	IgG2a, k	+	+	+	+	+	Lunney et al., 1983
9B	IgG2a, k	+	–	+	+	NT[b]	Lunney et al., 1983
40E	IgG2a, k	+	+	+	+	NT	Lunney et al., 1983
39D	IgG2a, k	+	+	+	+	NT	Lunney et al., 1983
40K	IgG2a, k	+	+	+	+	NT	Lunney et al., 1983
39G	IgG2a, k	+	+	+	+	NT	Lunney et al., 1983
40D	IgG2a, k	+	+	+	+	+	Lunney et al., 1983
7A	IgG2a, k	+	+	+	+	+	Lunney et al., 1983
74C	IgG3, k	+	+	+	+	+	Lunney et al., 1983
DC2.06	IgG2	+	+	+	+	+	Osborne et al., 1984
SLA-DQ reactivity							
1052-5-21	IgG3	+	–	+	–	NT	Sachs et al., unpublished
1038-13-25	IgM	–	+	–	+	NT	Sachs et al., unpublished
TH16B	IgG2a, k	+	+	+	+	+	Davis et al., 1984, 1987
H42A	IgG2a, k	+	+	+	+	+	Davis et al., 1984, 1987
TH21A	IgG2a, k	+	+	+	+	+	Davis et al., 1984, 1987
TH22	IgG2a, k	+	+	+	+	+	Davis et al., 1984, 1987
RH1A	IgG2a, k	+	+	+	+	+	Davis et al., 1984, 1987
Unspecified SLA class II reactivity							
227		–	–	NT	+	NT	Lunney et al., 1988
Genox 3.53		+	–	NT	+	NT	Lunney et al., 1988
H4	IgG3	NT	NT	NT	NT	+	Lewin et al., 1985
TH4B	IgM	+	+	NT	+	+	Davis et al., 1984, 1987
TH14B		NT	NT	NT	NT	+	Davis et al., 1984, 1987
TH81B		NT	NT	NT	NT	+	Davis et al., 1984, 1987

[a] Analyses were performed as described in the legend to Table 1.
[b] NT = not tested.

genes (Vaske et al., 1994 a,b). Autosomal Mendelian segregation was observed in three-generation reference families, and analyses of unrelated pigs from different breeds identified several breed-specific bands. An alternate linkage location of the TAP-2 locus is outside the SLA region (Archibald et al., 1995).

3. Serological Reagents

Class II specific alloantisera are available for certain SLA haplotypes but there has never been an international comparison test to assign specificities. MAb reactive with swine class II Mhc antigens are listed in Table 2 (reviewed by Lunney et al. 1988; 1994). These have been split into groups based on their class II locus reactivity, i.e., those reactive with DR or DQ antigens and those that are not yet specified. The first mAbs to be assigned a swine locus reactivity were the anti-mouse-I-E mAbs that showed strong cross-reactions with both human and swine class II antigens (Lunney et al., 1983). Two-dimensional gel electrophoresis verified that these mAbs recognized a subset of swine class II antigens from cells of NIH minipigs of each SLA haplotype and, because of the known homology between the mouse I-E and human DR antigens, these mAbs were designated as reacting with swine DR antigens (Lunney et al., 1983). This was confirmed by Hirsch et al. (1992) using cell lines transfected with specific cloned swine class II genes. Sequential immunoprecipitation results (Figure 3) proved that the anti-DR mAb 40D recognized an independent subset of swine class II antigens from that precipitated by mAb TH16a; therefore, TH16 mAb reacted with the SLA-DQ antigens, as has been confirmed with SLA-DQ transfected cells. It should be noted, though, that cross locus reactivity of anti-class II mAb has been reported in other species, i.e., that anti-HLA-DR mAb can react with certain HLA-DQ antigens (Shaw et al., 1985). Thus, the locus reactivity of each mAb cannot be assumed unless reactivity has been confirmed on cells of a particular SLA haplotype.

Originally, when class II SLA specific mAbs were analyzed for reactivity with polymorphic determinants, all were negative, i.e., they were monomorphic; almost none exhibited differential reactions on cells from NIH minipigs of different SLA haplotypes. There were, however, a few mAbs that appeared to recognize polymorphic determinants, e.g., 9B, 227, and Genox 3.53, but on further analysis these "allelic" differences appeared to be due to low reactivity with certain swine class II alleles. Only recently have Sachs and colleagues (unpublished) been successful at producing anti-DQ (and possibly anti-DR) mAbs that recognize allelic porcine class II specificities. Since most of the mAbs listed in Table 2 react with monomorphic determinants of class II antigens, they have limited applicability for use in functional assays. In summary, antigens encoded by the two SLA class II loci have been identified and characterized, although only two mAbs that recognize unique, polymorphic determinants of swine class II antigens have just recently been reported.

4. *Molecular Genotyping*

SLA class II expression and sequence also has been analyzed through gene amplification using locus-specific oligonucleotide primers (Gustafsson et al., 1990 a,b; Emery et al., 1993; Shia et al., 1993; Våge et al., 1993). Recent investigations using the polymerase chain reaction (PCR)-RFLP to specifically amplify regions of SLA class II genes revealed the presence of at least 16 DQB and 9 DRB alleles (Shia et al., 1995; Våge et al., 1994). In these studies oligonucleotide primers were designed for PCR amplification of the second exon which encodes the β1 domain. The β1 domain contains the putative antigen recognition site (ARS) (Brown et al., 1988) which binds the processed antigens for presentation to T cells and thus would be expected to be highly polymorphic. By digesting the PCR products with a pair of restriction enzymes, all 16 DQB and 7 of 9 DRB alleles could be distinguished by this method (Figure 4; Shia et al., 1995).

SLA β1 domain PCR products amplified using genomic DNA as a template were sequenced, and nucleic acid sequence information was obtained for five DQB and

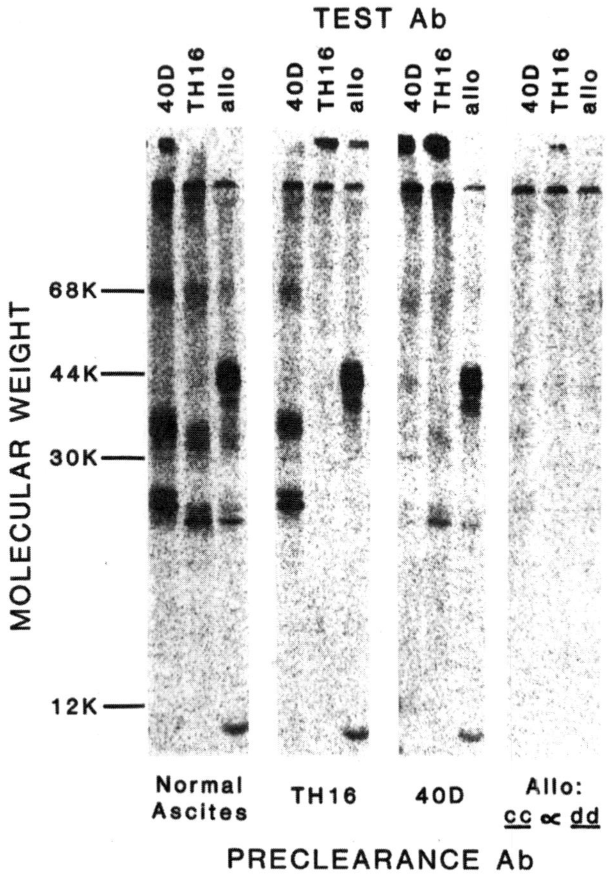

Figure 3 Sequential immunoprecipitation of SLA class II reactive mAb distinguishes SLA-DR antigens from SLA-DQ antigens. [^{35}S]-methionine labeled peripheral blood lymphocytes were lyzed with 1% Nonidet-P40 and the extract precleared of nonspecific precipitating material with normal pig serum. Aliquots of the lyzate were then precipitated with excess amounts of either anti-SLA-DR mAb, 40D, anti-SLA-DQ mAb, TH16, or alloantisera specific for the *d* haplotype. Aliquots of the precleared lyzates were then precipitated with the same mAb to prove that there had been total removal of that antigen, with the reciprocal mAb to verify the independence of the 2 class II antigen subsets or with the alloantiserum to verify that these were two independent SLA class II encoded products. Note that the class I antigens were not precleared by either mAb.

nine DRB alleles (Shia et al., 1995). Similarly, Våge et al. (1994) obtained genomic sequence for DRB in unrelated Norwegian Landrace pigs and identified five DRB alleles. Thus, including the NIH minipigs, sequence information at the ARS domain for SLA class II genes exists for 7 DQB and 13 DRB alleles. For the 93 amino acids which constitute the β1 domain, the deduced amino acid sequences for DQB alleles (Figure 5) showed an average of 12.0 ± 2.1 polymorphic residues compared to the SLADQB*C allele from NIH minipigs. Similarly, DRB alleles (Figure 6) showed an average of 13.1 ± 3.5 polymorphic residues compared to the SLADRB*C allele for the 94 amino acids which constitute the β1 domain.

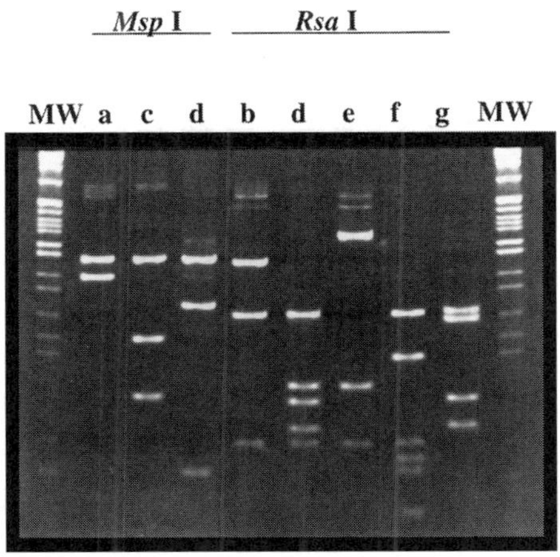

Figure 4 SLA-DRB PCR-RFLP pattern types. Genomic DNA from genetically diverse animals was isolated and amplified using DRB-β1 domain-specific oligonucleotides. Amplified products were then subjected to *Msp*I or *Rsa*I restriction endonuclease digestion with fractionation using an 8% polyacrylamide gel. Allelic variants are designated in lanes a through g.

```
                     10           20           30           40           50
                      |            |            |            |            |
                     *  * *                    * *          **
SLADQB*C        RDSPQDFVFQ FKGECYFYNG TQRVRGVARY IYNQEEHLRF DSDVGEFRAV

SLADQB*D        --------V- ---------- ----WS-D-- ------F--- ---M--Y---
SLADQB*01       ---------- --F----F-- -----L---- ------L--- ------Y---
SLADQB*02       ---------- --F----F-- -----L-T-- ------YA-- ------Y---
SLADQB*03       ---------- --F----F-- -----L-T-- ------YA-- ------Y---
SLADQB*04       ---------- --F----F-- -----L-T-- ------L--- ------Y---
SLADQB*05       --------V- ---------- ---------W V-----YV-- ----------

                     60           70           80           90
                      |            |            |            |
                   *    *      *   *  *   *    *     *          *
SLADQB*C        TPLGRPEADS WNSQKDVLEQ MRAEVDRVCK HNYQIEEGTT LQR

SLADQB*D        ------D--Y L-G--EA--- K---L-T--- ---------- ---
SLADQB*01       ---------- --G------- T---L-T--- ---------- ---
SLADQB*02       ---------- --G------- T---L-T--- ---------- ---
SLADQB*03       ------A--Y ------I--- T---L-T--- ---------- ---
SLADQB*04       ---------- --G------- T---L-T--- ----M----- ---
SLADQB*05       ------D--Y --G--E---- K---L-T--- ---------- ---
```

Figure 5 Amino acid sequence comparison of SLADQB alleles. SLA-DQB amino acid sequences at the β1 domain encoding the putative antigen recognition sequence were deduced from cDNA sequences. Data for alleles *S01, *S02,*S03, *S04, and *S05 were obtained from RT-PCR amplified SLA-DQB products (Shia et al., 1995), and cDNA from alleles *C and *D were obtained from Gustafsson et al., 1990b. Asterisks (*) indicate the position of the ARS residues. Amino acid numbering is as published for SLA-DQB*C beginning with amino acid 1 of the mature protein.

It is important to note, however, that all amino acids were not equally substituted. Of the 16 amino acids that compose the ARS, 10 residues in DQB and 15 residues in DRB alleles were found to be variable. In contrast, non-ARS residues had substitution rates of 13% for DQB and 20% for DRB, clearly demonstrating an accumulation of

```
                            10         20         30         40         50
                             |          |          |          |          |
                            *  * *               **       **
SLADRB*C            RDIPPHFLHQ LKFECHFFNG TERVRLLQRN CYNGEEYVRF DSDVGEYRAV

SLADRB*D            --T------L ---------- -------E Q Y-----FL-- ----------
SLADRB*S01          --IAQ--FFM G--------- -----Y--KY L-----F--- ---L--F---
SLADRB*S02          --T-----FL G--------- --Q----E-Q Y-----F--- ----------
SLADRB*S03          --T-----FL G--------- --Q----E-Q Y-----F--- ----------
SLADRB*S04                  FM E--------- -------D-H Y-----IL-- ----------
SLADRB*S05                  FL G--------- --Q----E-Q Y-----F--- ----------
SLADRB*S06                  FL V-D------- --------KQ Y-----H--- ----------
SLADRB*S07                  YL ---------- -------E-Q Y--------- ----------
SLADRB*S08                  FL V-D------- -------E-Q Y-----H--- ----------
SLADRB*S09                  LL V-S------- -----FME-H Y-----H--- -----D----
DRB1*03                  --FFM G-S------- -----Y-LKY L-----F--- ---L----E-
DRB1*05                  ----L ---------- -------E-Q Y-----FL-- ----------

                            60         70         80         90
                             |          |          |          |
                          *   *      *  * *  *   *   *   *
SLADRB*C            TELGRPDAKY RNSQKDLLEQ RRAAVDTYCR HNYRILDTFL VPRR

SLADRB*D            ---------D W--------- ---E------ ---------- ----
SLADRB*S01          ---------- W------M-- K--V------ ---------- ----
SLADRB*S02          ---------N Y--------- ---E------ ---------- ----
SLADRB*S03          ---------N Y--------- ---E------ ----TS---- ----
SLADRB*S04          ------E--D W--R--I--- ---E------
SLADRB*S05          ---------N Y--------- ---E------
SLADRB*S06          ---------- W--------- M---------
SLADRB*S07          ------V--D W--------- ---E------
SLADRB*S08          ---------- W-----I--- K--E---V--
SLADRB*S09          ------S--- W-----I--E K--E------
DRB1*03             ---------- W--------- --EK------ ---GVS-S
DRB1*05             ---------D W--------- ---E------ --------
```

Figure 6 Amino acid sequence comparison of SLADRB alleles. SLA-DRB amino acid sequences at the β1 domain encoding the putative antigen recognition sequence were deduced from cDNA sequences. Data for alleles *S01 to *S09 (Shia et al., 1995) and for alleles DRB1*03 and DRB1*05 (Våge et al., 1994) were obtained from PCR amplified SLA-DRB products. Alleles *C and *D were obtained from Gustafsson et al., 1990a. Asterisks (*) indicate the position of the ARS residues. Amino acid numbering is as published for SLA-DRB*C beginning with amino acid 1 of the mature protein.

polymorphism within the ARS. In addition, the ARS residues had a greater rate of nonconservative substitutions. Similar findings have been observed for class II alleles from other species (Sigurdardottir et al., 1992).

A porcine DRB pseudolocus designated DRB2 was detected by PCR amplification of genomic DNA using primers specific for the expressed DRB locus (Våge et al., 1994). The pseudogene contained a single base deletion in codon 53 and was detected in all 10 unrelated Norwegian Landrace tested. Although DNA sequence analysis identified three DRB2 alleles, the amino acid variation at the β1 domain was much less than the expressed DRB1 locus, reflecting the lack of overdominant selection at this pseudolocus. In contrast, a separate study using different primers to amplify SLA-DRB genes did not detect the pseudolocus using RT-PCR to amplify either mRNA transcripts or genomic DNA (Shia et al., 1995). This discrepancy likely is due to the specificity of the PCR primers used in each study. RT-PCR studies also suggested that only one SLA-DQB is expressed (Shia et al., 1995). In summary, it is still a question as to how many SLA class II loci exist or are expressed by immunocytes; the number of SLA class II genes may indeed vary depending on the actual SLA haplotype analyzed.

5. *Breed Frequencies for SLA Class II Alleles*

The PCR-RFLP method for genotyping DRB and DQB loci has been used to show that all ten of the breeds tested expressed multiple PCR-RFLP pattern types and approximately 20% of the animals were heterozygous for at least one restriction pattern (Shia et al., 1995). Using a pair of restriction enzymes, several PCR-RFLP pattern types were observed at high frequencies, but this may be a factor of the disparate numbers of animals tested for each breed. Importantly, despite the small number of animals tested for some breeds, no breed-specific PCR-RFLP patterns were observed. In addition, the PCR-RFLP method for genotyping DQB and DRB loci (Shia et al., 1995) was used to type a Meishan × Yorkshire reference family which contains 300 F2 progeny. RFLP patterns showed that genotypes were inherited in a Mendelian fashion and no recombinants were observed between DRB and DQB loci. These findings suggest a close linkage of the DRB and DQB loci on swine chromosome 7.

6. *Cross-Species Comparisons of Class II Alleles*

A phylogenetic comparison of the nucleic acid sequences for the β1 domains of class II Mhc molecules (Figures 7, 8, and 9) showed that individual SLA-DRB and SLA-DQB alleles are more closely related to each other than they are to specific alleles in other species. This is in contrast to several bovine DQB alleles which appear to be more closely related to ovine DQB alleles. In addition, SLA-DRB alleles are more closely related to ovine and bovine DRB alleles, whereas SLA-DQB alleles were more closely related to HLA-DQB1 alleles. These data suggest that human DRB alleles have undergone a higher rate of mutation compared to porcine alleles. It must be noted, however, that only a few HLA-DRB and HLA-DQB alleles were included in this analysis and the inclusion of more alleles might demonstrate slightly different relatedness patterns.

Comparison of SLA-DRB and SLA-DQB nucleotide sequences for the β1 domain showed that DQB alleles were 87% similar with each other and were 72 to 75% similar to HLA-DQB1 alleles. This is consistent with the 84 to 100% similarity observed between HLA-DQB1 alleles. Similarly, SLA-DRB alleles shared 85% similarity and were 62 to 72% similar to HLA-DRB alleles. SLA-DQB and SLA-DRB alleles shared only 61% similarity. Figure 9 shows the comparison of porcine and human DRB alleles.

C. SLA Class III Region

1. *Genes*

The SLA class III region has been extensively analyzed by Chardon and colleagues. The steroid 21-hydroxylase (CYP21) and tumor necrosis factor (TNF) genes

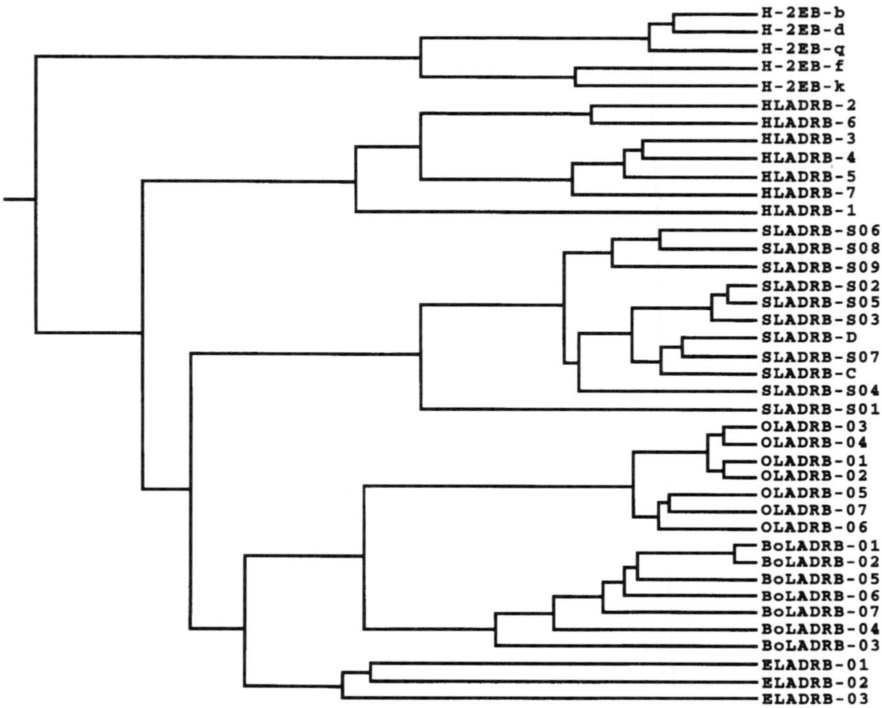

Figure 7 Dendogram of murine, human, porcine, ovine, bovine, and equine representative DRB alleles. The dendogram was constructed using the Pileup program of Genetics Computer Group. Distances along the vertical axis are proportional to the difference between sequences. Horizontal distances have no significance. Although the dendogram is not a phylogenetical tree, vertical branch lengths are proportional to the similarity between sequences.

have been cloned and, in addition to the complement Bf, C2 and C4 genes have each been mapped to the SLA complex (Chardon et al., 1988a, 1991; Geffrotin et al., 1990, 1991; Lie et al., 1988; Kuhnert et al., 1991; Choi et al., 1991). Three heat shock protein (HSP70) genes map to this region (Rouhenen-Lehto et al., 1993). More recently, the opposite strand gene (OSG), now designated Tenascin-X (TNX) because of its role as an extracellular matrix glycoprotein, was mapped to this region (Burghelle-Mayeur et al., 1992; Geffrotin et al., 1995).

Studies in humans and mice have shown that polymorphisms at several class III loci are associated with specific immune parameters. For example, a polymorphism at the first intron of the human tumor necrosis factor-β (TNF-β) gene correlates with a reduced level of TNF-β production (Bettinotti et al., 1993), and a polymorphism within the TNF-α promoter region is strongly associated with HLA haplotypes correlating with a wide range of autoimmune diseases (Wilson et al., 1993). Although many of the human class III homologs have been identified in pigs, little information is presently available concerning the presence of multiple alleles or their possible association with specific traits. However, Geffrotin et al. (1991) used RFLP analysis of 31 unrelated, SLA-typed pigs to demonstrate the existence of at least 6 allelic

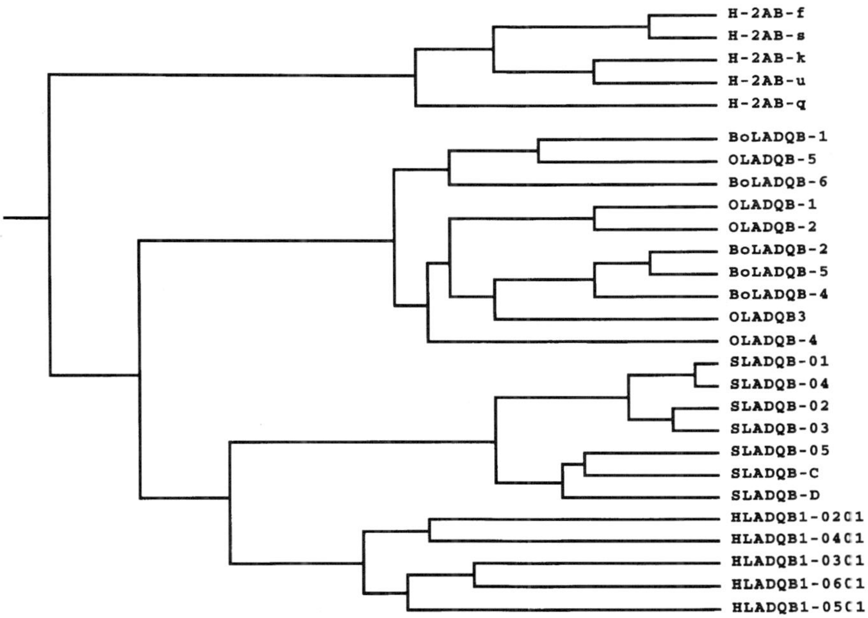

Figure 8 Dendogram of murine, human, porcine, ovine and bovine, representative DQB alleles. The dendogram was constructed using the Pileup program of Genetics Computer Group. Distances along the vertical axis are proportional to the difference between sequences. Horizontal distances have no significance. Although the dendogram is not a phylogenetical tree, vertical branch lengths are proportional to the similarity between sequences.

patterns for the CYP21 gene. Nunes et al. (1994) mapped three TNF genes into the class III region and later Looft et al. (1993) reported the isolation of a microsatellite complex downstream from the porcine TNF-α gene. DNA from four breeds of SLA class I-typed pigs yielded four microsatellite polymorphisms, whereas TNF-α RFLP typing identified only two alleles. Linkage disequilibrium between the microsatellite alleles and class I was observed. Importantly, breed-specific polymorphisms were observed, allowing distinction between the Meishan and Large White breeds. However, as with RFLP analyses, it is unknown whether the microsatellite alleles have any effect on TNF-α regulation or are associated with specific traits.

The BAT1 region was shown to be very rich in repeated sequences at the 5′-region and within introns (Peelman et al., 1995). It is uncertain whether the repeats serve a functional property, although they may be the site of a recombination hot spot (Peelman et al., 1995). RFLP analysis using eight different enzymes revealed several bands depending on the enzyme, but the bands were always monomorphic. Northern blots of RNA prepared from various tissues and probed with overlapping cosmid clones from the SLA class III region showed that most tissues expressed variable levels of several distinct mRNA bands and that peripheral blood leukocytes expressed most of the mRNA species observed (Nunes et al., 1994). This gene has now been determined to encode a nuclear RNA helicase that is evolutionarily highly conserved (Peelman et al., 1995).

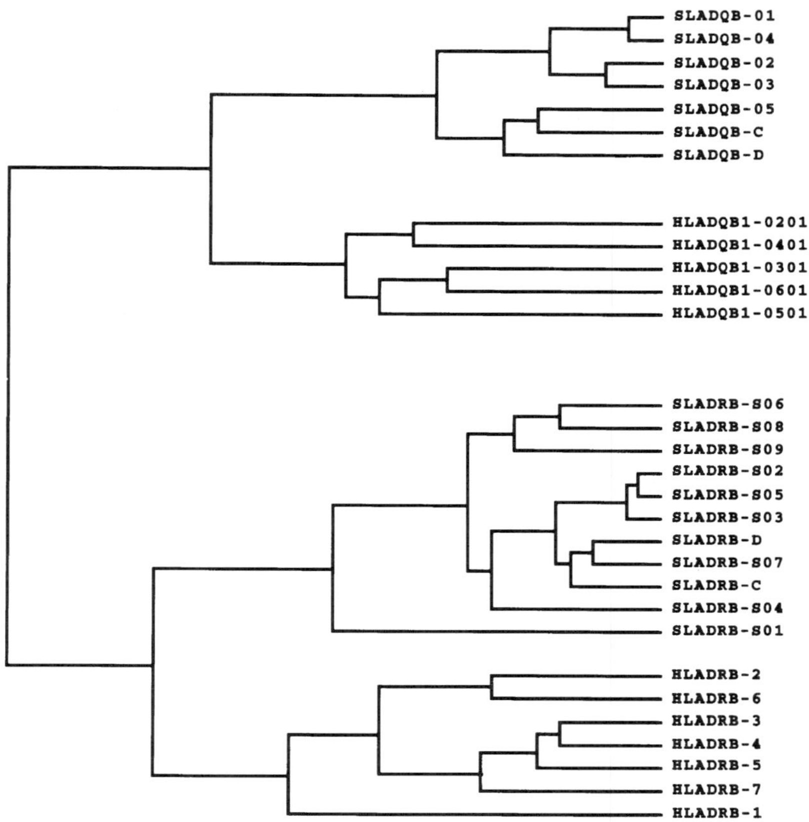

Figure 9 Dendogram comparison of human and porcine DRB and DQB alleles. The dendogram was constructed using the Pileup program of Genetics Computer Group. Distances along the vertical axis are proportional to the difference between sequences. Horizontal distances have no significance. Although the dendogram is not a phylogenetical tree, vertical branch lengths are proportional to the similarity between sequences.

2. Regional Map

A lambda clone was found to contain the porcine Bf gene and 6 exons of the complement factor 2 gene, demonstrating their close linkage within the swine SLA (Peelman et al., 1996). Overlapping cosmid clones containing genomic DNA from the swine SLA class III region have been obtained and hybridized with homologous and heterologous probes to identify porcine sequences (Nunes et al., 1994). The TNF cluster was found to contain lymphotoxin-β, BAT1, BAT2, BAT3, and NG, a sequence related to DNA binding factors which regulate proinflammatory cytokine gene expression; however, a pig B144-like locus has not been identified. In addition, a pig sequence corresponding to the human G7b locus was not detected between the HSP70 and the BAT6 genes. Nonetheless, the data highlight a remarkable conservation of the class III region between humans and pigs and suggests that allelic polymorphisms in class III genes of the pig may similarly impact host immune processes. After extensive investigations using overlapping cosmid clones, evidence suggests that there is only one gene each for CYP21, C2 and C4, even though other species have two or more CYP21 and C4 genes (Geffrotin et al., 1990, 1991; Xu et al., 1993; Chardon et al., personal communication). The order of genes within this region has recently been confirmed by Wu et al. (1995) using PFGE based techniques.

III. Mhc ASSOCIATIONS WITH SPECIFIC TRAITS

Numerous studies in humans have shown significant associations between specific Mhc alleles and predisposition for a wide variety of autoimmune and infectious diseases, including multiple sclerosis (Cullen et al., 1991), ankylosing spondylitis (Hammer et al., 1990), and rheumatoid arthritis (Altmann et al., 1991). These studies show that the inheritance of a particular allele increases the risk for various autoimmune diseases. Similar studies have linked disease susceptibility or immune status in domestic animal species to the inheritance of specific Mhc haplotypes (Lewin et al., 1991; Lunney and Grimm, 1994). In addition, several studies have suggested that some SLA haplotypes may influence carcass composition and growth rate performance. Thus, SLA haplotype impacts directly on both porcine health and development. However, the association of specific SLA haplotypes with a given phenotype is often frustrated by the findings that associations often differ with breed or age of the animals tested. Moreover, the magnitude of estimates of Mhc effects depends on the method of data analysis (Mallard et al., 1991).

A. SLA and Immune Response Associations

Studies assessing SLA expression after virus inoculation have implicated modulation of expression of SLA class I and class II antigens during infection as likely influencing both humoral and cellular anti-viral immune responses (Gonzalez Juarrero et al., 1992 a,b; Chinsakchai and Molitor, 1994). The potential role of interferon-gamma in this protective immune response is implied by the observation of up-regulation of SLA antigen expression only in animals which are likely to survive African swine fever virus infection (Gonzalez Juarrero et al., 1992 a,b). The protective role of class I restricted anti-viral specific cytotoxic T cells has yet to be confirmed (Martins et al., 1993; Pauly et al., submitted); high levels of natural killer-type cells may also influence viral recovery (Martins and Leitao, 1994).

T cells from responder pigs, determined to mediate enhanced destruction of encysted *Trichinella* larvae, did not express more SLA class I or II antigens when compared to T cells from nonresponder pigs (Dillender and Lunney, 1993). Similarly, the selection of pigs for high antibody response to hen egg white lysozyme and lymphocyte proliferation did not alter monocyte SLA-DQ or SLA-DR expression (Groves et al., 1993).

Several studies have examined the association of specific SLA haplotype on immune parameters (Table 3), in part for use as markers for indirect selection of disease-resistant animals. However, it is difficult to distinguish between the SLA

Table 3 Immune Parameters Associated with Specific SLA Haplotypes

Immune parameter	Breed	SLA association	Ref.
Serum Immunoglobulins			
IgG	NIH minipigs	Higher in SLA class II *d* haplotype	Mallard et al., 1989a
IgM	NIH minipigs	None significant	Mallard et al., 1989a
IgA	German breeds	Higher in H4	Buschmann et al., 1985
Specific antibody production			
Anti-(T,G)-A—L	NIH minipigs	Higher in H10, lower in H4	Lunney et al., 1986
	NIH minipigs	Higher in SLA class II *d* haplotype	Mallard et al., 1989b
Anti-lysozyme	NIH minipigs	Higher in H10 and H4	Lunney et al., 1986
	Large Whites	Higher in H12, lower in H10	Vaiman et al., 1978
Anti-sheep RBC	NIH minipigs	Higher in SLA class II *d* haplotype	Mallard et al., 1989b
Anti-*B. bronchiseptica*	Various breeds	Higher in H10	Rothschild et al., 1984a
Anti-*Salmonella*	NIH minipigs	Higher in *aa*, *ac* haplotypes	Lumsden et al., 1993
Anti-PRV			Rothschild et al., 1984b
Antibody avidity	NIH minipigs	None significant	Appleyard et al., 1992
DTH response	NIH minipigs	Higher in SLA class II *d* haplotype	Mallard et al., 1989c
Bacterial phagocytosis	NIH minipigs	Higher in *ad*, *aa* haplotypes	Lacey et al., 1989
Bactericidal activity	NIH minipigs	Lower in *ad*, *aa* haplotypes	Lacey et al., 1989
T. spiralis clearance	NIH minipigs	Higher in *aa* haplotype	Madden et al., 1990, 1993
Response to *T. spiralis*	NIH minipigs	Lower burden in *cc* haplotype	Lunney and Murrell, 1988
Macrophage O_2 production	Inbred Yorkshires	None with class II haplotype	Groves et al., 1993
Melanoblastoma appearance	Hybrid minipigs	Associated with SLA class I L9.10	Hruban et al., 1994
	Sinclair Swine	Associated with SLA *B* haplotype	Tissot et al., 1989
Lymphocyte proliferation	NIH minipigs	Lower in SLA class II *d* haplotype	Lumsden et al., 1993
Serum hemolytic complement	NIH minipigs	Not significant	Mallard et al., 1989b
	Large Whites	Higher in H10, lower in H12	Vaiman et al., 1978

and non-SLA effects, thus requiring multivariable analyses to estimate the contribution of the SLA haplotype on the parameter measured. SLA haplotype affects complement-dependent hemolytic activity (Lie et al., 1988a), presumably due to class III loci encoding complement factors. Indeed, complement factors 2, 4, and Bf map to the class III region of the SLA and a polymorphic band for complement factor 4 cosegregated with one SLA haplotype (Kirszenbaum et al., 1985). Mallard et al. (1989a) observed that genetic factors accounted for 56 to 62% of the variation observed for serum IgG levels in 8-week-old NIH miniature pigs, and that pigs of the *dd*, *dg*, and *gg* haplotypes were associated with higher serum IgG levels. Pigs of the *dd*, *dg*, and *gg* haplotypes also produced more antibody against sheep red blood cells, hen egg white lysozyme, and a purified tuberculin protein derivative (Mallard et al., 1989c). Pigs of these haplotypes all have class II genes of the *d*-haplotype, confirming the likely importance of class II genes in regulating immune responses. Importantly, heritability estimates for serum IgG concentration were similar to those reported in other species and suggests that selection for higher serum IgG is feasible. However, effects of the sire and dam were greater than effects of SLA haplotype, and serum IgM levels were not associated with SLA. In addition, SLA had no significant effect on antibody avidity (Appleyard et al., 1992). Thus, the impact of SLA haplotype on antibody production appears to be very limited.

Antibody titers to *Bordetella bronchiseptica* and to a pseudorabies vaccine were higher in pigs that reacted with anti-SLA^{a} haplotype alloantisera, but again dam and breed effects were larger (Rothschild et al., 1984a). Moreover, SLA did not associate with higher agglutinating antibody titers, either prevaccination or following vaccination with *Salmonella typhimurium* (Lumsden et al., 1993). Although the effect of SLA on lymphocyte proliferation was significant only after secondary vaccination, pigs of the class II^{d} haplotype also had significantly lower prevaccination lymphocyte proliferation and at 6 days after a second vaccination.

When monocyte bactericidal activity was measured, specific SLA haplotypes were significantly associated with more efficient killing of *Staphylococcus aureus* or *S. typhimurium*, but associations of specific haplotypes to killing changed as a function of the age of the pig (Lacey et al., 1989). Age and litter, but not sire, also had significant effects on bactericidal activity. Although a dominant non-Mhc locus designated *Ity* or *Bcg* significantly influences intracellular killing of pathogens by macrophages, the model of Lacey et al. (1989) accounted for 78 to 88% of the phenotypic variation for bactericidal activity. Thus, SLA haplotypes may help determine the relative bactericidal performance of host macrophages, either directly or via close linkage with additional loci.

Studies with outbred swine have implicated an antibody-mediated protective immunity against newborn *Trichinella spiralis* larvae (Marti and Murrell, 1986). Homozygous *cc* NIH miniature pigs developed 50% fewer muscle larvae compared to homozygous *dd* or *aa* pigs (Lunney and Murrell, 1988), an effect that correlated with earlier serum antibody reactivity and heightened lymphocyte blastogenesis to *T. spiralis* antigens. Although pigs of all three SLA genotypes were protected by prior exposure, homozygous *aa* pigs exhibited a significant reduction in encysted muscle larvae after *T. spiralis* rechallenge compared to homozygous *cc* or *dd* pigs (Madden et al., 1990). Madden et al. (1993) also observed that one copy of the *a* haplotype was associated with reduced larval burden, but was not sufficient for manifestation of the responder phenotype. These data indicated that a second locus appears to be required, but the mechanisms mediating enhanced secondary anti-encysted muscle larvae reactivity are unknown. No correlation was observed between enhanced larval clearance and *T. spiralis*-induced lymphocyte blastogenesis, cell expression of class I or II molecules, or parasite fecundity (Madden et al., 1993).

Taken together, these studies clearly show that specific SLA haplotypes can affect various host defense parameters. Indeed, SLA haplotype *B* in Sinclair Swine is associated with the expression of cutaneous malignant melanoma (Tissot et al., 1989). Only a single dose of the *B* haplotype is required for full penetration of the tumor initiation locus, and homozygous *B* swine develop more primary tumors than heterozygotes. Presently, the mechanism by which this SLA haplotype acts on tumor initiation in unknown. Hruban et al. (1994) also observed an association between an SLA class I allele and melanoblastoma appearance in hybrid minipigs.

B. Production Traits Associated with SLA Haplotypes

Association between Mhc and several economic traits have been indicated in several species. As shown in Table 4, the SLA region is associated with genes that affect body composition of growing pigs. The SLA haplotype H23, for example, was found to be strongly correlated with carcass leanness and weakly, but significantly, associated with ham development in female Landrace pigs (reviewed in Vaiman et al., 1988). The H4 haplotype was associated with lower fat content of the carcass in the Large White breed, but had the opposite effect on Landrace carcasses. Significant associations between SLA haplotypes and the birth and weaning weights of piglets also has been reported (Rothschild et al., 1986a), but studies using miniature pigs with defined SLA genotypes showed that litter, but not SLA, had significant effects on both birth and weaning weight (Mallard et al., 1991). In Durocs, a specific SLA class I RFLP band was associated with a decreased selection index and lower average daily weight gain, whereas two other RFLP fragments were associated with increased backfat thickness (Jung et al., 1989a). The same study reported the association of two RFLP bands with an increase in the selection index in Hampshires. Vaiman et al. (1988) similarly observed that SLA haplotype H6 in the Large White breed significantly decreased average daily weight gain. However, these studies could not account for the genetic influence of the sires and dams and individual haplotypes could not be assigned. Recently, Liu et al. (1995) reexamined several growth and production traits for associations with five chromosome 7 specific microsatellite markers. Their data identified an SLA associated QTL for weight and one for levels of backfat.

Reports also have suggested that Mhc genes are associated with reproductive performance. In pigs, matings between certain SLA haplotypes were associated with a decreased number of piglets (Vaiman and Renard, 1980). In miniature pigs, the

Table 4 Production Traits Associated with Specific SLA Haplotypes

Production trait	Breed	SLA association	Ref.
Growth Index	Durocs	Decreased with one class I RFLP	Jung et al., 1989a
	Hampshires	Increased with two class I RFLPs	Jung et al., 1989a
Average daily weight gain	Large White	Decreased in H6 haplotype	Vaiman et al., 1988
Carcass leaness	French Landrace	Improved in H23 haplotype	Vaiman et al., 1988
	Large White	Improved in H4 haplotype	Vaiman et al., 1988
Ham development	French Landrace	Greater in H23 haplotype	Vaiman et al., 1988
	Belgian Landrace	Decreased in H1 haplotype	Vaiman et al., 1988
Backfat thickness	Large White	Decreased in H4 haplotype	Vaiman et al., 1988
	Belgian Landrace	Increased in H1 haplotype	Vaiman et al., 1988
Litter size at birth	Large White	Positive association with H6	Renard et al., 1985
	NIH minipigs	Effect of sow and sire haplotypes	Mallard et al., 1987
Litter size at weaning	NIH minipigs	Effect of sow haplotype	Mallard et al., 1987
Ovulation rate	Synthetic line	Increased with H10 haplotype	Rothschild et al., 1985
		Decreased with H4 haplotype	Rothschild et al., 1985
	NIH minipigs	No effect of *ad*, *ac*, or *dd* genotypes	Mallard et al., 1987
Birth weight	NIH minipigs	No effect of seven genotypes	Mallard et al., 1991
	Large White	Increased with H12 haplotype	Rothschild et al., 1986a
Weaning weight	NIH minipigs	No effect of seven genotypes	Mallard et al., 1991
	Large White	Increased with H12 haplotype	Rothschild et al., 1986a
Male genital tract development	Large White	Increased in H15, H16 haplotypes	Rothschild et al., 1986b
		Decreased in H4 haplotype	Rothschild et al., 1986b

largest litters with good postnatal survival were from *dd* genotype sows and sires (Mallard et al., 1989c). Interestingly, *dd*-weaned pigs also had superior antibody response to sheep erythrocytes, chicken lysozyme, and tuberculin test antigens (Mallard et al., 1987). Thus, SLA haplotypes may impact piglet survival due to enhanced immune parameters. In general, though, it appears that litter size at birth and weaning are only slightly influenced by SLA haplotype (reviewed in Vaiman et al., 1988). One exception may be for sows carrying the SLA H6 haplotype which were found to consistently produce larger litters. This haplotype, however, appears to negatively influence growth performance in males (Table 4).

The presence of SLA haplotype(s) serologically related to the miniature pig *a* haplotype was associated with an increased ovulation rate (Rothschild et al., 1985). Results from a French herd showed that sows with the haplotype H10, which also resembles the miniature pig *a* haplotype, had higher ovulation rates. Testis weight in boars has been found to correlate with the ovulation rate of closely related females (Schnickel et al., 1983), and two haplotypes had significant positive effects on the development of the male genital tract (Rothschild et al., 1986b).

Mating failures and litter size were not influenced by the SLA (Riezebos et al., 1991; reviewed by Vaiman et al., 1988). However, a small but significant association between sows homozygous for SLA haplotype H16 and an increased number of mummified births compared has been observed (Riezebos et al., 1991). Interestingly, matings involving identical haplotypes occasionally gave smaller litters than expected, and litter size reduction was combined with a significant reduction in the expected number of SLA homozygous piglets (Vaiman et al., 1988). A higher mortality in homozygous SLA H2 haplotype embryos was determined to be the most likely mechanism accounting for loss of homozygous piglets and accounted completely for the reduced litter size. The increased percentage of embryo loss for H2 homozygotes was not apparent until day 30 of development. It must be noted, however, that, although the differences between expected haplotype frequencies and observed haplotype frequencies can be attributed to the loss of SLA homozygotes, the data show that only specific haplotypes are detrimental (Mallard et al., 1987).

IV. FUTURE PERSPECTIVES

Although tremendous progress has been realized during the past 5 years with respect to the molecular organization and molecules of the SLA, several issues still remain to be resolved. These include determining (1) the number of class I and class II loci which are encoded and expressed for the major representative SLA haplotypes; (2) delineating the tissue expression patterns of class I loci and whether their expression is associated with $\alpha\beta$ or $\gamma\delta$ T cell receptor utilization in swine; (3) identification of recombination "hot spots" or suppression within and between class I, III, and class II SLA regions; and (4) continued development of both serological and molecular reagents for characterization of class I and class II loci and molecules. In particular, progress toward the development of immunologic and molecular reagents to type for class I molecules has been slow and has limited studies in determining associations between SLA haplotypes with disease and other economically important traits. Nonetheless, the data summarized in Tables 3 and 4 indicate that the selection of animals bearing specific SLA haplotypes may improve disease resistance, reproductive performance, and production traits in domestic swine breeds.

ACKNOWLEDGMENTS

Work presented in this report has been supported by the National Pork Producers Council, The Biotechnology Research and Development Corporation, The University of Minnesota Agriculture Experiment Station, The University of Illinois Agriculture Experiment Station, the Agricultural Research Service-U.S.D.A., and the National Research Initiatives, Competitive Research Grants Program, U.S.D.A. (93-37205-9345). The authors also wish to express their appreciation to Drs. D. Sachs, P. Chardon, M. Vaiman, B. Mallard, M. Amoss, and C. Beattie for sharing unpublished information and personal communications.

REFERENCES

Altmann, D. M., Sansom, D., and Marsh, S. G. E., What is the basis for HLA-DQ associations with autoimmune disease?, *Immunol. Today*, 12, 267, 1991.

Appleyard, G. D., Mallard, B. A., Kennedy, B. W., and Wilkie, B. N., Antibody avidity in swine lymphocyte antigen-defined miniature pigs, *Can. J. Vet. Res.*, 56, 303, 1992.

Archibald, A. L., Haley, C. S., Brown, J. F., Couperwhite, S., McQueen, H. A., Nicholson, D., Coppieters, W., van de Weghe, A., Stratil, A., and Wintero, A. K., The PigMaP consortium linkage map of the pig (*Sus scrofa*), *Mammal., Genome*, 6, 157, 1995.

Bettinotti, M. P., Hartung, K., Deicher, H., Messer, G., Keller, E., Weiss, E. H., and Albert, E. D., Polymorphism at the tumor necrosis factor beta gene in systemic lupus erythematosis: TNFB-Mhc haplotypes, *Immunogenetics*, 37, 449, 1993.

Brown, J. H., Jardetzky, T., Saper, M. A., Samraoui, B., Bjorkman, P. J., and Wiley, D. C., A hypothetical model of the foreign antigen-binding site of class II histocompatibility molecules, *Nature (London)*, 332, 845, 1988.

Burghelle-Mayeur, C., Geffrotin, C., and Vaiman, M., Sequences of the swine 21-hydroxylase gene (CYP21) and a portion of the opposite strand overlapping gene of unknown function previously described in human, *Biochim. Biophys. Acta*, 1171, 153, 1992.

Buschmann, H., Krausslich, H., Hermann, H., Meyer, J., and Kleinschmidt, A., Quantitative immunological paramenters in pigs — experiences with the evaluation of an immunocompetence profile, *Z. Tierzuhtg. Zuchtgsbiol.*, 102, 189, 1985.

Campbell, R. D. and Trowsdale, J., Map of the human Mhc, *Immunol. Today*, 14, 349, 1993.

Chardon, P., Geffrotin, C., and Vaiman, M., Genetic organization of the SLA complex, in *The Molecular Biology of the Major Histocompatibility Complex of Domestic Animal Species*, Warner, C. M., Rothschild, M. F., and Lamont, S. J., Eds., Iowa State University Press, Ames, 1988, 63.

Chardon, P., Peelman, L., Nunes, M., Renard, C., and Vaiman, M., The RNA-helicase-related porcine BAT1 sequence, *Anim. Genet.*, in press.

Chardon, P., Renard, C., Kirszenbaum, M., Geffrotin, C., Cohen, D., and Vaiman, M., Molecular genetic analyzes of the major histocompatibility complex in pig families and recombinants, *J. Immunogenet.,* 12, 139, 1985a.

Chardon, P., Renard, C., and Vaiman, M., Characterization of class II histocompatibility antigens in pigs, *Anim. Blood Groups Biochem. Genet.*, 12, 59, 1981.

Chardon, P., Vaiman, M., Kirszenbaum, M., Geffrotin, C., Renard, C., and Cohen, D., Restriction fragment length polymorphism of the major histocompatibility complex of the pig, *Immunogenetics,* 21, 161, 1985b.

Chinsackchai, S. and Molitor, T. W., Immunobiology of pseudorabies infection in swine, *Vet. Immunol. Immunopathol.*, 43, 107, 1994.

Choi, C. S., Molitor, T. W., Lin, G. F., and Murtaugh, M. P., Complete nucleotide sequence of a cDNA encoding porcine tumor necrosis factor-alpha, *Anim. Biotechnol.*, 2, 97, 1991.

Cullen, C. G., Middleton, D., Savage, D. A., and Hawkins, S., HLA-DR and -DQ DNA genotyping in multiple sclerosis patients in Northern Ireland, *Hum. Immunol.*, 30, 1, 1991.

Davis, W. C., Marusic, S., Lewin, H. A., Splitter, G. A., Perryman, L. E., McGuire, T. C., and Gorham, J. R., The development and analysis of species specific and cross reactive monoclonal antibodies to leukocyte differentiation antigens and antigens of the major histocompatibility complex for use in the study of the immune system in cattle and other species, *Vet. Immunol. Immunopathol.*, 15, 337, 1987.

Davis, W. C., Perryman, L. E., and McGuire, T. C., The identification and analysis of major functional populations of differentiated cells, in *Hybridoma Technology in Agriculture and Veterinary Research*, Stern, N. J. and Gamble, H. R., Eds., NJ Rowman and Allanheld, Totowa, NJ, 1984, 121.

Dillender, M. J. and Lunney, J. K., Characteristics of T lymphocyte cell lines established from NIH minipigs challenge inoculated with *Trichinella spiralis*, *Vet. Immunol. Immunopathol.*, 35, 301, 1993.

Edfors-Lilja, I., Ellegren, H., Wintero, A. K., Ruohonen-Lehto, M., Fredholm, M., Gustafsson, U., Juneja, R. K., and Andersson, L., A large linkage group on pig chromosome 7 including the Mhc class I, class II (DQB), and class III (TNFB) genes, *Immunogenetics,* 38, 363, 1993.

Ehrlich, R., Lifshitz, R., Pescovitz, M. D., Rudikoff, S., and Singer, D. S., Tissue specific expression and structure of a divergent member of a class I Mhc gene family, *J. Immunol.*, 139, 593, 1987.

Emery, D. W., Shafer, G. E., Karson, E. M., Sachs, D. H., and LeGuern, C., Retrovirus-mediated transfer and expression of an allogeneic major histocompatibility complex class II DRB cDNA in swine bone marrow cultures, *Blood*, 81, 2460, 1993.

Flanagan, M., Jung, Y. C., Rothschild, M. F., and Warner, C. M., RFLP analysis of SLA class I genotypes in Duroc swine, *Immunogenetics,* 27, 465, 1988.

Gaycken, U., Shabahang, M., Meyer, J. N., and Glodek, P., Molecular characterization of the porcine Mhc class I region, *Anim. Genet.*, 25, 357, 1994.

Geffrotin, C., Chardon, P., de Andres-Cara, D. F., Feil, R., Renard, C., and Vaiman, M., The swine steroid 21-hydroxylase gene (CYP21): cloning and mapping within the swine leucocyte antigen complex, *Anim. Genet.,* 21, 1, 1990.

Geffrotin, C., Garido, J., Cleriot, C., and Vaiman, M., Study of the transcription of the major histocompatibility complex-class III tenascin-X gene: tissue distribution in 87-day-old fetuses and adults, *Anim. Genet.,* 25, Suppl. 2, 18, 1994.

Geffrotin, C., Renard, C., Chardon, P., and Vaiman, M., Marked genetic polymorphism of the swine steroid 21-hydroxylase gene, and its location between the SLA class I and class II regions, *Anim. Genet.,* 22, 311, 1991.

Gonzalez-Juarrero, M., Lunney, J. K., Sanchez-Vizcaino, J. M., and Mebus, C., Modulation of splenic macrophages, and swine leukocyte antigen (SLA) and viral antigen expression following African swine fever virus (ASFV) inoculation, *Arch. Virol.*, 123, 145, 1992a.

Gonzalez-Juarrero, M., Mebus, C. A., Pan, R., Revilla, Y., Alonso, J. M., and Lunney, J. K., Swine leukocyte antigen and macrophage marker expression on both African swine fever virus-infected and non-infected primary porcine macrophage cultures, [published erratum appears in *Vet. Immunol. Immunopathol.*, 34(3–4), 391, 1992], *Vet. Immunol. Immunopathol.*, 32, 243, 1992b.

Groves, T. C., Wilkie, B. N., Kennedy, B. W., and Mallard, B. A., Effect of selection of swine for high and low immune responsiveness on monocyte superoxide anion production and class II Mhc antigen expression, *Vet. Immunol. Immunopathol.*, 36, 347, 1993.

Gustafsson, K., Germana, S., Hirsch, F., Pratt, K., LeGuern, C., and Sachs, D. H., Structure of miniature swine class II DRB genes: conservation of hypervariable amino acid residues between distantly related mammalian species, *Proc. Natl. Acad. Sci. U.S.A.*, 87, 9798, 1990a.

Gustafsson, K., LeGuern, C., Hirsch, F., Germana, S., Pratt, K., and Sachs, D. H., Class II genes of miniature swine. IV. Characterization and expression of two allelic class II DQB cDNA clones, *J. Immunol.*, 145, 1946, 1990b.

Hammer, R. E., Majka, S. D., Richardson, J. A., Tang, J.-P., and Traurog, J. D., Spontaneous inflammatory disease in transgenic rates expressing HLA-B27 and human β_2m: an animal model of HLA-B27-associated human disorders, *Cell*, 63, 1099, 1990.

Hammerburg, C. and Schurig, G., Characterization of monoclonal antibodies directed against swine leukocytes, *Vet. Immunol. Immunopathol.*, 11, 107, 1986.

Hirsch, F., Germana, S., Gustafsson, K., Pratt, K., Sachs, D. H., and Leguern, C., Structure and expression of class II alpha genes in miniature swine, *J. Immunol.*, 149, 841, 1992.

Hirsch, F., Sachs, D. H., Gustafsson, K., Pratt, K., Germana, S., and LeGuern, C., Class II genes of miniature swine. III. Characterization of an expressed pig class II gene homologous to HLA-DQA, *Immunogenetics,* 31, 52, 1990.

Hradecky, J., Hruban, V., Hojny, J., Pazdera, J., and Stanck, R., Development of a semi-inbred line of Landrace pigs. I. Breeding performance and immunogenetic characteristics, *Lab. Anim.*, 19, 279, 1985.

Hruban, V., Horak, V., and Fortyn, K., Presence of specific Mhc haplotype in melanoblastoma-bearing minipigs, *Anim. Genet.,* 25 (Suppl. 2), C10, 1994.

Ivanoska, D., Sun, D. C., and Lunney, J. K., Production of monoclonal antibodies reactive with polymorphic and monomorphic determinants of SLA class I gene products, *Immunogenetics,* 33, 220, 1991.

Jung, Y. C., Rothschild, M. F., Flanagan, M. P., Christian, L. L., and Warner, C. M., Association of restriction fragment length polymorphisms of swine leucocyte antigen class I genes with production traits of Duroc and Hampshire boars, *Anim. Genet.,* 20, 79, 1989b.

Jung, Y. C., Rothschild, M. E., Flanagan, M. P., Pollack, E., and Warner, C. M., Restriction fragment length polymorphisms in the swine major histocompatibility complex, *Theor. Appl. Genet.,* 77, 271, 1989a.

Kenmochi, T., Mullen, Y., Miyamoto, M., and Stein, E., Swine as an allotransplantation model, *Vet. Immunol. Immunopathol.*, 43, 177, 1994.

Kirszenbaum, M., Renard, C., Geffrotin, C., Chardon, P., and Vaiman, M., Evidence for mapping pig C4 gene(s) within the pig major histocompatibility complex (SLA), *Anim. Blood Grps. Biochem Genet.*, 16, 65, 1985.

Kuhnert, P., Wuthrich, C., Peterhans, E., and Pauli, U., The porcine tumor necrosis factor-encoding genes: sequence and comparative analysis, *Gene*, 102, 171, 1991.

Kristensen, B., Renard, C., Ostergard, H., and Lunney, J. K., Reactivity of murine monoclonal anti-SLA class I antibodies on cells from outbred swine, *Anim. Genet.,* 23, S41, 1993.

Lacey, C., Wilkie, B. N., Kennedy, B. W., and Mallard, B. A., Genetic and other effects on bacterial phagocytosis and killing by cultured peripheral blood monocytes of SLA-defined miniature pigs, *Anim. Genet.,* 20, 371, 1989.

Letesson, J.-J., Coppe, P., Lostrie, N., Creimers, R., and Depelchin, A., Production and characterization of monoclonal antibodies raised agasinst BoLA class I antigens, *Vet. Immunol. Immunopathol.*, 13, 213, 1986.

Lewin, H. A., Calvert, C. C., and Bernoco, D., Cross-reactivity of a monoclonal antibody with bovine, equine, ovine, and porcine peripheral blood B lymphocytes, *Am. J. Vet. Res.*, 46, 785, 1985.

Lewin, H. A., Clamp, P. A., Beever, J. E., and Schook, L. B., Mapping genes for resistance to infectious diseases in animals, in *Gene Mapping Techniques and Applications*, Schook, L. B., Lewin, H. A., and McLaren, D. G., Eds., Marcel Dekker, New York, 1991, 283.

Lie, W. R., Rothschild, M. F., and Warner, C. M., Mapping of C2, Bf and C4 genes to the swine major histocompatibility complex (swine leukocyte antigen), *J. Immunol.,* 139, 3388, 1987.

Lie, W.-R., Rothschild, M. F., and Warner, C. M., Preparation and characterization of murine monoclonal antibodies against swine lymphocyte antigens, *Immunology*, 64, 599, 1988b.

Liu, H. C., Rothschild, M. F., and Tuggle, C. K., Analysis of pig chromosome 7 genetic markers for growth and carcass performance traits, *J. Anim. Sci.*, in press.

Looft, C., Nunes, M., Chardon, P., Vaiman, M., and Kalm, E., Presence of microsatellite-SINE complex structures downstream of the swine tumor necrosis factor alpha gene, *Anim. Genet.,* 23 (Suppl. 1), 40, 1993.

Lumsden, J. S., Kennedy, B. W., Mallard, B. A., and Wilkie, B. N., The influence of the swine major histocompatibility genes on antibody and cell-mediated immune responses to immunization with an aromatic-dependent mutant of *Salmonella typhimurium*, *Can. J. Vet. Res.*, 57, 14, 1993.

Lunney, J. K., Characterization of swine leukocyte differentiation antigens, *Immunol. Today*, 14, 147, 1993.

Lunney, J. K., Current status of the swine leukocyte antigen complex, *Vet. Immunol. Immunopathol.*, 43, 19, 1994.

Lunney, J. K. and Grimm, D. R., Major histocompatibility complex: current state of knowledge and its use in and impact on livestock improvement, *Proceedings of the 5th World Congress on Genetics Applied to Livestock Production*, 20, 230, 1994.

Lunney, J. K. and Murrell, K. D., Immunogenetic analysis of *Trichinella spiralis* infections of swine, *Vet. Parasitol.*, 29, 179, 1988.

Lunney, J. K., Osborne, B. A., Sharrow, S. O., Devaux, C., Pierres, M., and Sachs, D. H., Sharing of Ia antigens between species. IV. Interspecies cross-reactivity of monoclonal antibodies directed against polymorphic mouse Ia determinants, *J. Immunol.*, 130, 2786, 1983.

Lunney, J. K. and Pescovitz, M. D., Phenotypic and functional characterization of pig lymphocyte populations, *Vet. Immunol. Immunopathol.*, 17, 135, 1987.

Lunney, J. K., Pescovitz, M. D., and Sachs, D. H., The swine major histocompatibility complex: its structure and function, in *Swine in Biomedical Research*, Tumbleson, M. E., Ed., Plenum Press, New York, 1986, 1821.

Lunney, J. K. and Sachs, D. H., Transplantation in miniature swine. IV. Chemical characterization of the SLA and Ia-like antigens, *J. Immunol.,* 120, 607, 1978.

Lunney, J. K., Sun, D. C., Ivanoska, D., Pescovitz, M. D., and Davis, W. C., SLA monoclonal antibodies, in *The Molecular Biology of the Major Histocompatibility Complex of Domestic Animal Species*, Warner, C. M., Rothschild, M. F., and Lamont, S. J., Eds., Iowa State University Press, Ames, 1988, 97.

Madden, K. B., Moeller, R. F., Jr., Douglass, L. W., Goldman, T., and Lunney, J. K., *Trichinella spiralis*: genetic basis and kinetics of the anti-encysted muscle larval response in miniature swine, *Exp. Parasitol.*, 77, 23, 1993.

Madden, K. B., Murrell, K. D., and Lunney, J. K., *Trichinella spiralis*: major histocompatibility complex-associated elimination of encysted muscle larvae in swine, *Exp. Parasitol.*, 70, 443, 1990.

Mallard, B. A., Kennedy, B. W., and Wilkie, B. N., The effect of swine leukocyte antigen haplotype on birth and weaning weights in miniature pigs and the role of statistical analysis in this estimation, *J. Anim. Sci.*, 69, 559, 1991.

Mallard, B. A., Wiley, D. C., Croy, B. A., Kennedy, B. W., and Friendship, R., The influence of the swine major histocompatibility complex (SLA) on reproductive traits in miniature swine, *J. Reproduc. Immunol.*, 12, 201, 1987.

Mallard, B. A., Wilkie, B. N., and Kennedy, B. W., The influence of the swine major histocompatibility genes (SLA) on variation in serum immunoglobulin (Ig) concentration, *Vet. Immunol. Immunopathol.*, 21, 139, 1989a.

Mallard, B. A., Wilkie, B. N., and Kennedy, B. W., Influence of major histocompatibility genes on serum hemolytic complement activity in miniature swine, *Am. J. Vet. Res.*, 50, 359, 1989b.

Mallard, B. A., Wilkie, B. N., and Kennedy, B. W., Genetic and other effects on antibody and cell mediated immune response in swine leucocyte antigen (SLA)-defined miniature pigs, *Anim. Genet.*, 20, 167, 1989c.

Marti, H. P. and Murrell, K. D., *Trichinella spiralis*: antifecundity and antinewborn larvae immunity in swine, *Exp. Parasitol.*, 62, 370, 1986.

Martins, C. L. V., Lawman, M. J., Scholl, T., Mebus, C. A., and Lunney, J. K., African swine fever virus specific porcine cytotoxic T cell activity, *Arch. Vir.*, 129, 211, 1993.

Martins, C. L. V. and Leitao, A., Porcine immune responses to African swine fever virus (ASFV) infection, *Vet. Immunol. Immunopathol.*, 43, 99, 1994.

Nunes, M., Peelman, L., Vaiman, M., Bourgeaux, N., and Chardon, P., Characterization of six new loci within the swine major histocompatibility complex class III region, *Mamm. Genome*, 5, 616, 1994.

Nunes, M., Yerle, M., Dezeure, F., Gellin, J., Chardon, P., and Vaiman, M., Isolation of four HSP70 genes in the pig and localization on chromosomes 7 and 14, *Mamm. Genome*, 4, 247, 1993.

Osborne, B. A., Lunney, J. K., Pennington, L. R., Sachs, D. H., and Rudikoff, S., Two dimensional gel analysis of swine histocompatibility antigens, *J. Immunol.*, 131, 2939, 1984.

Peelman, L. J., Mattheeuws, M., Van Zeveren, A., van de Weghe, A., and Bouquet, Y., Conservation of the *RD-BF-C2* organization in the pig Mhc class-III region: mapping and cloning of the pig *RD* gene, *Anim. Genet.*, 27, 35, 1996.

Peelman, L. J., Chardon, P., Nunes, M., Geffrotin, C., Vaiman, M., Van Zeveren, A., Coppieters, W., van de Weghe, A., Bouquet, Y., Choy, W. W., Strominger, J. L., and Spies, T., The BAT1 gene in the Mhc encodes an evolutionarily conserved putative nuclear RNA helicase of the DEAD family, *Genomics*, 26, 210, 1995.

Peelman, L., van de Weghe, A., Coppieters, W., van Zeveren, A., and Bouquet, Y., Cloning, characterization and mapping of swine lymphocyte antigen class III genes, *Anim. Genet.*, 23(Suppl. 1), 40, 1993.

Pescovitz, M. D., Lunney, J. K., and Sachs, D. H., Preparation and characterization of monoclonal antibodies reactive with PBL, *J. Immunol.*, 133, 368, 1984.

Pescovitz, M. D., Popitz, F., Sachs, D. H., and Lunney, J. K., Expression of Ia antigens on resting porcine T cells. A marker of functional T cells, in *Advances in Gene Technology: Molecular Biology of the Immune System*, Streilein, J. W., Ed., ICSU Press, 1985, 271.

Pratt, K., Sachs, D. H., Germana, S., El-Gamil, M., Hirsch, F., Gustafsson, K., and LeGuern, C., Class II genes of miniature swine. II. Molecular identification and characterization of B (b) genes from the SLA^c haplotype, *Immunogenetics,* 31, 1, 1990.

Rabin, M., Fries, R., Singer, D., and Ruddle, F. H., Assignment of the porcine major histocompatibility complex to chromosome 7 by *in situ* hybridization, *Cytogen Cell Genet.*, 39, 206, 1985.

Renard, C., Bolet, G., Dando, P., and Vaiman, M., Relations d'un marquer genetique le complex majeur d'histocompatibilitie avec la prolificite des truies etal mortalite des porcelets, *J. Rech. Porcine en France,* 17, 105, 1985.

Renard, C., Kristensen, B., Gautschi, C., Hruban, V., Fredholm, M., and Vaiman M., Joint report of the first international comparison test on swine lymphocyte alloantigens (SLA), *Anim. Genet.,* 19, 63, 1988.

Riezebos, A. Y. D., Molenaar, B. A. J., Nieuw-Dalland, B. V., and Renard, C., Association between Mhc and litter size traits in a commercial pig line, *Anim. Genet.,* 22, 150, 1991.

Rothschild, M. F., Chen, H. L., Christian, L. L., Lie, W. R., Venier, L., Cooper, M., Briggs, C., and Warner, C. M., Breed and swine lymphocyte antigen haplotype differences in agglutination titers following vaccination with *B. bronchiseptica*, *J. Anim. Sci.*, 59, 643, 1984a.

Rothschild, M. F., Hill, H. T., Christian, L. L., and Warner, C. M., Genetic differences in serum-neutralization titers of pigs after vaccination with pseudorabies modified live-virus vaccine, *Am. J. Vet. Res.*, 45, 1216, 1984b.

Rothschild, M. F., Renard, C., Bolet, G., Dando, P., and Vaiman, M., Effect of swine lymphocyte antigen haplotypes on birth and weaning weights in pigs, *Anim. Genet.,* 17, 267, 1986.

Rothschild, M. F., Renard, C., Sellier, P., Bonneau, M., and Vaiman, M., Swine lymphocyte antigen (SLA) haplotype effects on male genital tract development and androstenone level, in *Third World Congress on Genetics Applied to Livestock Production*, Dickerson, G. E. and Johnson, R. K., Eds., Lincoln, NE, 1986b, 197.

Rothschild, M. F., Zimmerman, D. R., Johnson, R. K., Venier, L., and Warner, C. M., SLA haplotype differences in lines of pigs which differ in ovulation rate, *Anim. Blood Grp. Biochem. Gen.*, 15, 155, 1985.

Rouhonen-Lehto, M. K., Rothschild, M. F., and Larson, R. G., Restriction fragment length polymorphisms at the heat shock protein HSP70 gene(s) in pigs, *Anim. Genet.,* 24, 67, 1993.

Sachs, D.H., The pig as a potential xenograft donor, *Vet. Immunol. Immunopathol.*, 43, 185, 1994.

Sachs, D. H., Germana, S., El-Gamil, M., Gustafsson, K., Hirsch, F., and Pratt, K., Class II genes of miniature swine, *Immunogenetics,* 28, 22, 1988.

Sachs, D. H., Leight, G., Cone, J., Schwarz, S., Stuart, L., and Rosenberg, S., I. Fixation of the major histocompatibility complex, *Transplantation*, 22, 559, 1976.

Schinckel, A., Johnson, R. K., Pumfrey, R. A., and Zimmerman, D. R., Testicular growth in boars of different genetic lines and its relationship to reproductive performance, *J. Anim. Sci.*, 56, 1065, 1983.

Shaw, S., Ziegler, A., and DeMars, R., Specificity of monoclonal antibodies directed against human and murine class II histocompatibility antigens as analyzed by binding to HLA-deletion mutant cell lines, *Hum. Immunol.*, 12, 191, 1985.

Shia, Y.-C., Bradshaw, M., Rutherford, M. S., Lewin, H. A., and Schook, L. B., PCR-based genotyping for characterization of SLA-DQB and SLA-DRB alleles in domestic pigs, *Anim. Genet.*, 26, 91, 1995.

Shia, Y.-C., Gautschi, C., Ling, M.-S., Beever, J. E., McLaren, D. G., Lewin, H. A., and Schook, L. B., RFLP analysis of SLA haplotypes in Swiss Large White and American Hampshire pigs using SLA class I and class II probes, *Anim. Biotechnol.*, 2, 75, 1991.

Shia, Y.-C., Lewin, H. A., and Schook, L. B., PCR-based typing of the swine major histocompatibility complex (SLA) class II genes, *Anim. Genet.*, 24(Suppl. 1), 42, 1993.

Sigurdardottir, S., Borsch, C., Gustafsson, K., and Andersson, L., Gene duplications and sequence polymorphism of bovine class II DQB genes, *Immunogenetics*, 35, 205, 1992.

Singer, D. S., Camerini-Otero, R. D., Satz, M. L., Osborne, B., Sachs, D., and Rudikoff, S., Characterization of a porcine genomic clone encoding a major histocompatibility antigen: expression in mouse L cells, *Proc. Natl. Acad. Sci. U.S.A.*, 79, 1403, 1982.

Singer, D. S., Ehrlich, R., Golding, H., Satz, L., Parent, L., and Rudikoff, S., Structure and expression of class I Mhc genes in the miniature swine, in *The Molecular Biology of the Major Histocompatibility Complex of Domestic Animal Species*, Warner, C M., Rothschild, M. F., and Lamont, S. J., Eds., Iowa State University Press, Ames, 1988, 53.

Tissot, R. G., Beattie, C. W., and Amoss, M. S., Jr., The swine leucocyte antigen (SLA) complex and Sinclair swine cutaneous malignant melanoma, *Anim. Genet.*, 20, 51, 1989.

Trowsdale, J., Ragoussis, J., and Campbell, R. D., Map of the human Mhc, *Immunol. Today*, 12, 443, 1991.

Våge, D. I., Olsaker, I., Lingaas, F., and Lie, O., High degree of genetic diversity revealed in the first domain exon of a porcine DRB gene, *Anim. Genet.*, 24(Suppl. 1), 42, 1993.

Våge, D. I., Olsaker, I., Lingaas, F., and Lie, O., Isolation and sequence determination of porcine class II DRB alleles amplified by PCR, *Anim. Genet.*, 25, 73, 1994.

Vaiman, M., Chardon, P., and Cohen, D., DNA polymorphism in the major histocompatibility complex of man and various farm animals, *Anim. Genet.*, 17, 113, 1986.

Vaiman, M., Chardon, P., and Renard, C., Genetic organization of the pig SLA complex. Studies on nine recombinants and biochemical and lyostrip analysis, *Immunogenetics*, 9, 353, 1979.

Vaiman, M., Metzger, J., Renard, C., and Vila, J., Immune response gene(s) controlling the humoral anti-lysozyme response (Ir-Lys) linked to the major histocompatibility complex SL-A in the pig, *Immunogenetics*, 7, 231, 1978.

Vaiman, M. and Renard, C., Deficit of piglets homozygous for the SLA histocompatibility complex in families, *Anim. Blood Groups Biochem. Genet.*, 11, Suppl. 1, 57, 1980.

Vaiman, M., Renard, C., and Bourgeaux, N., SLA, the major histocompatibility complex of swine: its influence on physiological and pathological traits, in *The Molecular Biology of the Major Histocompatibility Complex of Domestic Animal Species*, Warner. C. M., Rothschild, M. F., and Lamont, S. J., Eds., Iowa State University Press, Ames, 1988, 23.

Vaiman, M., Renard, C., Lafage, P., Ameteau, J., and Nizza, P., Evidence of a histocompatibility system in swine (SLA), *Transplantation*, 10, 155, 1970.

Vaske, D. A., Liu, H. C., Larson, R. G., Warner, C. M., and Rothschild, M. F., Rapid communication: *Taq*I restriction fragment length polymorphism at the porcine transporter associated with antigen processing 2 (TAP2) locus, *J. Anim. Sci.*, 72, 798, 1994a.

Vaske, D. A., Ruohonen-Lehto, M. K., Larson, R. G., Warner, C. M., and Rothschild, M. F., Rapid communication: restriction fragment length polymorphisms at the porcine transporter associated with antigen processing 1 (TAP1) locus, *J. Anim. Sci.*, 72, 255, 1994b.

Viza, D., Sugar, J. R., and Binns, R. M., Lymphocyte stimulation in pigs. Evidence for the existence of a single major histocompatibility locus: PL-A, *Nature (London)*, 227, 949, 1970.

Warner, C. M., Genetic manipulation of the major histocompatibility complex, *J. Anim. Sci.*, 63, 279, 1986.

Watanabe, M., Suzuki, T., Tanaguchi, T., and Shinohara, N., Monoclonal anti-Ia murine alloantibodies crossreactive with the Ia-homologues of other mammalian species including humans, *Transplantation*, 36, 712, 1983.

White, D. and Lunney, J., Transplantation in pigs, *Transplant. Proc.*, 11, 1170, 1979.

Wilson, A. G., de Vries, N., Pociot, F., di Giovine, F. S., van der Putte, L. B. A., and Duff, G. W., An allelic polymorphism within the human tumor necrosis factor a promoter region is strongly associated with HLA A1, B8, and DR3 alleles, *J. Exp. Med.*, 177, 557, 1993.

Wu, L., Rothschild, M. F., and Warner, C. M., Mapping of the SLA complex class III region by pulsed field gel electrophoresis, *Mammal. Genome,* 6, 607, 1995.

Xu, Y., Rothschild, M. F., and Warner, C. M., Mapping of the SLA complex of miniature swine: mapping of the SLA gene complex by pulsed field gel electrophoresis, *Mamm. Genome*, 2, 2, 1992.

CHAPTER 10

The Equine Major Histocompatibility Complex

Eliane Marti, Gabor Szalai, Douglas F. Antczak, Ernest Bailey, Heinz Gerber, and Sandor Lazary

CONTENTS

0-8493-8032-4/96/$0.00+$.50

I. INTRODUCTION

Almost 20 years ago several investigators independently began studying equine leukocyte antigen polymorphisms using sera from primi- and multiparous mares or alloantisera developed for classical blood group typing. At that time the role of the major histocompatibility complex (Mhc) in transplantation biology was evident from studies in other species; the high degree of polymorphism and the complexity of the Mhc molecules and genes were just beginning to be uncovered. The first disease associations such as that of HLA-B27 with ankylosing spondylitis in humans[1] or B haplotypes with susceptibility/resistance to Marek's disease in chickens[2] were also reported. Motivation for research on the equine Mhc was manifold: the use of the high polymorphism for parentage testing or for population studies, the possible physiological or pathological influence of the Mhc, and the role of the Mhc in fertility.

For the following reasons the horse seemed to be an excellent species for such studies. First, horses are a relatively long lived species due to the manner of their use. Also, the owners usually handle and care for each animal individually and record their fitness and disease conditions. Finally, immunological processes are involved in various pathological conditions of the horse, i.e., summer dermatitis, where a genetic influence has been long suspected.

II. RECOGNITION OF THE EQUINE Mhc

A. Nomenclature Used for the Equine Mhc

The genetic system of the antigen presenting structures located within the Mhc of the horse is called the equine leukocyte antigen (ELA) system. The participants of the international workshops on ELA agreed to follow rules for nomenclature of the Mhc class I and class II loci adopted for use with the Mhc of humans, which is called HLA. Thus, the series of class I allelic antigens first described were denoted as locus A products. Class I antigens which were shown later to be encoded on other loci were assigned as locus B and C products, respectively. The internationally accepted ELA specificities are presented with the locus assignment and the number of the allelic antigen, i.e., A1, A2, etc. Antigens recognized in the workshops but of which the relationship to other alleles is not clarified have a prefix of W (workshop). Locally defined ELA antigens are designated with the abbreviation of the laboratory, i.e., Be for Berne, Ith for Ithaca, etc.

The allelic class II antigens are denoted as products of the D region. Using DNA based molecular techniques genes similar to the human homologs of DQA, DQB, DRA, and DRB have been shown to be present in the horse, and these loci are called the same as in man.

Recently a new nomenclature was proposed for the Mhc of non-primate mammals.[3] An abbreviation of the scientific name of the species would be used, i.e., for the horse (*Equus caballus*), the Mhc would be designed "Eqca". This nomenclature has not yet been widely adopted.

B. Determination of the Allelic Structures

Sera from over 90% of primiparous mares contain Mhc class I-specific alloantibodies, as detected in the lymphocyte microcytotoxicity test developed by Terasaki and McClelland.[4] Raw sera or sera after absorptions with leukocytes were tested on cell panels from selected horses. The pattern of these reactions allowed the definition of 6 to 12 antigen specificities in various laboratories, each using its own terminology.[5-13] In the 1980s, five comparison tests and workshops resulted in an internationally accepted terminology for the serologically detected leukocyte antigens of the horse.[14-18]

To date, two series of allelic ELA antigens and one separate structure can be determined serologically; the first contains 17 alleles encoded by locus A which are internationally accepted (A1 to A10, A14, A15, A19, W16, W17, W18, and W20). Further specificities are recognized locally without international acceptance, such as Be-22, -24, -25, -26, and -108. The second series are all encoded by D region loci and contain three internationally accepted specificities (W13, W22, and W23) and again there are at least two locally defined allelic antigens (BeVIII and Be200). A serum cluster defined the ELA W21 antigen, which was shown to be encoded by yet another locus using lysostripping.[19] A further antigen coded by locus B was demonstrated by Bernoco and co-workers[20] based on population, family, and molecular data.

Sera from pregnant mares were used to define the locus A and W21 alloantigens, while all D loci specific alloantisera were prepared by immunizations with lymphocytes between horses which displayed identical locus A specificities but which stimulated one another in mixed lymphocyte reaction (MLR).

C. Molecular Weights of the ELA Antigens

The determination of the molecular weight of immunoprecipitated antigens representing the two ELA series justified the locus designations A and D. The 42- to 44-kilodalton (kDa) glycoproteins represent locus A products and the dimers with approximately 28- and 33-kDa chains are locus D products.[17,21-24]

D. Topography

The cellular distribution of Mhc class I and class II molecules in horses appears to be similar to that described in other species, with a few exceptions. First, polymorphic Mhc class I antigens are expressed in high density on a subpopulation of trophoblast cells (discussed in detail below). Second, Mhc class II antigens appear to have a broader range of expression in horses compared to the more intensively studied species. In humans and mice class II Mhc molecules are expressed on B cells in a constitutive manner, but they are not expressed on resting T cells. Activation of human, but not mouse, T cells causes these cells to express class II molecules.[25] The situation in the horse is quite different: virtually all T cells express class II molecules, although at a lower density than do B-cells.[21] Both $CD4^+$ and $CD8^+$ T cells express Mhc class II antigens, and in young horses the levels of Mhc class II antigens appear to be reduced compared to adults.[26] It is possible that T cells express the products of fewer class II loci than do B cells,[24] but this has not been firmly established. Horses carrying the Mhc class II allele W13 express lower levels of Mhc class II antigens on both B and T cells than do horses carrying other Mhc class II types.[26] Mhc class I and class II antigens can be selectively upregulated in epithelial cells, for example, in the glands of the uterus during pregnancy.[27]

Horse Mhc class I antigens are expressed on platelets, while Mhc class II antigens are not.

E. Inheritance of ELA, Linkage Relationships

One of the early questions about the ELA system was whether or not it was also one of the eight previously discovered horse blood group systems.[28] Since both the murine and the chicken Mhc were initially identified as blood group systems and the horse possessed several complex blood group systems, this seemed a reasonable assumption. Nevertheless, absorption studies clearly demonstrated that the equine lymphocyte alloantigens were not expressed on red blood cells.[9,11] Family studies, however, demonstrated linkage between the A blood group system and the ELA system.[9,29] The distance between the loci was demonstrated to be 1.6 cM and, within the American Standardbred breed, linkage disequilibrium between the A and ELA loci was observed.[30] The linkage association found among pacing Standardbred horses was not observed among trotting Standardbred horses or Thoroughbreds; consequently the origin of the linkage disequilibrium was thought due to gametic associations rather than biological or physical causes. Guerin and co-workers[29] investigated a family in which recombination had been observed between the blood group A and ELA system. Using Southern blotting and mixed lymphocyte culture studies they demonstrated that the A blood group locus lays outside the region including both class I and class II Mhc genes. Little is known about the biochemical nature of the A blood group system. Blood group systems linked to the Mhc have been identified in several species including mice, humans, cattle, chickens and pigs. Whether or not this system is homologous to any of those is unknown.

Family studies using Southern blotting demonstrated linkage of complement component C4 and 21-hydroxylase to the equine Mhc, as observed in other species.[31]

RFLPs were identified using a human cDNA probe for complement component C4 and genomic horse DNA digested with BamHI or EcoRI. Linkage was demonstrated between ELA-A and C4 loci in a three-generation family of Arabian horses, where the ELA was typed using serological methods and C4 was typed using Southern blotting. No RFLP was uncovered for 21-hydroxylase in this study, although the human cDNA probe did hybridize with horse DNA. Kay and co-workers[31] used restriction enzymes which cut genomic DNA infrequently, generating large fragments. They demonstrated that the probes for 21-hydroxylase and C4 hybridized to some of the same large fragments, indicating they occurred closely together on the same chromosome.

Another locus found linked to the ELA-A locus was equine soluble class I substance (ESCI). Soluble class I molecules in the serum of horses were discovered by Lew and colleagues,[32] using Western blot technology. These molecules were found in all members of the order Perissodactyla, including horses, donkeys, zebras, tapirs, and rhinoceros, with two exceptions. Soluble class I molecules were not detected in the sera of some domestic horses and in none of the four Przewalski horses tested. Soluble class I molecules were not detected in any other species tested, including members of the closely related order Ariodactyla. The gene encoding the equine soluble class I molecule was demonstrated to be linked to the Mhc of the horse in subsequent studies.[33]

Before the discovery of the equine molecule, soluble class I molecules had only been detected in members of the genus *Mus*. The equine soluble class I molecule has many similarities to the Q10 soluble class I molecule of mice, which is produced by cells in the liver and which is encoded by a gene located in the murine Mhc.[34]

The ESCI molecule has a lower molecular weight than the membrane-bound form (36 vs. 44 kDa). It would thus probably lack about 30 to 40 amino acids from the carboxy-terminus of the standard class I molecule that encode the transmembrane and cytoplasmic portions of the molecule. Equine soluble class I is bound to beta-2 microglobulin, and exists as a high molecular weight complex of 200 to 300 kDa in serum. The molecule is very abundant, with an estimated serum concentration of 30 μg/ml.

Soluble class I molecules could be produced by a variety of mechanisms, including mutations or frame shifts that lead to truncation, alternative splicing that produces secreted forms of a class I molecule, or by processing of the membrane-bound form to release the extracellular portion of the molecule. In the mouse, a frame shift in exon V produces a stop codon which results in the production of a truncated molecule with no transmembrane or cytoplasmic domain.[34] Detailed studies of the genes encoding the ESCI molecule are required to determine which mechanism operates in the horse.

Although all wild mice appear to carry the Q10 molecule, some inbred strains of mice do not.[34] The function of this molecule is unknown, but the existence of a null allele that circulates at high frequency in mouse and horse populations strongly suggests that the gene is not essential for survival, at least not in laboratory mice and horses.

As observed with the A blood group system, considerable linkage disequilibrium was apparent within breeds between the ELA-A and ESCI loci.[35]

The ELA loci are not closely linked to the *Tcp-1* locus, contrary to the situation in mice.[36] Bailey[37] suggested such linkage may exist based on the observation of excess transmission of the ELA-A10 allele in a large Standardbred horse family and possible homology of this phenomena to that caused by Mhc-linked genes in the mouse, including the *Tcp-1* locus of mice. Langemeier and co-workers[36] studied the equine *Tcp-1* homolog using Southern blotting and excluded linkage between the ELA-A locus and the equine *Tcp-1* homolog up to 25 cM (Lod Score = –2.0), thus demonstrating that these genes could not be responsible for the reported transmission ratio distortion.

Human blood clotting factor F13A is linked to the telomeric side of the HLA complex, and also found to be polymorphic. Similar physical properties have been uncovered for the F13A homolog in the horse, and the possible locus order F13A-A-ELA is suggested.[38]

III. MOLECULAR APPROACHES TO ELA

Molecular approaches to the study of the ELA system have made it possible to develop and address fundamental questions about the genetics and organization of the Mhc in the horse. As noted elsewhere, serology was very effective in identifying the ELA-A locus. Linkage disequilibrium between Mhc loci, as well as the difficulty of making antisera to antigens at other loci than ELA-A, made it difficult to progress beyond this point rapidly using serology. The application of molecular techniques made it possible to map ELA to a chromosome region, to demonstrate the complexity of the system, and to begin typing genes using molecular rather than serological techniques.

A. Physical Mapping

The equine Mhc was localized to equine chromosome 20q14-q22 by *in situ* hybridzation. Both Ansari et al.[39] and Mäkinen et al.[40] used heterologous class I probes, derived from porcine and human clones, respectively.

B. Southern Blotting Studies

Vaiman and co-workers[41] published the first Southern blots on the ELA system. They used human cDNA probes for Mhc class I, class II (DRB and DQA), and complement C4. They noted that, compared to cattle and swine, hybridization of horse DNA with class I and class II probes revealed extensive polymorphism. A two allele polymorphism was observed for the C4 locus.

Alexander and co-workers[42] used Southern blotting to demonstrate that the ELA typing results correlated with characterization of the system using molecular techniques. They compared the RFLP patterns among 14 standardbred horses selected based on their homozygosity for the ELA-A alleles A1, A3, A4, A5, A6 and A10. Murine cDNA for class I, class II Aα, Eα, Aβ, and Eβ were used as probes. The Southern blot resulting from use of the class I probe revealed 12 to 14 nonpolymorphic fragments and 10 to 19 bands which were polymorphic among these horses. No two horses had identical patterns, although those sharing ELA types were more similar than those not sharing types. Alexander and co-workers suggested that the horse may have as many as 30 class I loci, like the mouse, and that some of those loci may be homologous to the Qa/Tla system of the mouse, exhibiting limited polymorphism. Their studies using class II probes identified extensive polymorphism using the Aβ, Eβ and Aα probes. Again, horses with the same ELA type were more similar than those with different ELA types. Extensive cross hybridization was observed between the Eβ and Aβ probes. No variation was observed among the horses using the Eα probe, consistent with observations among mice in which Eα is nonpolymorphic.

Guerin et al.[29,43] used family studies involving ELA typed horses and Southern blotting with human class I and class II probes to demonstrate that the ELA types segregated with the RFLP patterns.

Hänni et al.[44] extended these studies on class II genes using human Mhc probes. Extensive polymorphism was reported with the use of the probes for DQA, DPB, DQB, and DRB, each exhibiting approximately 20 polymorphic bands. DRA exhibited several polymorphic bands; this contrasts with the observation of Alexander et al.,[42] in which no variation was detected when using a probe for the murine homolog, Eα. Hänni and co-workers[44] were unable to detect an equine homolog to the DPA gene. These workers compared the RFLP results to serological results and identified RFLPs using DQB and DRB probes which were associated with the ELA W13 specificity.

C. Sequencing Mhc Genes

Southern blotting was very useful to identify the existence in the horse of Mhc loci homologous to those found in other species. However, RFLPs only reflect sequence variation which results in gain or loss of specific sequences resulting in cutting sites for restriction enzymes. Furthermore, RFLPs for a gene do not necessarily reflect sequence variation in the coding sequence and may be a result of sequence variation of intron DNA or in DNA flanking the entire gene sequence. Advances in DNA sequencing and development of the polymerase chain reaction have led to sequencing of equine Mhc genes.

1. *Class I Genes*

Two full-length sequences of horse Mhc class I molecules have been determined from clones isolated from a cDNA library.[45] These clones show strong homology to one another and to Mhc class I genes of other species. The horse used for construction of the library was homozygous for expressed Mhc antigens, and it is likely that the two clones represent the products of different polymorphic Mhc class I genes. Another complete sequence of an equine class I gene has been reported to EMBL/Genbank.[46] Ellis et al.[47] screened a cDNA library from lymphocytes of an ELA-A heterozygote, finding clones for seven different class I genes. Consequently, they suggested that the horse has at least four class I loci which are expressed.

Ellis and Martin[48] also reported the complete sequence of B-2-microglobulin, which, while not encoded by the equine Mhc, is an integral part of the functioning class I molecule.

2. *Class II Genes*

a. *DRA*

A sequence for DRA was reported using a cDNA clone from a partially processed transcript.[49] This clone lacked the leader sequence and exon 1 but included part of intron 1 and all of intron 2. Computer analysis of possible translation of the gene revealed open reading frames in all exons and indicated a fully functional gene rather than a pseudogene. Northern blotting studies demonstrated the presence of normal sized transcripts (1 kb) and the absence of DRA transcripts homologous in size to this clone (2 kb), thus indicating that this clone represented a rare processing event. The exon and intron sequences showed a strong similarity to sequences from the human DRA gene.

Investigations of the second exon of this DRA gene using DNA sequencing and allele specific PCR led to the discovery of three alleles that are common among Arabian and Thoroughbred horses.[50] This observation is in contrast to the low or absent polymorphism of this locus among mice and humans. One of the alleles, ELA-DRA(2)1, was found among other equid species while the other two alleles, ELA-DRA(2)2 and ELA-DRA(2)3, were found only among horses. The DNA substitutions leading to these alleles did not involve palindromic sites for restriction enzymes and therefore could not be responsible for the RFLPs found during Southern blotting studies.[44]

b. *DRB*

Gustaffson and Andersson[51] reported three sequences for equine DRB second exon to EMBL/Genbank. They noted a striking similarity between horse and human alleles in sharing 11 of 16 residues in the antigen recognition site.

c. *DQA*

The full-length cDNA of the horse DQA has been sequenced.[52] The predicted protein sequence consists of 232 amino acids, as seen in other species.

The second exon of the DQA locus from six horses was amplified by PCR and sequenced.[53] The animals represented different ELA class II haplotypes and thus also were allelic at the DQB locus. Little DNA variation was found in the second exon of the DQA gene. Three Swiss halfbred-riding horses (ELA W13, W22, Be200) one Freiberger (ELA undefinied), and one Arabian (ELA undefined) displayed identical DNA sequences. However, the sequence from an American Throroughbred revealed strikingly different DNA sequence at various positions. Probably this sequence is from the same locus because (1) either the mutations occur at the antigen-binding sites or (2) in the translated amino acid sequence the replaced residues have similar chemical and structural characteristics (such as L to V or E to D). It is still to be investigated whether the two alleles reflect breed differences only. These results contradict previous RFLP studies at the DQα locus, where a high degree of polymorphism was observed.

d. DQB

Gyllensten and co-workers[54] used polymerase chain reaction with primers corresponding to conserved sequences among species to amplifiy and sequence the second exon of DQB from a variety of species, including the horse. They found two sequences for the horse, presumably alleles at the ELA-DQB locus. Szalai and co-workers[55] extended this work, sequencing the second DQB exon from 16 horses selected for homozygosity at the ELA-A locus and for class II genes. Extensive variation was observed and 13 different alleles were uncovered. All of the ELA class II serological specificities were associated with a group of DQB alleles which share the same five amino acids at positions 56 to 60 of the DQB chain but differ at other residues. Unlike other species where there is an extensive amino acid motif sharing between alleles, amino acid associations were found at different positions of the DQB molecule. Moreover, these residues were conserved in alleles which correspond to the same class II serological specificity. Horse-specific amino acids have been identified. Comparing the horse DNA sequences to other mammalian DQB sequences, striking differences were observed in case of the alleles associated with the ELA class II W13 and W23 serotypes, indicating that these alleles are selected for a specific function such as presentation of an antigenic peptide derived from an equine pathogen.

The full-length cDNA of equine DQB has been sequenced[56] and the amino acid composition predicted. The mature peptide seems to be 237 residues long, which is novel among higher mammals, as up to now only rodents displayed the additional 8 amino acids in the cytoplasmic region.

A method for typing DQB was developed by Mashima et al.[57] based on PCR amplification of the second exon followed by restriction enzyme digestion to detect two different alleles. The relationship of these alleles to those reported by Szalai et al.[55] was not investigated.

IV. MIXED LYMPHOCYTE REACTION

It is generally accepted that in one-way mixed lymphocyte culture (MLC) the degree of reactivity of the cells is mainly related to the incompatibility of Mhc class II loci antigens.[58,59] The reactivity of lymphocytes from horses in one-way MLC was found to be similar as observed in other species.[5,11,29] In family studies the reactivity segregated in the great majority of cases with the Mhc class I alleles. Consequently, lymphocytes from fullsib offspring inheriting identical parental haplotypes do not react against one another. This observation was the first evidence that the established serological system of the horse recognizes Mhc gene products. Horse cells homozygous for ELA DW13 or DW22 when used as stimulatory "typing cells" did not induce reactions from unrelated responder cells carrying the same class II allele as the typing cells. In cases where genetic crossing over events took place, the MLC reactivity was linked to the D loci antigens, and the class I incompatibility did not influence the degree of the reaction.[60]

V. BONE MARROW TRANSPLANTATION

The naturally occurring severe combined immunodeficiency disease (SCID) is an important problem for breeders of Arabian horses, to which breed the disease is largely confined, as well as a good model for a similar disease of humans. Affected horses and humans lack both T and B lymphocytes. They are unable to make immune responses and usually die within 2 months of birth without treatment. The disease in horses is caused by an autosomal recessive gene that is found at high frequency in Arabians. Attempts to treat the condition have included allogeneic bone marrow transplantation, the aim of which is to engraft bone marrow stem cells from a normal donor animal into the SCID recipient. These stem cells have the capacity to differentiate in the recipient into a functioning immune system of T and B cells. This procedure, however, usually leads to a lethal graft-versus-host (GVH) disease when immunocompetent donor lymphocytes recognize and begin to respond to the foreign histocompatibility antigens of the recipient SCID horses.[61]

Studies of Mhc antigens of Arabians[12] permitted typing of horse families carrying SCID genes and the selection of donor-recipient combinations matched for Mhc antigens. Bone marrow transplantation between Mhc-matched full sibling horses resulted in correction of SCID in one case.[62] This recipient developed virtually normal immune function and lived for over 1 year with no signs of GVH disease and was alive at least 6 years later.[63] This is remarkable considering that the bone marrow cells came from a normal immunocompetent yearling horse. Mature T cells in the bone marrow inoculum might have been expected to cause GVH disease when stimulated by minor histocompatibility antigens of the host. The use of a sex-matched fullsibling as the donor probably aided in this regard, by decreasing the number of genetic differences between donor and recipient encoding minor histocompatibility antigens. It would be useful to compare skin graft rejection in Mhc-matched horses that are either unrelated or full siblings. Such research could help determine the number of minor histocompatibility antigen loci in horses.

VI. ANTIGEN FREQUENCY VARIATIONS IN BREEDS

One major impetus to studying the equine major histocompatibility complex has been to uncover genes influencing diseases which may be linked to the Mhc. Population studies have been the most effective approach for identifying associations in outbred populations, specifically human populations; consequently, identifying the distribution of alleles among horse populations and breeds has been an important aspect of ELA studies. As new alleles were identified their distributions among horse populations were investigated, including studies of Thoroughbred and Standardbred horses,[13,30] Arabian horses,[10,12] Belgian Saddlebreds, Belgian Draft horses, Belgian Trotters, Shetland ponies,[64] and Swiss Warmblood horses.[65] These studies demonstrated extensive genetic variation of the ELA system within and between breeds. However, the results were difficult to compare between reports due to differences in terminology and differences in the reagents used for typing the horses at each laboratory. Therefore, at the Third ELA workshop a major effort was made to compare the distribution of genes among diverse horse populations using a standard set of typing reagents.[16] A total of 12 laboratories participated to type 1179 horses of 12 different breeds for 10 ELA antigens (ELA-A1 through ELA-A10) and the ELY-2.1 antigen recognized at previous workshops.[14,15] The results of this workshop demonstrated clearly the wide variation in gene frequency between breeds and also demonstrated that those ten ELA antigens were most likely products of a single locus, since no horse was found among the diverse breeds which possessed more than two of these antigens.

VII. THE EQUINE Mhc AND REPRODUCTION

For many years transplantation biologists and reproductive physiologists have been fascinated by immunological aspects of mammalian pregnancy (see Reference 66 for a review). The question of how the mammalian fetus survives and thrives in the potentially hostile immunological environment of its mother's uterus has not been completely answered. Most studies point to at least three main components: the lack of expression of Mhc antigens by the trophoblast, the paucity of molecules unique to the trophoblast that might be antigenic to the mother, and the possibility of local immunosuppression in the uterus.

The contribution of the horse to these studies has come by virtue of cellular and humoral immune responses surrounding the unique, fetal-derived trophoblast cells of the endometrial cups. The endometrial cups form between days 36 and 40 through the migration of the invasive trophoblast cells of the chorionic girdle into the endometrium. The endometrial cup cells are the source of equine chorionic gonadotropin (eCG), which is detected in maternal blood between days 40 and 120 of pregnancy. The endometrial cup cells attract an accumulation of maternal leukocytes (mostly lymphocytes) that appear to wall off and lead to the death of the cup cells and to their sloughing into the uterine lumen.[67] No such accumulations of maternal leukocytes are found in the endometrium along the border of allantochorion, the noninvasive and major component of the equine trophoblast. The endometrial cup reaction represents the most striking example of a maternal cell-mediated immune response to the fetal-placental unit, although the immunological details of this response have not yet been elucidated.

In parallel to this indication of a cellular maternal anti-fetal immune response in equine pregnancy is the evidence for a humoral response to the fetus. Blood samples from virtually all foaling mares were found to contain cytotoxic antibody directed against paternally inherited alloantigens expressed on lymphocytes from the foal and the mating stallion.[7] These antibodies are first detected shortly after the time of endometrial cup formation,[68] and they are directed primarily, but not exclusively, against paternal class I Mhc antigens.[22,68] Maternal antibodies to fetal Mhc antigens are generated as a result of donkey pregnancy, although apparently at a lower frequency than in horse pregnancy.[69]

The relationship between the formation of the endometrial cups and the subsequent attraction of maternal leukocytes to the cup region, on the one hand, and the sensitization of the mare to paternal Mhc antigens, as evidenced by the appearance of cytotoxic antibody in her serum, on the other, suggested that the endometrial cups might express high levels of paternal Mhc antigens. Efforts to demonstrate class I Mhc antigens on the endometrial cup cells were largely unsuccessful, however.[70]

The explanation of this apparent paradox came with the discovery that only the chorionic girdle cells, and not the mature endometrial cup cells or the noninvasive trophoblast cells of the allantochorion, express class I Mhc antigens.[27,70] The Mhc class I antigens of horse chorionic girdle cells are expressed on the cell surface and they have the same molecular weight as Mhc class I molecules expressed on horse lymphocytes.[27] The expression of the Mhc class I antigens on differentiating endometrial cup cells is lost over a 5- to 10-day period between days 40 and 50 of gestation. This loss of Mhc expression appears to be developmentally regulated and not determined by uterine factors acting on the chorionic girdle cells after they invade the endometrium.[71]

Both paternal and maternal polymorphic Mhc class I antigens are expressed on the chorionic girdle, demonstrating that these genes are not influenced by gametic imprinting.[72] When injected intravenously into horses, isolated Mhc class I antigen-positive chorionic girdle cells can induce the formation of cytotoxic alloantibodies and prime an animal for strong secondary antibody responses to Mhc class I antigens.[72]

The mechanisms which result in the selective expression of the class I antigens on the developing chorionic girdle cells, but not on the noninvasive trophoblast of the allantochorion, and the subsequent down regulation of the class I antigens on the mature endometrial cup cells, have not yet been determined. Neither is it known whether the expression of the class I antigens is required for chorionic girdle cell differentiation or invasion into the endometrium.

There is some evidence that maternal antibody responses to the ELA-A3 class I allele are weaker than those to other A locus alleles,[70,73] but this observation has not been followed at the cellular level of the endometrial cup reaction.

It is important to ask whether the immunogenetic relationship between mother and fetus has any effect on fetal survival. It is possible that Mhc differences between mother and fetus may be beneficial to pregnancy and fetal development. If so, this would be a powerful way to maintain the Mhc polymorphism in natural breeding populations. Studies of Mhc antigens in feral horses and wild equids such as zebra might be elucidating in this respect. In experimental studies, the intensity of the endometrial cup reaction did not appear to be diminished in Mhc compatible pregnancies, even though no maternal antibodies to paternal lymphocyte antigens were generated by the Mhc-compatible pregnancies.[74] It is possible that the cellular leukocyte response to the cup cells in these Mhc-compatible pregnancies was directed against the so-called minor histocompatibility antigens, which are considered to be the degradation products of normal cellular proteins presented in association with Mhc class I molecules. Alternatively, the endometrial cup reaction may be inflammatory in nature, and not immunological.

In field investigations in domestic horses, Park et al.[75] found no differences between the distribution of ELA antigens among Standardbred horses with decreased reproductive efficiency and among aborting mares compared to mares of normal fertility on the same farm. In a larger study of Standardbred horses, MacCluer and colleagues[76] demonstrated no effect of ELA alleles on gestation length, but small, statistically significant effects of the ELA types of the dam, the sire, and of sire-dam ELA incompatibility on fertility rate. The effects differed between trotting and pacing Standardbreds, possibly reflecting an influence by genes linked to the ELA-A locus or due to differences in genetic background. Bailey[37] demonstrated that, in one extended Standardbred family, stallions carrying the ELA-A10 allele showed a small, but statistically significant, segregation distortion of this Mhc type in their offspring. Heterozygous stallions transmitted their A10 allele to their offspring more often than they transmitted their other ELA allele (57% of the time). This distortion of expected 50:50 transmission was not seen in mares from the same family, nor from stallions or mares of unrelated Standardbred families carrying the ELA-A10 allele. This observation suggests that the effect may be determined by a genetic region in or near the Mhc of the horse distinct from the ELA-A10 allele itself.

VIII. ASSOCIATION BETWEEN Mhc AND ANTIBODY TITER AFTER VACCINATION

In an early report a relationship between the immune response to tetanus toxoid-toxin and the presence of specific alloantigens on the leukocyte membrane of unrelated horses was identified.[77] After the ELA system was described, the influence of the ELA region on antibody production was examined further. Antibody titers of ELA typed foals (N = 184, representing various breeds) were measured after the first vaccination and after a boosting injection. The vaccine contained multivalent viral antigens such as equine herpes virus type 1 (EHV-1), influenza A/equi 1 and 2, and traces of ovalbumin. The results showed that the age of the foals, the specific environment of the stud, and for some vaccine antigens the sex of the animals influenced the immune response significantly. Possible ELA association was investigated by analysis of variance. When all animals were taken into account, evidence for a weak influence ($0.1 > p > 0.01$) of certain ELA antigens was demonstrated. Further analysis showed that Thoroughbreds carrying the ELA class I antigen A10 displayed lower antibody titers against EHV-1 antigen(s) than foals lacking this ELA specificity ($p < 0.01$). Among Warmblood foals, a similar, strong association ($p < 0.01$) was found between the presence of class II DBeVIII specificity and higher antibody titers against influenza A equi 2 virus hemagglutination antigen. The other, rather weak associations ($0.1 > p > 0.01$) should be further investigated.[78] Titrations against specific epitopes of the individual antigens might allow clarification of the specific influence of ELA on the immune reactions of the horse.

IX. DISEASE ASSOCIATIONS

A. Investigated Diseases

After the first report on the association of Hodgkin's disease with HLA antigens in humans[79] extensive work began to shed light on other Mhc-influenced diseases. These studies resulted in a number of HLA-disease associations, which are reproducible and statistically beyond any doubt.[80]

In horses, possible associations between susceptibility for diseases and the ELA complex have been investigated, as summarized in Table 1. For laryngeal hemiplegia[81] and chronic bronchitis[82] a non-Mhc-based genetic element can be shown. Meredith et al.[83] studied laminitis among Standardbreds and Thoroughbreds, but failed to demonstrate increased frequency of any ELA antigens in diseased animals vs. controls.

However, as we see below, a strong association between sarcoid tumors and Mhc antigen has been described. Also, to some extent, susceptibility to insect bite dermal hypersensitivity seems to be linked to the Mhc.

1. Equine Sarcoid

The equine sarcoid is the most common skin tumor in the horse. These fibroblastic, wart-like lesions are caused by a virus closely related or identical to the bovine papilloma virus (BPV) type 1 or 2.[84-86]

The association between class I or class II antigens and susceptibility to sarcoids has been described in different breeds and also from different parts of the world.[65,83,87-89] In French, Swiss, Irish, and Swedish Warmbloods as well as in Thoroughbreds the class II antigen DW13 occurs with increased frequency in sarcoid-affected horses compared to controls (corrected $P < 0.01$). The antigen DW13 is commonly linked to the class I alleles A3, A5 and A15. Therefore, depending on the investigated subpopulation, these class I alleles can also display some weak association with susceptibility. In breeds where the class II allele DW13 does not occur, sarcoids are either very rare, i.e., among Standardbreds[83,90] or the susceptibility to develop sarcoids is linked to yet undefined class II specificities but to defined class I alleles, as, for example, ABe108 in Freibergers (Swiss draft horse breed) or A1 in Arabians. Despite the strong association of DW13 with sarcoid susceptibility in Warmblood horses, the etiological fraction varies between 0.28 and 0.72 within Warmblood breeds; in other words, part of the affected horses do not display the DW13 antigen. To investigate whether there is a systematic segregation of a specific haplotype among diseased animals, halfsibling groups sired by ELA heterozygous stallions were the most informative. Gerber et al.[88] demonstrated, that, within such a halfsibling group nearly all sarcoid-affected offspring inherited one particular paternal ELA haplotype. As shown in Table 2, haplotypes cosegregating with the sarcoid affection in families are not necessarily those found associated with the disease in population studies. Moreover, as we already saw above, sarcoid susceptibility is associated with different class I or class II antigens depending on the breed or the family studied. These facts suggest that the class I or class II genes might

Table 1 **Summary of Investigations on ELA Associations with Various Conditions in Unrelated Horses, in Halfsiblings, and in Multiple-Case Families**

Condition	Country	Breed	Material	No.patients/controls	ELA associations	Ref.
Sarcoid	Switzerland	Swiss WB	Unrelated	55/248	DW13**, A5**	65
	Switzerland	Swiss WB	Unrelated	130/941	DW13***	89
	Switzerland	Irish WB	Unrelated	38/128	DW13**	89
	Switzerland	French WB	Unrelated	34/120	A3**, DW13****	89
	Switzerland	Freiberger	Unrelated	64/453	ABe108**	89
	Switzerland	Swiss WB	Halfsiblings	47/195	DW13****	88, 89
	U.S.	Thoroughbred	Unrelated	18/99	A3****, DW13****	83
	Sweden	Swedish WB	Unrelated	69/62	A3*, DW13**	87
	Sweden	Swedish WB	Families	12/14	DW13**	87
Sweet itch	Norway	Icelandic	Unrelated	136/167	DW22**	99
	Switzerland	Swiss WB	Families	9/15	DW23**	98
Chronic bronchitis	Switzerland	Swiss WB	Halfsiblings	40/113	None	82
Laminitis	U.S.	Thoroughbred	Unrelated	30/99	None	83
	U.S.	Standardbred	Unrelated	25/106	None	83
Hemiplegia laryngis	Switzerland	Swiss WB	Halfsiblings	22/25	None	81

Note: The significance values in unrelated groups were reported as corrected p (p × number of investigated ELA alleles). Significance levels: * $0.05 > p > 0.01$; ** $0.01 > p > 0.001$; ***$0.001 > p > 0.0001$; **** $p < 0.0001$.

Table 2 Distribution of Transmitted Paternal ELA Haplotypes in Informative Halfsiblings with and without Sarcoids

	Paternal haplotypes		No. of sarcoid affected offspring with paternal haplotype		No. of healthy offspring with paternal haplotype		
Stallion No.	a	b	a	b	a	b	RR
1	A3, DW13	A1, D und[a]	5	1	1	1	
2	A3, DW13	A3, DW23	11	1	42	40	
3	A3, DW13	A18, D und	4	0	7	8	
4	A3, DW13	A19, D und	4	0	5	7	11.0
5	A3, DW13	A20, D und	8	1	11	8	($p < 0.0001$)
6	A5, DW13	A15, DW23	4	0	3	2	
7	A15, DW13	A15, D und	7	1	28	32	
Total			43	4	97	98	
8	A5, DW23	A2, DBe VIII	3	1	5	3	
9	A5, DW23	A2, DW22	5	1	4	3	2.3
10	A5, DW23	A19, DW23	13	4	14	9	($p = 0.4$)
Total			21	6	23	15	
11	A15, DW13	A5, D und	7	1	6	6	
12	A3, DW13	A5, D und	4	3	11	11	1.74
13	A3, DW13	A5, D und	4	4	5	4	(p = 0.38)
Total			15	8	22	21	

[a] Undefined.

only act as genetic markers for the disease, and a putative "sarcoid-susceptibility" gene might be located between these loci.

Recently, similar associations between class II Mhc genes and Papilloma virus-induced tumors have been described in other species. Wank and Thomssen[91] showed that women carrying the HLA-DQW3 class II Mhc allele have an increased risk of developing papilloma virus-induced squamous cell carcinoma of the cervix. In a more recent study, Apple et al.[92] found a papilloma virus type-specific association between HLA DR-DQ haplotypes and the same disease. In rabbits, class II Mhc genes are also associated with the development of tumors induced by Shope Papilloma virus.[93] These comparable findings in three different species suggest a similar virus-host interactions in oncogenic transformation, which have a papilloma virus etiology and are influenced by Mhc class II or closely linked genes.

2. Insect Bite Dermal Hypersensitivity

Insect bite dermal hypersensitivity, also called sweet itch, summer eczema, or summer dermatitis is an allergy to bites from insects mostly of the genera *Culicoides* and *Simulium*. A hereditary predisposition to the disease has been described by many authors.[94-98]

An association between ELA and susceptibility to summer dermatitis has been described in two different breeds. Halldorsdottir et al.[99] studied the repartition of the Mhc antigens in 303 Icelandic horses; 136 were affected with summer dermatitis. The class II antigen DW22 (former Be8) occurred more often in affected animals than in controls (corrected $P < 0.01$), with a relative risk of 2.53. The etiological fraction was low (0.186) because the frequency of this antigen in the population is also low. In other studies[89,98] ELA typing of four Swiss Warmblood multiple-case families descended from the same healthy sire gave the following results: 8 of the 9 affected offspring carried the paternal haplotype A15,DW23; on the other hand, only 3 of the 15 nonaffected offspring displayed this paternal haplotype ($p < 0.005$). In the case of a halfsibling group by another stallion, summer dermatitis affected offspring also displayed the class II ELA DW23 specificity in an increased frequency. It still remains to be investigated whether this association can also be found in population studies within Warmblood breeds.

X. CONCLUSIONS

Investigations of the Mhc of the horse have been important for several reasons. First, as a component of comparative evolutionary studies of the Mhc, the ELA region has revealed a soluble class I molecule that is apparently shared only by rodents and an expression pattern of class II molecules on T lymphocytes that is different from that described for humans or mice. The emergence of DNA sequences of horse Mhc genes will increase the value of cross species comparisons from the viewpoints of both structural Mhc genes and their noncoding control regions. With the development of molecular techniques, serological methods are giving way to

PCR-based methods for typing ELA genes. At this time the number of class I and class II loci present and expresssed is unknown for the horse. However, the horse homologs of DRA, DRB, DQA, and DQB have been investigated with some surprising results, specifically the extent of polymorphisms of DRA and the limited polymorphism of DQA. These findings represent a distinct contrast with the situation for the murine and human Mhc. Investigation of these differences among species should lead to a better understanding of the evolution and function of the entire gene complex.

Second, studies of the equine Mhc have added a unique perspective to our understanding of the function of the Mhc molecules. For example, research on reproductive immunology of the mare has demonstrated that strong transplantation reactions can be mounted to Mhc class I antigens presented by trophoblast cells. This shows that the uterus is an immunological competent mucosal surface during pregnancy. In the case of equine sarcoid, where a susceptibility to papilloma virus infection and subsequent tumor development is linked to the Mhc class II region, similar findings in humans and rabbits infected with related viruses may lead to the elucidation of a common mechanism resulting in tumor formation and progression.

Finally, knowledge of the Mhc of the horse is important in equine medicine. The diagnosis and treatment of conditions, such as summer dermatitis and sarcoid, which have significant associations with particular ELA antigens, can already be considered in a different light, based upon the demonstration of a strong genetic component in the development of those diseases. Other ELA-disease associations will no doubt follow as reagents and techniques become more refined and sensitive.

REFERENCES

1. Brewerton, D. A., Caffrey, M., Hart, F. D., James, D. C. O., Nicholls, A., and Sturrock, R. D., Ankylosing spondylitis and HL-A27, *Lancet*, II, 904, 1973.
2. Briles, W. E., Stone, H. A., and Cole, R. K., Marek's disease: effects of B histocompatibility alloalleles in resistant and susceptible chicken lines, *Science*, 195, 193, 1977.
3. Klein, J., Bontrop, R. E., Dawkins, R. L., Erlich, H. A., Gyllensten, U. B., Heise, E. R., Jones, P. P., Parham, P., Wakeland, E. K., and Watkins, D. I., Nomenclature for the major histocompatibility complexes of different species: a proposal, *Immunogenetics*, 31, 217, 1990.
4. Terasaki, P. I. and McClelland, J. D., Microdroplet assay of human serum cytotoxins, *Nature*, 204, 998, 1964.
5. Lazary, S., de Weck, A. L., Straub, R., and Gerber, H., Characterization of lymphocyte alloantigens permitting identification of horse families (preliminary results), in *Proc. of the First International Symposium on Equine Hematology*, Kitchen, H. and Krehbiel, J. D., Eds., Am. Assoc. of Equine Practitioners, Golden, CO, 1975, 132.
6. Schmid, D. O., Cwik, S., and Emerich, M., Lymphocyte antigens of the horse, in *Proc. of the First International Symposium on Equine Hematology*, Kitchen, H. and Krehbiel, J. D., Eds., Am. Assoc. of Equine Practitioners, Golden, CO, 1975, 120.
7. Bright, S., Antczak, D. F., and Ricketts, S., Studies on equine leukocyte antigens, in *Equine Infectious Diseases IV*, Bryans, J. T. and Gerber, H. G., Eds., Veterinary Public. Inc., Princeton, NJ, 1978, 229.

8. de Weck, A. L., Lazary, S., Bullen, S., Gerber, H., and Meister, U., Determination of leukocyte histocompatibility antigens in horses by serological techniques, in *Equine Infectious Diseases IV,* Bryans, J. T. and Gerber, H. G., Eds., Veterinary Publ. Inc., Princeton, NJ, 1978, 221.
9. Bailey, E., Stormont, C., Suzuki, Y., and Trommerhausen-Smith, A., Linkage of loci controlling alloantigens on red blood cells and lymphocytes in the horse, *Science*, 204, 1317, 1979.
10. Bailey, E., Identification and genetics of horse lymphocyte alloantigens, *Immunogenetics*, 11, 499, 1980.
11. Lazary, S., de Weck, A. L., Bullen, S., Straub, R., and Gerber, H., Equine leukocyte antigen system. I. Serological studies, *Transplantation*, 30, 203, 1980.
12. Mottironi, V. D., Perryman, L. E., Pollara, B., Mickey, M. R., Swift, R., and McGrath, P., Major histocompatibility locus in the arabian horse, *Transplantation*, 31, 290, 1981.
13. Antczak, D. F., Bright, S. M., Remick, L. H., and Bauman, B. E., Lymphocyte alloantigens of the horse. I. Serologic and genetic studies, *Tissue Antigens*, 20, 172, 1982.
14. Bull, R., Ed., Joint Report of the First International Workshop on Lymphocyte Alloantigens of the Horse, *Animal Blood Groups and Biochemical Genetics*, 14, 119, 1983.
15. Bailey, E., Antczak, D. F., Bernoco, D., Bull, R. W., Fister, R., Guerin, G., Lazary, S., Matthews, S., McClure, J., Meyer, J., Mottironi, V. D., and Templeton, J., Joint Report of the Second International Workshop on Lymphocyte Alloantigens of the Horse, *Anim. Blood Groups Biochem. Genet.*, 15, 123, 1984.
16. Antczak, D. F., Bailey, E., Barger, B., Bernoco, D., Bull, R. W., Byrns, G., Guerin, G., Lazary, S., McClure, J., Mottironi, V. D., Symons, R., Templeton, J., and Varewyck, H., Joint Report of the Third International Workshop on Lymphocyte Alloantigens of the Horse, *Anim. Genet.*, 17, 363, 1986.
17. Bernoco, D., Antczak, D. F., Bailey, E., Bell, K., Bull, R. W., Byrns, G., Guerin, G., Lazary, S., McClure, J., Templeton, J., and Varewyck, H., Joint Report of the Fourth International Workshop on Lymphocyte Alloantigens of the Horse, *Anim. Genet.*, 18, 81, 1987.
18. Lazary, S., Antczak, D. F., Bailey, E., Bell, T. K., Bernoco, D., Byrns, G., and McClure, J. J., Joint Report of the Fifth International Workshop on Lymphocyte Alloantigens of the Horse, *Anim. Genet.*, 19, 447, 1988.
19. Hirni, H., Lazary, S., and Gerber, H., Serological and genetic study of the ELA W21 specificity, *Anim. Genet.*, 19, 391, 1988.
20. Bernoco, D., Byrns, G., Bailey, E. and Lew, A. M., Evidence of a second polymorphic ELA class 1 (ELA-B) locus and gene order for three loci of the equine major histocompatibility complex, *Anim. Genet.,* 18, 103, 1987.
21. Crepaldi, T., Crump, A., Newman, M., Ferrone, S., and Antczak, D. F., Equine T lymphocytes express class II Mhc antigens, *J. Immunogenet.*, 13, 349, 1986.
22. Donaldson, W. L., Crump, A. L., Zhang, C. H., Kornbluth, J., Kamoun, M., Davis, W., and Antczak, D. F., At least two loci encode polymorphic class I Mhc antigens in the horse, *Anim. Genet.,* 19, 379, 1988.
23. Hesford, F., Lazary, S., Curty-Hänni, K., and Gerber, H., Biochemical evidence that equine leucocyte antigens W13, W22 and W23 are present on horse major histocompatibility complex class II molecules, *Anim. Genet.,* 20, 395, 1988.
24. Monos, D. S., Wolf, B., Radka, S. F., Rifat, S., Donawick, W. J., Soma, L. R., Zmijewski, C. M., and Kamoun, M., Equine class II Mhc antigens: identification of two sets of epitopes using anti-human monoclonal antibodies, *Tissue Antigens*, 34, 111, 1989.
25. Klein, J., *Natural History of the Major Histocompatibility Complex*, John Wiley & Sons, New York, 1986.

26. Barbis, D. P., Bainbridge, D., Crump, A. L., Zhang, C. H., and Antczak, D. F., Variation in expression of Mhc class II antigens on horse lymphocytes determined by Mhc haplotype, *Vet. Immunol. Immunopathol.*, 42, 103, 1994.
27. Donaldson, W. L., Zhang, C. H., Oriol, J. G., and Antczak, D. F., Invasive equine trophloblast expresses conventional class I major histocompatibility complex antigens, *Development*, 110, 63, 1990.
28. Stormont, C. and Suzuki, Y., Genetic systems of blood groups in horses, *Genetics*, 50, 915, 1964.
29. Guerin, G., Varewyck, H., Bertaud, M., and Chasset, P., Analysis of a horse family with a crossing-over between the ELA complex and the A blood group system, *Anim. Genet.*, 19, 1, 1988.
30. Bailey, E., Linkage disequilibrium between the ELA and the A blood group systems in standardbred horses, *Anim. Blood Groups Biochem. Genet.*, 14, 37, 1983.
31. Kay, P. H., Dawkins, R. L., Bowling, A. T., and Bernoco, D., Heterogeneity and linkage of equine C4 and 21-hydroxylase genes, *J. Immunogenet.*, 14, 247, 1987.
32. Lew, A. M., Valas, R. B., Maloy, W. L., and Coligan, J. E., A soluble class I molecule analogous to mouse Q10 in the horse and related species, *Immunogenetics*, 23, 277, 1986.
33. Lew, A. M., Bailey, E., Valas, R. B., and Coligan, J., The gene encoding the equine soluble class I molecule is linked to the horse Mhc, *Immunogenetics*, 24, 128, 1986.
34. Lew, A. M., Maloy, W. L., and Coligan, J. E., Characteristics of the expression of the murine soluble class I molecule (Q10), *J. Immunol.*, 136, 254, 1986.
35. Bailey, E., Woodward, J. G., Albright, D. G., and Alexander, A. J., Use of RFLP marker genes to study serological and physiological traits of the horse, in *Molecular Biology of the Major Histocompatibility Complex of Domestic Animal Species*, Warner, C., Lamont, S., and Rothschild, M., Eds., Iowa State University Press, Ames, 1988, 135.
36. Langemeier, J. L., Bailey, E., and Henney, P. J., Linkage studies between the Tep-1, Tep-10 and Mhc-Eqca-A loci in the horse, *Immunogenetics*, 38, 359, 1993.
37. Bailey, E., Segregation distortion within the equine Mhc: analogy to a mouse T/t complex trait, *Immunogenetics*, 24, 225, 1986.
38. Weitkamp, L. R., Bailey, E., MacCluer, J. W., and Guttormsen, S. A., Polymorphism for equine F13; linkage of F13A with ELA and A, *Anim. Genet.*, 18(suppl. 105), 1989.
39. Ansari, H. A., Hediger, R., Fries, R., and Stranzinger, G., Chromosomal localization of the major histocompatibility complex of the horse (ELA) by *in situ* hybridization, *Immunogenetics*, 28, 362, 1988.
40. Mäkinen, A., Chowdhary, B., Mahdy, E., Andersson, L., and Gustavsson, I., Localization of the equine major histocompatibility complex (ELA) to chromosome 20 by *in situ* hybridization, *Hereditas*, 110, 93, 1989.
41. Vaiman, M., Chardon, P., and Cohen, D., DNA polymorphism in the major histocompatibility complex of man and various farm animals, *Anim. Genet.*, 17, 113, 1986.
42. Alexander, A. J., Bailey, E., and Woodward, J. G., Analysis of the equine lymphocyte antigen system by Southern blot hybridization, *Immunogenetics*, 34, 136, 1987.
43. Guerin, G., Bertaud, M., Chardon, P., Geffrotin, C., Vaiman, M., and Cohen, D., Molecular genetic analysis of the major histocompatibility complex in an ELA typed horse family, *Anim. Genet.*, 18, 323, 1987.
44. Hänni, H., Hesford, F., Lazary, S., and Gerber, H., Restriction fragment length polymorphisms of horse class II Mhc genes observed using various human alpha- and beta-chain cDNA probes, *Anim. Genet.*, 19, 395, 1988.

45. Barbis, D. P., Maher, J. K., Stanek, J., Klaunberg, B. A., and Antczak, D. F., Horse cDNA clones encoding two Mhc class I genes, *Immunogenetics*, 40, 163, 1994.
46. Ellis, S. A. and Martin, A. J. E., Caballus mRNA for Mhc class I heavy chain, *EMBL/Genbank*, X71809, 1993.
47. Ellis, S. A., Martin, A. J., and Morrison, W. J., Analysis of equine Mhc class I, *Anim. Genet.*, 25(Suppl. 2), 19, 1994.
48. Ellis, S. A. and Martin, J. A., Nucleotide sequence of horse beta 2-microglobin cDNA, *Immunogenetics*, 38, 383, 1993.
49. Albright, D., Bailey, E., and Woodward, J. G., Nucleotide sequence of a cDNA clone of the horse (*Equus caballus*) DRA gene, *Immunogenetics*, 34, 136, 1991.
50. Bailey, E., Variation within the antigen-binding site of the major histocompatibility complex DRA gene of domestic horses, in *Equine Infectious Diseases VII*, Nakajima, H. and Plowright, W., Eds., R. & W. Publications, Newmarket, UK, 1994, 123.
51. Gustafsson, K. T. and Andersson, L., Structure and polymorphism of equine Mhc class II DRB genes: convergent evolution on the antigen binding site, *EMBL/Genbank*, L25644, L25645, L25646, 1993.
52. Szalai, G., Antczak, D. F., Gerber, H., and Lazary, S., Molecular cloning and characterization of equine DQA cDNA, *Immunogenetics*, 40, 457, 1994.
53. Szalai, G., Gerber, H., and Lazary, S., DNA sequence analysis of the second exon of the equine DQα chain, XXIV International Conference on Animal Genetics, Prague, August 23–27, 1994, in *Anim. Genet.*, 25(Suppl. 2), 22, 1994.
54. Gyllensten, U. B., Lashkari, D., and Erlich, H. A., Allelic diversification of the class II DQB locus of the mammalian major histocompatibility complex, *Proc. Natl. Acad. Sci. U.S.A.*, 87, 1835, 1990.
55. Szalai, G., Bailey, E., Gerber, H., and Lazary, S., DNA sequence analysis of serologically detected ELA class II haplotypes in the equine DQB locus, *Anim. Genet.*, 24, 187, 1993.
56. Szalai, G., Antczak, D. F., Gerber, H., and Lazary, S., Molecular cloning and characterization of equine DQB cDNA, *Immunogenetics*, 40, 458, 1994.
57. Mashima, S., Hirota, K., and Mukoyama, H., Genomic organization of horse major histocompatibility complex, *Anim. Genet.*, 25(Suppl. 2), 20, 1994.
58. Bach, F. H. and Hirschhorn, K., Lymphocyte interaction: a potential histocompatibility test *in vitro*, *Science*, 142, 813, 1964.
59. Bach, F. H., Albertini, R. J., Klostermann, H., and Day, E., Evaluation of compatibility at the major histocompatibility locus in man, *Histocompatibility Testing*, Munksgaard, Copenhagen, 1967, 55.
60. Lazary, S., Dubath, M.-L., Luder, C., and Gerber, H., Equine leukocyte antigen system. IV. Recombination within the major histocompatibility complex (Mhc), *J. Immunogenet.*, 13, 315, 1986.
61. Perryman, L. E. and Liu, I. K. M., Graft versus host reactions in foals with combined immunodeficiency, *Am. J. Vet. Res.*, 41, 187, 1980.
62. Bue, C. M., Davis, W. C., Magnuson, N. S., Mottironi, V. D., Ochs, H. D., Wyatt, C. R., and Perryman, L. E., Correction of equine severe combined immunodeficiency by bone marrow transplantation, *Transplantation*, 42, 14, 1986.
63. Perryman, L. E., personal communication, 1993.
64. Varewyck, H., Bouquet, Y., Lazary, S., Guerin, G., Weghe, A. van de, and Zeveren, A. van, Contribution to serology and antigen distribution, *Anim. Blood Groups Biochem. Genet.*, 16, 217, 1985.
65. Lazary, S., Gerber, H., Glatt, P. A., and Straub, R., Equine leucocyte antigens in sarcoid-affected horses, *Equine Vet. J.*, 17, 283, 1985.

66. Antczak, D. F. and Allen, W. R., Maternal immunological recognition of pregnancy in equids, *J. Reprod. Fertil.*, 37(suppl.), 69, 1989.
67. Allen, W. R., Immunological aspects of the equine endometrial cup reaction, in *Immunobiology of Trophoblast*, Edwards, E. G., Howe, C. W. S., and Johnson, M. H., Eds., Cambridge University Press, 1975, 217.
68. Antczak, D. F. and Allen, W. R., Invasive throphoblast in the genus Equus, *Annales d'Immunologie — Institut Pasteur*, 135D, 325, 341, 1984.
69. Kydd, J., Miller, J., Antczak, D. F., and Allen, W. R., Maternal antifetal cytoxic antibody responses of equids during pregnancy, *J. Reprod. Fert.*, 32(suppl.), 361, 1982.
70. Crump, A., Donaldson, W. L., Miller, J., Kydd, J., Allen, W. R., and Antczak, D. F., Expression of major histocompatibility complex (Mhc) antigens on equine trophoblast, *J. Reprod. Fert.*, 35(suppl.), 379, 1987.
71. Donaldson, W. L., Oriol, J. G., Plavin, A., and Antczak, D. F., Developmental regulation of class I major histocompatibility complex antigen expression by equine trophoblastic cells, *Differentiation*, 52, 69, 1992.
72. Donaldson, W. L., Oriol, J. G., Pelkaus, C. L., and Antczak, D. F., Paternal and maternal major histocompatibility complex class I antigens are expressed co-dominantly by equine trophoblast, *Placenta*, 15, 123, 1994.
73. Antczak, D. F., Miller, J. M., and Remick, L. H., Lymphocyte alloantigens of the horse. II. Antibodies to ELA antigens produced during equine pregnancy, *J. Reprod. Immunol.*, 6, 283, 1984.
74. Allen, W. R., Kydd, J., Miller, J., and Antczak, D.F., Immunological studies on fetomaternal relationships in equine pregnancy, in *Proceedings of the 38th Easter School, University of Nottingham,* Crighton, D. B., Ed., Butterworths, London, 1984, 183.
75. Park, C. A., Hines, H. C., and Threlfall, W. R., Short Communication, Equine lymphocyte antigens and reproduction in the Standardbred mare, *Anim. Genet.,* 20, 99, 1989.
76. MacCluer, J. W., Bailey, E., Weitkamp, L. R., and Blangero, J., ELA and fertility in American Standardbred horses, *Anim. Genet.,* 19, 359, 1988.
77. Lazary, S., Kuslys, A., Bullen, S., De Weck, A. L., and Gerber, H., Leukocyte antigens (ELA) in the antibody production to tetanus toxoid and toxin in the horse, in *Equine Infectious Diseases IV, J. of Equine Medicine and Surgery, Suppl. #1*, Bryans, J. T. and Gerber, H., Eds., Veterinary Publications, Princeton, NJ, 1978, 237.
78. Bodo, G., Marti, E., Gaillard, C., Weiss, M., Bruckner, L., Gerber, H., and Lazary, S., Association of the immune response with the major histocompatibility complex in the horse, in *Equine Infectious Diseases VII,* Nakajima, H. and Plowright, W., Eds., R. & W. Publications, Newmarket, UK, in press.
79. Amiel, J. L., Study of the leukocyte phenotypes in Hodgkin's disease, *Histocompatibility Testing 1967,* Curtoni, E. S., Mattiuz, P. L., and Tosi, M. R., Eds., Munksgaard, Copenhagen, 1967, 79.
80. Tiwari, J. L. and Terasaki, P. I., *HLA and Disease Associations,* Springer-Verlag, New York, 1985.
81. Poncet, P. A., Montavon, S., Gaillard, C., Barrelet, F., Straub, R., and Gerber, H., A preliminary report on the possible genetic basis of laryngeal hemiplegia, *Equine Vet. J.*, 21, 137, 1989.
82. Marti, E., Gerber, H., Essich, G., Ouehla, J., and Lazary, S., The genetic basis of equine allergic diseases. I. Chronic hypersensitivity bronchitis, *Equine Vet. J.*, 23, 457, 1991.

83. Meredith, D., Elser, A. H., Wolf, B., Soma, L. R., Donawick, W. J., and Lazary, S., Equine leukocyte antigens: relationships with sarcoid tumors and laminitis in two pure breeds, *Immunogenetics*, 23, 221, 1986.
84. Angelos, J., Marti, E., Lazary, S., and Carmichael, L. E., Characterization of BPV-like DNA in equine sarcoids, *Arch. Virol.*, 119, 95, 1991.
85. Reid, S. W. J. and Smith, K. T., The equine sarcoid: detection of papillomaviral DNA in sarcoid tumors by use of consensus primers and the polymerase chain reaction, in *Equine Infectious Diseases VI*, Plowright, W., Rossdale, P. D., and Wade, J. F., Eds., R. & W. Publications, Newmarket, 1991, 297.
86. Otten, N., Tscharner von, C., Lazary, S., Antczak, D. F., and Gerber, H., DNA of bovine papillomavirus type 1 and 2 in equine sarcoids: PCR detection and direct sequencing, *Arch. Virol.*, 132, 121, 1993.
87. Broström, H., Fahlbrink, E., Dubath, M.-L., and Lazary, S., Association between equine leucocyte antigens (ELA) and equine sarcoid tumors in the population of Swedish halfbreds and some of their families, *Vet. Immunol. Immunopathol.*, 19, 215, 1988.
88. Gerber, H., Dubath, M.-L., and Lazary S., Association between predisposition to equine sarcoid and Mhc in multiple-case families, in *Equine Infectious Diseases V*, Powell, D. G., Ed., The University Press of Kentucky, 1988, 272.
89. Lazary, S., Marti, E., Szalai,G., Gaillard, C., and Gerber, H., Studies on the frequency and associations of equine leucocyte antigens in sarcoid and summer dermatitis, *Anim. Genet.*, 25(suppl. 1), 75, 1994.
90. Mohammed, H. O., Rebhun, W. C., and Antczak, D. F., Factors associated with the risk of developing sarcoid tumors in horses, *Equine Vet. J.*, 24, 165, 1992.
91. Wank, R. and Thomssen, C., High risk of squamous cell carcinoma of the cervix for women with HLA-DQw3, *Nature (London)*, 352, 723, 1991.
92. Apple, R. J., Erlich, H. A., Klitz, W., Manos, M. M., Becker, T. M., and Wheeler, C. M., HLA DR-DQ associations with cervical carcinoma show papillomavirus-type specificity, *Nature Genet.*, 6, 157, 1994.
93. Han, R., Breitburd, F., Marche, P. N., and Orth, G., Linkage of regression and malignant conversion of rabbit viral papillomas to Mhc class II genes, *Nature (London)*, 396, 66, 1992.
94. Rick, R. F., Studies on allergic dermatitis (Queensland Itch) of the horse, I. and II., *Austr. Vet. J.*, 29, 177, 1953.
95. Ishihara, T. and Ueno, H., Studies on summer mange of the horse. II. Histopathological findings on infected skin, *Bull. Natl. Inst. Anim. Health (Tokyo)*, 32, 215, 1957.
96. McCaig, J., Recent thoughts on sweet itch, in *Veterinary Annual*, 15th ed., Wright and Sons, Bristol, 1975, 204.
97. Strothman, A., Beitrag zum Sommerekzem (Allergische Dermatitis) der Islandpferde, Literaturstudie und eigene Untersuchungen, Diss. med. vet. Hannover, 1982.
98. Marti, E. Gerber, H., and Lazary, S., On the genetic basis of equine allergic diseases. II. Insect bite dermal hypersensitivity, *Equine Vet. J.*, 24, 113, 1992.
99. Halldorsdottir, S., Lazary, S., Gunnarsson, E., and Larsen, H. J., Distribution of leucocyte antigens in Icelandic horses affected with summer eczema compared to non-affected horses, *Equine Vet. J.*, 23, 300, 1991.

CHAPTER 11

The Major Histocompatibilty Complex of the Rabbit

Nopporn Sittisombut

CONTENTS

0-8493-8032-4/96/$0.00+$.50

I. HISTORICAL PERSPECTIVES

The landmark discovery of immunological basis of allograft rejection in rabbits reported by Medawar during 1944 and 1945 is the beginning of Mhc research not only in this species but also in many others.[1] Although the Mhc systems are currently studied with regards to their primary role in the antigen presentation to T lymphocytes, it is in the context of graft rejection that much of the early studies of rabbit Mhc are conducted. Progress in the field has been reviewed;[2,3] in this section, the most pertinent literature that bear importance to our current view of the rabbit Mhc is again highlighted to render a whole picture.

A. Rabbit and the First Demonstration of Immunological Basis of Allograft Rejection

Modern scientific inquiry into the laws that govern the fate of tissues taken from one individual and grafted into another individual had started in two animal models, mouse and rabbit. While initial studies in mouse are mainly conducted using tumor tissues, it is in rabbits that normal tissue transplantation studies are successful.[4] Indeed, studies done in rabbit by Medawar demonstrate for the first time the role of immune system in the rejection of normal tissue grafts.[1,5] To commemorate the discovery in its 50th anniversary, the findings, so nicely summarized by Klein,[4] are reproduced as follows:

> Medawar worked with outbred rabbits rather than mice, and used skin rather than splenic tissue as grafts... He demonstrated that in the first week after transplantation auto- and allografts behave in exactly the same way: they heal in and establish a normal blood supply. In the second posttransplantation week, however, differences between the two types of grafts begin to emerge. While the autografts gradually return to the condition of normal skin, the allografts begin to show signs of degeneration. They become inflamed and edematous, and their dermis is infiltrated by mononuclear cells. The inflammatory phase is followed by obliteration of the graft's vascular system and disintegration of the epidermis and dermis. The necrotic graft is then replaced by ingrowth of host epidermis. The destruction (rejection) of the graft is usually completed two to three weeks of transplantation. A second graft from the same donor placed on the same recipient is rejected much more rapidly than the first one, usually within the first week after transplantation. This second-set reaction is sometimes so rapid that the second graft does not have time to heal in. Apparently, the first graft sensitizes the host, which then responds to the second graft in an accelerated fashion. The sensitization is specific: only second-set grafts from the same donor as the first-set grafts are rejected by an accelerated reaction: second grafts from unrelated allogeneic donors are rejected by a typical first-set (unaccelerated) reaction. The sensitization is systemic, with the second graft provoking a second-set reaction in almost any part of the body, regardless of the location of the first graft. These three characteristics of the rejection process (accelerated response to second grafts, specificity, and systemic character of sensitization) were considered by Medawar to be evidence that the allograft reaction was immunological by nature.

In a subsequent study, Medawar showed that, whereas preinjection of rabbits with red blood cells did not affect the rejection of subsequent skin grafts derived from the same donors, intradermal injection with white blood cells clearly immunized the recipients against subsequent skin grafts as all grafts from the white cell donor were rejected in an accelerated manner.[6] The results imply that the histocompatibility antigens are shared by white blood cells and skin. However, it is not until 2 decades later that Tissot and Cohen conclusively demonstrated the presence of rabbit leukocyte alloantigens that strongly affected the fate of skin allografts.[7]

Until recently, studies of rabbit Mhc were hampered greatly by the lack of inbred rabbit lines and lymphoid cell lines. A long generation time and the phenomenon of inbred depression of fecundity are responsible for much of the difficulties in the generation and maintainance of inbred rabbits.[8,9] Because rabbits reproduce at an average of one generation a year, the large scale breeding of rabbits needed to obtain an inbred line or a congenic line is an extremely time consuming and expensive task.[10] In addition, inbred depression occurs with an unusual severity. Although there are now a few inbred rabbit lines and lymphoid cell lines available,[11] the latter are not derived from inbred rabbit. These major problems preclude the possibility of studying complex histocompatibility systems through the classical histogenetic and somatic cell genetic routes.

B. Estimation of the Number of Transplantation Loci

In animal systems where inbred lines are available, the multiplicity of the genes that affect the survival of allografts can be readily estimated by testing the F2 generations of the cross between two inbred lines for acceptance of grafts from a parental line. The proportion of F2 animals that accepts grafts is then used to derive the number of genes involved.[4,12] Due to the lack of rabbit inbred lines during the 1940s, Medawar could only estimate the number of the histocompatibility genes by performing cross-skin grafting in every combination in a limited group of outbred rabbits.[5] When all of the skin allografts among 22 rabbits were rejected within 24 days, it was concluded that there were at least seven different histocompatibility antigens in the tested rabbit population. This approach is limited by the number of the animals that can be practically cross-grafted at the same time; higher estimates may be obtained if a larger number is employed. Chai obtained an estimate of 23 to 64 antigens by exchanging skin grafts among siblings at three different generations of four inbreeding lines.[13] When the observed percentage of the survived grafts was compared with that of the theoretical value calculated by using the ideal inbreeding colony,[14] the observed numbers, especially those derived from later generations, were found to be overestimates possibly due to the selection of heterozygotes during inbreeding process. Later, a better estimate of 19 unlinked histocompatibility loci, including a major histocompatibility loci, was obtained from the results of skin graft exchanges among F2 and F3 generations of the cross between two inbred lines.[10] This number remains the best estimate of rabbit histocompatibility loci, but it can be regarded only as the minimum estimate because the effect of linkage is not taken into consideration and also because the B and Y lines employed in this study are derived from the same hybrid population and may carry the same alleles at some loci.

Skin grafting experiments also suggest the presence of a major histocompatibility locus. While performing the cross-skin grafting experiments in which a recipient was grafted with skin from several outbred donors, Medawar noted great variation in both the intensity of rejection reaction and the graft survival time among rejected grafts.[5] Subsequently, Chai observed that, whereas reciprocal skin grafts between members of two inbreeding B and Y lines were quite uniformly rejected within 2 weeks, skin grafts that were exchanged among the F2 and F3 generations of the cross between the B and Y lines were rejected in a less uniform manner (range: 7 days to more than 1 year).[10,13] When the number of rejected skin grafts among the F2 and F3 generations is plotted against time of complete rejection, the distribution is multimodal. Mathematical analysis indicates that the observed number of the earliest rejected grafts fit the expected number that can be derived from the one-locus hypothetical model in which rejection is assumed to be due to allelic difference at a single genetic locus. This result suggests the presence of a histocompatibility locus which exerts the strongest effect on the graft-host interaction.

C. Serological Typing and Mixed Lymphocyte Reaction

Attempts to identify and characterize the product of the major histocompatibility locus on blood cells are performed by using alloantisera derived from rabbits which receive multiple allogeneic skin grafts or spleen cells. Employing hemaglutination test, Kapitchnikov et al. found hemagglutinating antibody in the recipients of skin grafts incompatible at the Hg red blood cell antigen.[15] Cohen et al. directly tested the effect of incompatibility at five red blood cell loci on skin graft survival and found that donor-recipient combinations that were different at only the Hg locus resulted in accelerated graft rejection (mean graft survival time: 6.1 to 6.4 days) when compared with rejection of compatible grafts (mean graft survival time: 9.3 days). Accelerated graft rejection in Hg-incompatible donor-recipient pairs suggests that the Hg locus is a histocompatibility locus.[16] However, compatibility at this locus does not allow grafts to survive longer than grafts that are exhanged between untyped rabbits, indicating that it is not a candidate for the major locus.

The effect of matching of leukocyte alloantigens on graft survival had been studied most extensively by Tissot and Cohen, who obtained a series of leukocytotoxic antisera by skin grafting animals within or between different inbreeding lines.[7] Segregation and pedigree analyses indicate that these antisera recognize antigens encoded by allelic forms of an autosomal locus, RL-A (RLA-A in the current nomenclature). The antigens appear to be complex in that leukocytes from any given animal could react differentially with four or more antisera. Typing of known homozygotes from partially inbred lines reveals seven distinct reactivity patterns representing seven allelic forms. The Mendelian segregation of these complex alleles is then confirmed by typing offsprings of the crosses between known heterozygotes. The importance of the RLA-A locus on survival of allografts is determined by exchanging skin grafts between 35 pairs of siblings of partially inbred lines. Forty-nine

RLA-A compatible grafts survive with the median survival time of 14 days (range: 8 to 80 days), whereas 21 incompatible grafts have the median survival time of only 8 days (range: 6 to 12 days). Skin grafts from unrelated rabbits are rejected with the median survival time of 9 days (range: 6 to 16 days). Thus, compatibility at the RLA-A locus clearly enhances graft survival. The influence of RLA-A locus is further enhanced by suppressing the effect of mismatch at minor loci with 6-mercaptopurine treatment; the mean survival time of the compatible grafts is extended to 22 and 27 days while incompatible grafts were all destroyed within 12 days. Up to now, 13 distinct RLA-A alleles have been documented (Table 1) and the RLA-A locus remains the best serologically characterized histocompatibility locus in rabbit.

The presence of a genetic locus controlling *in vitro* MLR response was also reported by Tissot and Cohen.[17] This MLC locus, now known as RLA-D, is closely linked to the RLA-A and He blood group loci in the rabbit linkage group VII. Compatibility at both of the RLA-A and RLA-D loci results in the longest graft survival time, although this effect is observed only among inbred siblings.[18] The results of typing assay employing the micro method of one-way MLR and homozygous typing cells identify six distinct alleles of the RLA-D locus (Table 1).[19,20] Together, RLA-A and RLA-D alleles constitute up to 17 distinct RLA haplotypes found among laboratory rabbits.[20]

Table 1 Assignment of Rabbit Mhc Haplotypes

RLA haplotype	RLA-A haplotype		RLA-D haplotype		Rabbit strain[d]	Ref.
	SD[a]	RFLP[b]	MLC[c]	RFLP[b]		
1	A1	A1	D1	D1	III/Dw, A	20, 30, 38
2	A2	A2	D2	D2	B	20, 30, 38
3	A3	A3	D3c	D3c	C	20, 30, 38
4	A4	nd	D2	nd	na	20
5	A5	A5	D4	D4	E	20, 30
6	A6	A6	D3f	D3a	F	20, 30, 38
7	A7	A7	D4	D7	na	20, 30
8	A8	A8	D3	D6	na	20, 30
9	A9	nd	D3	nd	na	20
10	A10	A7	D2	D2	na	20, 30
11	A11	A11	D3l	D3a	B/J	20, 30, 38
12	A12	nd	D5	D5	na	20, 38
13	A13	A13	D4	D8	na	20, 30
14	nd	nd	D10	D10	III/J	38
15	nd	A15	nd	D15	na	30
16	nd	A16	nd	D2	na	30
17	nd	A1	nd	D17	na	30

Note: nd = not done. na = not available.

[a] Alleles defined by serological typing.

[b] Haplotypes defined by RFLP analysis.

[c] Alleles defined by mixed lymphocyte culture test.

[d] III/Dw, B/J and III/J are inbred strains; the others (A, B, C, E and F) are RLA homozygous rabbits maintained by selective breeding.

Although the significance of the major histocompatibility locus on allograft survival has been clearly shown, the minor loci as a group also exert strong effect on graft survival. When skin grafts are exchanged among three groups of animals that are identical for both the RLA-A and RLA-D antigens, grafts that are exchanged among inbred siblings survive longer than those among non-inbred siblings and unrelated animals.[18] Because the effect of the major locus is eliminated by matching of RLA-A antigens and MLR reactivity, reduction of graft survival time in unrelated groups reflects the effect of disparity at minor histocompatibility loci. Weak MLR reactivities between RLA-A and RLA-D matched cells[19] may also reflect disparity between the minor histocompatibility antigens. These loci have not yet been characterized structurally.

II. CLASS I RLA ANTIGENS AND GENES

A. Class I RLA Antigen and β_2-Microglobulin

Several attempts were made in the late 1960s and early 1970s to isolate histocompatibility antigens from rabbit lymphoid cells,[21,22] but the first successful isolation of rabbit Mhc antigen by Kimball et al.[23] was possible only after the efficient method for isolating membrane proteins had been developed and the basic structure of Mhc molecules elucidated in mouse and human.[24] The finding that β_2-microglobulin associates noncovalently with class I molecules allows the use of anti-β_2-microglobulin to purify class I molecules from cell lysates.[25]

In rabbits, β_2-microglobulin can be readily purified from urine of rabbits with induced kidney diseases. The complete primary structure of rabbit β_2-microglobulin reveals 72.4% sequence identity with the human counterpart.[26] The most extensive region of identity is between residues 8 and 20. Rabbit β_2-microglobulin is most homologous to the CH3 domain of rabbit Ig molecules. There appears to be minimal polymorphism of rabbit β_2-microglobulin, although this has not been studied extensively.

By using anti-β_2-microglobulin affinity chromatography, Kimball et al. purified class I Mhc molecules from NP-40-solubilized membrane glycoproteins of RL-5, a T-lymphocyte cell line derived from an inbred B/J (RLA-A11) rabbit.[23] After separating β_2-microglobulin by gel filtration in formic acid, isoelectric focusing analysis of the 43-kD protein peak reveals, unexpectedly, only one major protein. N-terminal amino acid analysis of this 43-kD glycoprotein discloses 89 and 82% amino acid sequence identity with HLA-B7 and H-2K^b molecules, respectively. Later, Wilkinson et al. employed two anti-class I murine monoclonal antibodies, which were derived by immunizing mice with the glycoprotein fraction of the RL-5 cells, to further characterize class I molecules from RL-5 cells.[27] Both antibodies immunoprecipitate the same 43-kD class I glycoprotein from RL-5 cells as the anti-β_2-microglobulin, again confirming the earlier result. However, sequential immunoprecipitation and immunoelectron microscopic study with the anti-rabbit class I monoclonal antibody and anti-β_2-microglobulin reveals two species of class I molecules in the RL-5 cell line: one species is associated with β_2-microglobulin on the cell surface, the other is not associated with β_2-microglobulin and is located in the cytoplasm and along the inner surface of cell membrane. These two species do not differ in either the N-terminal amino acid sequence or electrophoretic mobility by two-dimensional gel electrophoretic analysis. It appears that RL-5 cells overexpress class I heavy chain and, due to the relative paucity of β_2-microglobulin, some of these molecules that fail to associate with β_2-microglobulin remain sequestered inside the cells.[3]

B. Class I RLA Genes

The observation that the RL-5 cell line predominantly expressed only one species of class I molecules is quite distinct from the situation in human and mouse in which several class I gene products are documented. To detect other class I molecules, Tykocinski et al. prepared a cDNA library from the RL-5 cell line and identified more than 100 clones that cross-hybridized with the H-2L^d mouse Mhc class I probe.[28] Restriction enzyme analysis of 31 randomly picked clones reveals a major group of abundantly expressed (26/31) clones and three other groups of rare and structurally atypical clones. The largest group, represented by pR9, encodes predominant class I protein identified earlier from the same cell line. The second group consists of two clones (pR14 and pR26) which contain an unusual inframe stop codon only 18 bases 3′ to the first ATG codon. The translation from the 5′ most AUG codon would yield a very short peptide, while translation from the second AUG codon may be inefficient and yields class I molecule with slightly shortened signal sequence.

Representing the third group, a unique pR11 clone also contains an inframe stop codon in exon 5 in otherwise typical class I message.[3] This mutation results in translated product which lacks the cytoplasmic tail. Cloning and sequence analysis of corresponding genomic clone reveals the germ line origin of this mutation and exclude the possiblity of cloning artifacts.[3] The last cDNA clone, pR27, lacks some parts of the 5′ sequence and is very unusual in the organization.[29] This clone encodes a portion of $\alpha 1$ domain and typical $\alpha 2$ and $\alpha 3$ domains, but the typical transmembrane region is not present. Sequences 3′ to the $\alpha 3$ coding region is replaced with parts of intron 4 and exon 5 sequences; translation through the intron would result in additional 49 amino acid region which contains up to 12 charged residues. Based on S1 nuclease analysis, the intron 4 sequence is not spliced out of the mRNA species in RL-5 cells.[29] The clone pR27 also contains a highly repetitive sequence at the 3′ end. Analysis of cDNA clones thus reveals that there are at least four transcriptionally active RLA class I genes in RL-5 cells and that one of these is predominantly expressed.

The multiplicity of class I genes in the rabbit genome is assessed further by Southern blot hybridization, employing a highly conserved exon 4 sequence derived from a class I genomic clone.[3,29,30] Up to eight hybridizing bands are detected in HindIII-digested genomic DNA of RL-5 cells and rabbits homozygous for two RLA haplotypes (Figure 1).[3,29] With BamHI digestion, from four to nine bands are found among rabbits carrying 10 RLA-A haplotypes (Table 2).[30] In addition, a single copy gene corresponding to the unusual pR27 cDNA is observed in all rabbits tested (Table 2). Thus, with an approximate number of 8 to 13 genes, the RLA class I gene

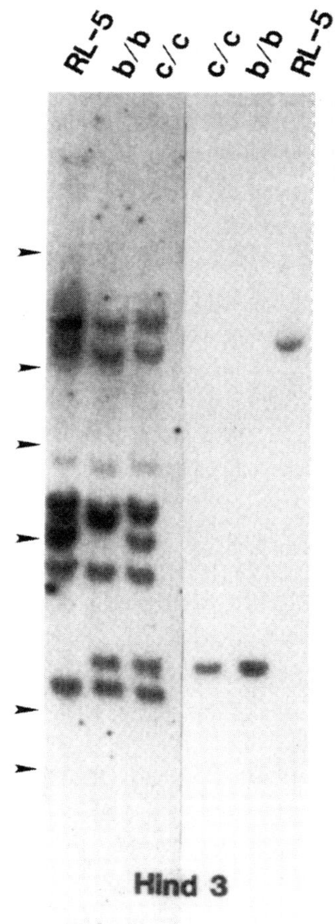

Figure 1 Southern blot analysis of rabbit genomic DNA with RLA class I exon 4 and pR27 probes. Genomic DNA from the RL-5 cells and RLA-2 (b/b) and RLA-3 (c/c) homozygous rabbits was digested with HindIII and probed with a RLA class I exon 4 probe (left-hand panel) and pR27 3′ probe (right hand panel). Size markers are HindIII-digested λ DNA. (From Rebierre, M.-C., Marche, P.N., and Kindt, T.J., *J. Immunol.,* 139, 2066, 1987. With permission.)

family appears to be less complex than those of several other species.[31] Currently, a number of these RLA class I genes has been cloned and partially characterized. Based on differential hybridization with multiple cDNA and oligonucleotide probes, up to four distinct subfamilies of these genes are recognized.[3] Except for the third subfamily which contains only one member, the other three subfamilies appear to be composed of multiple genes. At least one member of the first and second subfamilies and the sole member of the third subfamily are known to be expressed; their messages correspond to the cDNA clones pR9, pR11, and pR27, respectively.[3] These class I genes are present in isolated recombinant phage clones; how they relate

Table 2 List of RLA-A Haplotypes as Defined by RFLP Analysis[a]

RLA-A Haplotype	Probe (restriction enzyme)[b] RLA-A exon 4 (19-1-3′) (BamHI)							pR27-3′ (BglII)
A1	18[c]	11	8.5	3.7				5.6
A2	18	6.1	4.7	3.7				5.6
A3	18	11	6.1	3.7				7.2
A5	18	11	8.5[d]	6.7	6.1	4.7[d]	3.7	7.2
A6	18	11	10	6.7	6.1	4.7	3.7	5.6
A7	18	14	6.7	3.7				5.6
A8	25	18	11	6.1	3.7			7.2
A11	18	11	10	6.7	6.1	4.7	3.7	3
A13	18	11	8.5	4.7	3.7			4.4
A15	25	18	11	6.1	4.7			10
(A16)	18	14	11[e]	6.7	6.1[e]	3.7		5.6

[a] Adapted from Reference 30.
[b] Probes are listed in Table 4.
[c] Size of restriction fragment in kb.
[d] May be doublets.
[e] The status of haplotype 16 with respect to these fragments is not yet clear.

to each other topologically in the rabbit genome is not yet known. The possibility that rabbits also maintain class Ib Mhc genes will also be known from further analysis of these genes.

Genomic organization of the most abundantly expressed class I gene is determined by comparison of the pR9 cDNA clone with the corresponding genomic clone, 19-1.[32] Similar to the functional class I genes from other species, the coding sequences are organized in eight exons. They encode, respectively, the signal sequence, the first, second, and third extracellular domains, the transmembrane portion, two parts of the cytoplasmic tail, and the 3′ untranslated portion. As in other class I genes, the promoter region contains CAAT and TATA elements required for proper expression. It is striking that sequence similarity between the clone 19-1 and a human Mhc class I gene is observed in both coding sequences and noncoding regions all through the length of the genes.[32]

From available coding sequences of four class I genes from the RLA-A11 haplotype, structural differences among these class I genes appear to be concentrated in the $\alpha1$ and $\alpha2$ extracellular domains.[3,28,29] The $\alpha3$ domains are the most conserved. This has been thought to reflect the need to generate polymorphic peptide-binding regions in the $\alpha1$ and $\alpha2$ domains and at the same time to conserve the $\alpha3$ domain for interacting with the common subunit, β_2-microglobulin, in the assembly into functional class I molecules.[3]

C. Restriction Fragment Length Polymorphism

Allotypic variation of rabbit RLA class I molecules has not yet been assessed at the primary sequence level. However, an exon 4 probe (probe 19-1-3′) has been used to identify polymorphism of restriction enzyme sites within and around the class I genes.[30] The highly conserved nature of this exon allows simultaneous identification of multiple class I genes. With BamHI digestion, distinct patterns of hybridizing bands have been correlated with known RLA-A alleles defined previously by serology in nine out of ten haplotypes tested (Tables 1 and 2).[30] RFLP analysis also identifies two new RLA-A haplotypes present in an extended rabbit family.[30]

D. Expression

Expression of RLA class I genes has been studied mainly in the virally transformed RL-5 T cell line. As mentioned earlier, at least four class I genes are now known to be transcriptionally active in this line.[28,29] By employing the conserved exon 4 probe, four mRNA species with the size of 1.4, 1.9, 2.8, and 3.9 kb are detected.[29] In this line, pR27 probe hybridized with two large mRNA species. At the protein level, products of only one gene, 19-1, can be demonstrated with the use of anti-β_2-microglobulin and two anti-class I monoclonal antibodies.[23,27] It remains to be determined whether the predominant expression of a class I gene in the RL-5 line reflects the situation in the B/J rabbit and other rabbits in general or whether it is an unusual phenomenon resulting from the transformation of a T cell by a herpes virus.

In the rabbit thymus, two class I mRNA species, 1.4 and 2.8 kb in size, can be demonstrated with the exon 4 probe.[29] Only the 1.4-kb form is present in lymph node and liver tissue. Expression of the unsual gene corresponding to pR27 appears to be limited to the thymus; its mRNA is not found in brain, testis, lymph node, liver, or muscle.[29]

III. CLASS II RLA ANTIGENS AND GENES

Although studies of the rabbit Mhc with serological techniques and MLR tests provided suggestive evidence for two classes of Mhc molecules in rabbits, the recombinant animals that are crucial for the subdivision of the "locus" by classical genetics have not been identified. On the other hand, biochemical and molecular genetic studies are instrumental in identifying distinct molecular species corresponding to the Mhc class I and class II molecules described in other animal species and allowing the fine subdivision of the RLA locus accordingly.

A. Analysis of Class II RLA Antigens

By cross-immunization of rabbits with lymphoid cells from spleen, mesenteric lymph nodes, and Peyer's patches and by absorbing resulting alloantisera with donor sera to remove antibodies specific for immunoglobulins, Knight et al. obtained two alloantisera, anti-Ia1 and anti-Ia2, that recognized antigens primarily found on B-lymphocyte surface.[33] The antigens are composed of two proteins of molecular weight 28,000 and 30,000 to 32,000 and, by pedigree analysis, appear to be products of a polymorphic gene identical or closely linked to the RLA-D locus. Functional study is possible after an isolation of a monoclonal antibody, 2C4, which is specific for a monomorphic determinant on cell surface antigens recognized by both anti-Ia1 and anti-Ia2 alloantisera.[34] When 2C4 is included in the one-way MLR cultures, lymphocyte proliferation is inhibited by 80%. The antibody 2C4 also inhibits *in vitro* proliferative response of rabbit lymphocytes to bovine insulin, poly ($Glu^{50}Tyr^{50}$) and bacterial lipopolysaccharide. These results reveal that 2C4 recognizes cell surface proteins that are involved in the MLR and in the interaction between B and T lymphocytes and strongly suggest that the molecules recognized by the alloantisera and 2C4 are products of the RLA-D locus.

Analysis of immunoprecipitates of 2C4 and metabolically labeled rabbit spleen cell lysate by two-dimensional nonequilibrium pH gradient electrophoresis reveals multiple protein spots similar in distribution to the alpha, beta, and invariant components of class II Mhc molecules in other species.[35] Partial N-terminal amino acid sequence determination of solubilized alpha chain and beta chain spots identifies that most spots are homologous to HLA-DQ alpha and beta chains. An exceptional alpha spot is more homologous to the HLA-DR alpha chain.[35] Although not all spots can be analyzed, this analysis gives suggestive evidence for more than one type of RLA-D molecule.

B. Class II RLA Genes

All of the class II RLA genes identified so far have been isolated from cDNA libraries and genomic libraries by cross hybridization with known HLA or H-2 probes. Five distinct alpha chain genes and eight beta chain genes have been cloned.[36-39] Because previous information is not available on the subdivision of the RLA-D locus and also because of the fact that, except for the nonfunctional RLA-DFa gene, other RLA class II genes are highly related to corresponding HLA class II genes, current nomenclature of rabbit class II genes is based on that of the HLA system. From the linkage of genes within the cosmid and phage recombinant clones, the RLA class II region has been divided into three subregions: RLA-DP, -DQ, and -DR.[37]

1. RLA-DP Subregion

In this subregion, two alpha chain genes, RLA-DPα1 and -DPα2, and a beta chain gene, RLA-DPβ, are tightly linked within a 45 kb distance (Figure 2).[37] The original designation of these genes based on cross hybridization with human class II probes is confirmed by subsequent sequence analysis of the RLA-DPα1 and -DPβ genes which shows 80% and 86%, respectively, amino acid identity of the second extracellular domains of the rabbit genes with the human counterparts.[40] There is evidence from Southern blot hybridization of genomic DNA for an additional DPβ-like chain in the rabbit genome,[38] but it appears to lie outside a 35-kb stretch of DNA in the 5′ direction of the RLA-DPα2 gene that has been cloned from this subregion.[37] The organization of genes within the RLA-DP subregion slightly differs from the human cluster: in the rabbit the DPα1-DPβ duplex appears to have undergone an inversion event when compared with the RLA-DPα2, resulting in the

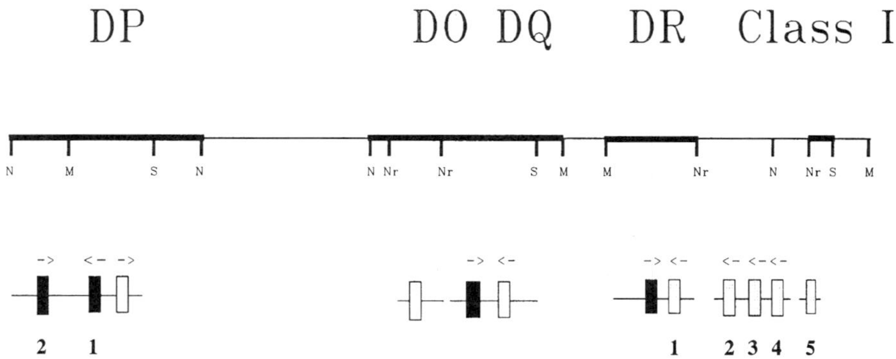

Figure 2 Schematic representation of the RLA chromosomal region. Upper panel, linkage of RLA class I and class II genes as determined by pulsed field gel electropheretic analysis; thick bars represent the smallest fragments that hybridize to the RLA-DPα probe (50- and 120-kb NotI + M1uI fragment), the RLA-DOβ and RLA-DQα probes (150-kb NotI + M1uI fragment), the RLA-DRα probe (47-kb M1uI + NruI fragment), and pR27 class I probe (18-kb NruI + Sa1I fragment). N, NotI; M, M1uI; Nr, NruI; S, Sa1I. Lower panel, linkage of RLA class II genes as determined by overlapping cosmid and phage recombinant clones; RLA-DP subregion, left panel; RLA-DQ subregion, middle panel; and RLA-DR subregion, right panel. Solid box, RLA class II alpha gene; clear box, class II beta gene; arrow represents transcriptional orientation of the gene in the direction 5′ → 3′; numbers represent RLA-DPα1 and -DPα2 (left panel) and RLA-DRβ1-5 (right panel). The RLA-DFα gene, HLA-DNA-like gene and an additional DPβ-like gene have not been mapped. Not drawn to scale. (Adapted from References 37, 38, and 44.)

following array, DPα2-DPα1-DPβ, instead of DPB2-DPA2-DPB1-DPA1 as found in human.

The RLA-DPα1 and RLA-DPβ genes from the RLA-D10 haplotype encode typical Mhc class II molecules.[40] However, unfavorable nucleotides surrounding the first AUG codon of the RLA-DPβ gene, at least in the RLA-D10 haplotype, have been thought to result in reduced translational efficiency of its mRNA. Although the coding sequences of the RLA-DPβ gene are most homologous to the HLA-DPB gene, this gene contains a complex deletion/insertion mutation in the second exon which is absent from all 32 known HLA-DPB alleles.[40,41] Because this mutation is also found at a similar site of the H-2Aβ1 gene of five haplotypes, it is assumed to have been arisen by a gene conversion-like event after speciation.[40] However, it is also possible that, according to the trans-species hypothesis of Mhc evolution,[42] rabbit RLA-DPβ allele may descend from one member of a pool of class II β genes, which already exist before mammalian diversification, while the HLA-DPB alleles derive from another β ancestor allele present in the same pool.

2. RLA-DQ Subregion

A tightly linked RLA-DQα and -DQβ duplex has been cloned (Figure 2).[36,37] Unlike the situation in human where two HLA-DQA-DQB duplexes have been found, there is no evidence for other DQ-like genes in the rabbits.[36,37,43] In accordance to the biochemical analysis of spleen cell with 2C4 monoclonal antibody, the RLA-DQα gene from the RL-5 cell line encodes functional class II alpha chain.[43] It also shares the highest level of sequence identity with the HLA-DQA gene.[42]

The RLA-DOB gene has been isolated recently also by cross hybridization with the corresponding HLA probe.[39] Evidence from pulse field gel electrophoretic analysis indicates that this gene is located on the same 150-kb NotI-MluI fragment as the RLA-DQα gene in the RLA-D1 haplotype, indicating linkage of this gene with the RLA-DQ subregion.[44] In the human genome, an isolated DOB gene is also found linked in the centromeric side of the HLA-DQ cluster.[45] The RLA-DOB gene shares 74% amino acid identity to HLA-DOB and 67% to the mouse H-2Ob gene product.[43]

3. RLA-DR Subregion

The RLA-DR subregion is the most complex class II gene cluster in rabbit. A functional RLA-DRα gene and a related, but nonfunctional, RLA-DFα gene have been cloned.[36,37,46,47] In addition, up to five RLA-DRβ genes have been detected and cloned from two genomic libraries.[37,38]

Similar to RLA-DPα and -DQα genes, the RLA-DRα gene encodes a typical Mhc class II alpha chain; it is most homologous to its human counterpart, the HLA-DRα gene, at all four exons than to other human or rabbit class II alpha chain genes.[46] In contrast, the RLA-DFα gene is a small remnant containing only exon 4 coding sequence.[47] The RLA-DFα gene is also unusual in that (1) the exon 4 sequence is preceded by a noncoding stretch derived from a part of intron 4 in the RLA-DRα gene and (2) the intron 4-exon 4 sequence is flanked on both sides by a 130-bp direct repeat with 85% sequence identity.[47] Like the RLA-DRα gene, the RLA-DFα gene has been detected in all rabbits tested as well as in hare and cottontail genomes, indicating that, following its possible generation by a gene duplication event in the rabbit ancestor, the RLA-DFα gene is well preserved through the evolution of lagomorphs.[47]

The evidence for five RLA-DRβ genes in the rabbit genome has been obtained from genomic blot hybridization and molecular cloning. Southern blot hybridization of the rabbit genome from six RLA-D haplotypes with several exon 3-containing RLA-DRβ probes reveals four to seven hybridizing bands depending on the restriction enzyme used (Figure 3, Table 3).[38] Characterization of these RLA-DRβ genes is achieved by cloning them from genomic libraries and identifying regions which are less conserved for differential hybridization.[37,38] It is evident that at high stringency of hybridization, the 3′ probe, which contains sequences 3′ to the conserved exon 3, derived from a phage or cosmid clone hybridizes only to itself and to its most homologous gene(s). From all cross hybridizations among eight cloned genes from two genomic libraries, only five unique RLA-DRβ genes, designated RLA-DRβ1-5, can be detected. With the exception of RLA-DRβ3 and -DRβ4 genes, the 3′ probe from any of the remaining RLA-DRβ genes detects only one distinct strongly hybridizing band in the genomic blot analysis of RLA-D homozygous rabbits, instead of four to seven bands with the exon 3-containing probes.[38] In an exceptional case, the 3′ probes derived from RLA-DRβ3 and -DRβ4 genes cross

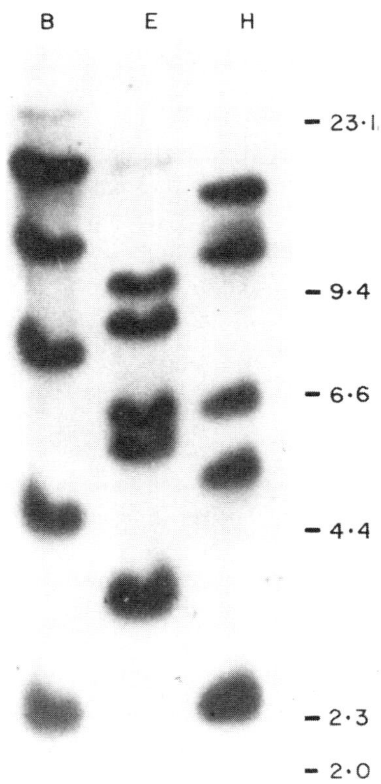

Figure 3 Southern blot analysis of rabbit genomic DNA with RLA-DRβ probe. Genomic DNA from a RLA-11 homozygous rabbit was digested wirh BamHI (B), EcoRI (E), and HindIII (H) and probed with the β3 exon from RLA-DRβ1 gene. Size markers in kb are indicated. (From Sittisombut, N., Tissot, R., and Knight, K. L., *J. Immunogenetics,* 16, 63, 1989. With permission.)

hybridize with each other in the genomic blot analysis; indicating closer sequence homology between these two genes.[38] Thus, cross hybridization experiment of cloned genes with less conserved probes reveals the presence of five distinct RLA-DRβ genes in the rabbit genome.

In two recombinant clones derived from two RLA-D haplotypes, the RLA-DRβ1 gene is found closely linked to the RLA-DRα gene in the 3′-3′ orientation.[36-38] In the second group of overlapping clones, the RLA-DRβ2, -DRβ3, -DRβ4 genes are found in tandem array in the 5′-3′ orientation.[38] The RLA-DRβ5 gene has not yet been linked with other clusters. Currently, it is not yet clear whether all of the RLA-DRβ genes encode functional class II beta molecules. It is possible that some of these genes may be structurally defective as has been observed previously in human.[48]

LeGuern et al. previously isolated five distinct class II α genes, but only four of these have been characterized fully.[36] The fifth gene, originally designated RLA-DPα2, is dissimilar to another RLA-DPα2 gene that is found in close linkage with the RLA-DPα1-DPβ duplex by cosmid cloning and by mapping with pulsed field

Table 3 List of RLA-D Haplotypes as Defined by RFLP Analysis

RLA-D Haplotype	Locus (restriction enzyme)							
	DPα1 (BamHI)	DPα2 (HindIII)	DPβ (EcoRI)	DQα (PvuII)	DQα (EcoRI)	DQβ (HindIII)	DRα (EcoRI)	DRβ[a] (BamHI)
D1	14[b]	8.3	[3.8,8.5][c,d]	8	2.8	3.3	6	2.4,5.2,7.4,9.2,27.5
D2	14	8.3	[3.8,8.5]	1.8	6.3	[5.5,6.1]	5.4	2.4,5.2,7.2,10.5,12,27.5
D3a	14	8.3	[2.7,8.5]	5.9	3.7	[2.8,8.4]	4.8	2.4,4.5,7.2,10.5,12,27.5
D3c	24[e]	7.6	[3.8,8.5]	5.9	3.7	[2.8,8.4]	4.8	2.4,4.5,7.2,10.5,12,27.5
D4	14	nd	nd	3	9.6	nd	6	nd
D5	24[e]	8.3	[3.8,8.5]	nd	7.8	16.5	5.6	2.4,5.9,7.2,12,27.5
D6	14	nd	nd	6.3	2.8	nd	4.8	nd
D7	24	nd	nd	4.7	6.3	nd	5.4	nd
D8	14	nd	nd	6.3	6.3	nd	6	nd
D10	14	8.3	[3.8,8.5]	nd	3.7	[2.8,8.4][f]	4.8	2.4,4.5,7.3,12,18.5
D15	4	nd	nd	3	3.3	nd	6	nd
D17	14	nd	nd	3.7	3.7	nd	6	nd
Ref.	30,38	38	38	30	30,38	38	30,38	38

[a] Pattern represents all five RLA-DRβ genes.
[b] Size of restriction fragment in kb.
[c] [ ] represents cleavage within an allelic form.
[d] Additional hybridizing band due to another DPβ-like gene is observed, see Reference 38.
[e] Reported as the >27.5-kb band in Reference 38.
[f] Reported as the [2.9, 8.4]-kb fragments in Reference 38, but are likely to be identical to the [2.8, 8.4] fragments of D3 haplotypes.

gel electrophoretic analysis.[37,45] This unusual clone is a candidate for the rabbit homolog of the HLA-DNA gene.[3,49] Whether rabbits also retain HLA-DM-like genes remains to be investigated.

C. Restriction Fragment Length Polymorphism

Polymorphism of the rabbit Mhc class II region has been studied only by RFLP analysis. Based on a limited set of restriction enzymes used, the most extensive allotypic variation of RFLP pattern is observed with the RLA-DQα gene (Table 3).[30] As a group, the RLA-DRβ genes also appear to be quite polymorphic and distinct RFLP patterns of these genes correlate with the RLA-D typing data in much the same way as the class I genes (Table 3). It is clear from the RFLP analysis that MLR reactivities in rabbits are likely to be governed by the RLA-DQ and RLA-DR subregion genes; rabbits with three distinct RLA-A haplotypes which do not cross stimulate in the MLR test are found to be identical in the RLA-DQ and RLA-DR subregions but they do differ by the RLA-DP RFLP patterns.[30,38]

Based on available data from the analysis of 13 RLA-A haplotypes and 12 RLA-D haplotypes, RFLP analysis is sufficiently sensitive to distinguish most serological defined and MLR-defined haplotypes tested. RFLP can now be regarded as a valuable tool for identifying RLA haplotypes. A minimum set of probe/enzyme combinations that is needed for this purpose includes (1) for RLA-A typing, the class I exon 4 probe (19-1-3′)/BamHI digestion, (2) for general RLA-D typing, the RLA-DQα probe/PvuII and EcoRI digestions, and (3) for subtyping of the RLA-D3 haplotype, the RLA-DPα1 probe/BamHI digestion (Table 4).[30,38] It should be noted that, although RFLP analysis has been useful in determining Mhc-disease association in rabbits,[50] it may not truly reflect coding sequence polymorphism at certain loci. Indeed, RFLP of the RLA-DOB gene has been documented despite the identity of β1 domain coding sequences in at least eight RLA haplotypes.[39] Similarly, Han et al. did not observe the correlation of RFLP of the RLA-DRα gene with variations in the coding sequence of the α1 domain of this molecule.[50]

D. Expression

Expression of the RLA class II genes has been determined mostly at the mRNA level. Messenger RNA corresponding to the RLA-DQα,-DQβ, -DRα, and -DRβ genes are readily detected in rabbit spleen cells by RNA blot analysis.[51,52] Comparatively much lower levels of the RLA-DPα1, -DPα2, and -DPβ mRNA are present in spleen. These class II messages can also be detected in other tissues including bone marrow, appendix, lymph node, and thymus.[51] Low levels of the RLA-DOβ mRNA transcripts appear to be restricted to appendix, spleen, and lymph node.[51]

Only the alpha and beta chains belonging to RLA-DQ subregion are identified clearly by N-terminal amino acid sequence analysis of immunoprecipitated glycoprotein spots from rabbit spleen.[35] As mentioned earlier, partial sequence of an additional alpha spot is compatible with an expressed RLA-DRα chain, but the possibility that the DRα-like spot is in fact an allelic form of the RLA-DQα chain cannot be entirely ruled out. Expression of the class II genes from the RLA-DR and

Table 4 List of Probes and Alleles/Patterns Detected by RFLP Analysis

Locus	Probe[a]	Region	Enzyme	No. of alleles/ haplotypes tested	Alleles/patterns (kb)	Ref.
Class I						
R9	19-1-3′(g)	Exons 6–8 (0.5 kb)[b]	BamHI	10/10	10 patterns of multiple fragments, see Table 2	30, 32
R27	R27-3′(c)	3′ UT (0.35 kb)	Bg1II	5/10	3, 4.4, 5.6, 7.2, 10	29, 30
Class II						
DPα1	DPα,DPα1 (g)	Exons 2,3[c] (0.6 kb)	BamHI	3/12	4, 14, 24[d]	30, 38, 51
DPα2	DPα2 (g)	Exons 3,4 (1.45 kb)	HindIII	2/6	7.6, 8.3	38
DPβ	DPβ (g)	Exon 3 (2 kb)	EcoRI	2/6	[2.7, 8.5],[e] [3.8, 8.5][e]	38
DQα	DQα (g)	Exons 3,4[c] (0.5 kb)	PvuII	6/10	1.8, 3, 4.7, 5.9, 6.3, 8	30, 38, 43
			EcoRI	6/12	2.8, 3.3, 3.7, 6.3, 7.8, 9.6	
DQβ	DQβ (g)	Exon 3 (2 kb)	HindIII	4/6	[2.8, 8.4],[e] 3.3, [5.5, 6.1],[e] 16.5	38
DRα	DRα (g)	Exons 3,4[c] (0.8 kb)	EcoRI	4/12	2.4,[f] 4.8, 5.4, 5.6, 6	30, 36, 38, 50
DRβ	7.1-3′ (c)	Exon 3 and other 3′ sequences (0.7 kb)	BamHI	5/6	5 patterns of multiple fragments, see Table 3	38
DFα	DN,RDF (g)	Intron 4 and exon 4 (0.5 kb)	BglII	4/10	2.7, 4.7, 8, 10	30, 47

[a] Probes were derived from either genomic clone (g) or cDNA clone (c).
[b] Size of probe in kb.
[c] Multiple similar probes have been employed but only the smallest probe is shown.
[d] Reported as the >27.5-kb band in Reference 38.
[e] Represents cleavage within an allelic form.
[f] Present in an outbred rabbit of unknown haplotype.

-DP subregions has now been tested directly by transfecting corresponding genes into a murine lymphoma cell line.[52] Stably transfected cells expressing the transcripts of RLA-DRa and -DRβ2 genes are reactive with a new monoclonal antibody, RDR34, which recognizes non-DQ class II molecules, indicating the surface expression of the RLA-DR glycoproteins derived from the α and β2 genes. Although the four remaining RLA-DRβ genes have not yet been tested by transfection, mRNA transcripts of RLA-DRβ1 and -DRβ3 or -DRβ4 genes can be found by northern blot hybridization of total RNA from rabbit spleen.[52] Expression of the RLA-DP genes has not been detected by the transfection experiment at either the RNA or protein level.[52]

E. Elements of the Promoter Region

Constitutive expression of the Mhc class II genes *in vivo* is limited to a few cell types.[53] In other cells, various inducers, such as IFNγ, IL-4, and IFNγ plus TNF, are required.[53,54] Transcriptional control of expression of both types is exerted through the interaction of trans-acting, DNA-binding proteins with corresponding cis-acting elements in the promoter region or intron of class II genes.[54,55] Among several cis-acting elements that are well conserved in human and murine class II genes are the W, X, and Y boxes.[54-56] Other elements, such as the octamer motive, are found in some but not all class II genes. Sequence analysis of the promoter region of RLA-DRα and RLA-DPβ genes reveals that these elements are conserved in the rabbit class II genes.[37,46] Based on sequence homologies, the RLA-DQα gene promoter appears to contain the X and Y boxes in addition to the octamer motive.[46] The significance of sequence homology has been assessed in the RLA-DPβ promoter by testing the interaction between the rabbit X and Y boxes with specific DNA-binding factors derived from a murine B cell line.[57] Both of the X and Y boxes in the rabbit DPβ chain gene can be recognized by murine factors in a specific manner.[57] The role of these elements in the transcriptional control has not yet been tested with functional assays in rabbit cells.

IV. GENOMIC ORGANIZATION OF RLA GENES

The organization of RLA genes in rabbit genome has been assessed in two haplotypes by pulsed field gel analysis of singly or doubly digested genomic DNA followed by hybridization with class I and II probes.[44] Similar to the class II region in human, the order of RLA class II subregions is DP-DOβ-DQ-DR (Figure 2). Within each subregion, the exact spatial orientation of the α-β genes is not yet clear. However, it is at least possible to deduce from the distribution of the NruI sites among cloned genes that, in the case that the RLA-DP subregion is centromeric to the RLA-DR subregion as is in human, both the RLA-DPβ and -DRβ1 genes are likely to be located telomeric to the respective α genes (Figure 2).[37,44] Based on the hybridization results derived with three RLA α probes and the RLA-DOβ probe, the size of rabbit class II region is estimated to be 630 and 1200 kb, respectively, in the RLA-1 and -2 haplotypes. The difference in size of the class II region between the two haplotypes is due to a 700-kb insertion between the RLA-DQ and -DR subregions.[44]

Due to the multiplicity of class I genes, the size of RLA class I region is not clearly known. A class I exon 4 probe reveals four MluI-digested bands with the size of 80, 120, 180, and 240 kb.[44] Quite unexpectedly, the 180-kb band also hybridizes to pR27 and RLA-DRα probes, indicating a very close proximity of the class I and II regions in rabbit genome. Moreover, because no physical linkage of the C4, CY21, C2, and HSP70 genes with the class I or class II regions can be detected in this analysis, it appears that the RLA class I and class II regions are connected to each other without intervening class III genes as in human and mouse.[44,45] Such organization is also distinct from those of chicken where there exist distinct Mhc regions separated on two linkage groups. Within each Mhc region, chicken class I and II genes intermingle with each others.[58,59]

V. DISEASE ASSOCIATION

Study of the association between RLA genes and disease in rabbits is possible following the development of RFLP analysis. Currently, the only example of disease-Mhc association in rabbit is the case of regression and malignant conversion of virus-induced papilloma.[50] Infection of cottontail rabbits and domestic rabbits with Shope cottontail rabbit papillomavirus leads to the development of skin warts which may regress spontaneously or persist for a very long time. Among persistors, a large fraction develops invasive carcinomas.[60] In domestic rabbits, the heterologous but preferred host due to a higher proportion of malignant conversion, papilloma regression occurs in 10 to 40% of all papilloma-bearing animals whereas epidermoid carcinomas develop in as much as 73%.[60,61] Histologically, papilloma regression is associated with intense leukocyte infiltration of the tumor tissue, most prominently at the epidermal basement membrane and subjacent dermis.[62,63] Activated T cells expressing RLA-DQ antigens constitute the majority of infiltrating cells. In contrast, much fewer leukocytes and T cells are observed distributed randomly in the dermis of progressing papilloma.[62,63] Paradoxically, enhanced antibody and T-cell proliferative responses to viral structural proteins are documented in rabbits with progressive lesions.[64,65] In progressor rabbits, viral antigen-specific T cell proliferation has been elicited from leukocytes derived from peripheral blood, lymph nodes, and spleen. Such higher systemic immunologic response in progressor rabbits have been attributed to the difference in cellular pattern of viral gene expression; expression of viral structural genes is limited to the uppermost epithelial layer in papilloma lesions but is widespread over the entire epithelial tissue in dysplastic lesions and in carcinomas.[66]

It has been observed earlier that the proportion of spontaneous regression and malignant conversion varies in domestic rabbits from different commercial suppliers.[62] RFLP analyzes of RLA class I and II genes and TCR β genes of papilloma-bearing outbred New Zealand rabbits reveal significant associations between early regression (within 1 to 6 weeks) with the 5.6-kb (EcoRI) RLA-DRα allele and the 3.0-kb (PvuII) RLA-DQα allele.[50] On the other hand, malignant conversion is associated with the 1.8-kb (PvuII) RLA-DQα allele. These findings are consistent with the variations in regression/progression frequency noted above. Most importantly, RLA linkage with papilloma progression is consistent with the linkage between HLA-DQw3 and cervical carcinoma in human.[67-69] At present, it is not yet clear how genetic predisposition is related to the final outcomes of papilloma development or to the contrasting histologic pictures and systemic immunologic responses in rabbits bearing regressing and progressing tumors. In this regard, rabbits will continue to play an important role as a model for studying the interaction between Mhc, B- and T-cell functions, and viral-induced tumor development.

VI. CONCLUSION AND FUTURE PERSPECTIVES

It is evident that much has been learned about rabbit Mhc during the last 2 decades following the demonstration of this fascinating family of genes in early 1970s. One of the most frequently recurring themes is the close similarity between RLA genes and their HLA counterparts. Thus far, resemblance of coding sequences, conservation of the major class II subregions, and similarity of genomic organization of class II genes are well documented. Based on these similarities, rabbits may help serve as an experimental model for studying Mhc-disease association for human diseases, especially with regards to class II genes as in the case of virus-induced papillomas. Furthermore, the contraction of RLA-class I genes may help simplify studies of class I gene function.

Clearly, there is much more to be studied. As has been summarized in a recent short review on Mhc-papilloma association, there is an urgent need to "...know more of rabbit Mhc genetics: how many class II loci are expressed, and the allelic variation and the mapped locations of each, and the haplotypes and the location and variability of other candidate genes including the tumor necrosis factor and peptide transporter loci".[70] The need is indeed more pressing as the potential of rabbit for the experimental studies of HIV and HTLV infections has now been realized.[71,72]

REFERENCES

1. Medawar, P. B., The behaviour and fate of skin autografts and skin homografts in rabbits, *J. Anat.*, 78, 176, 1944.
2. Cohen, C. and Tissot, R. G., Specialized research application. II. Serological genetics, in *The Biology of Laboratory Rabbits*, Weisbroth S. H., Flatt, R. E., and Kraus, A. L., Eds., Academic Press, New York, 1974, chap. 7.
3. Rebiere, M. C., Marche, P. N., LeGuern, C., and Kindt, T. J., Molecular studies of the rabbit major histocompatibility complex, in *The Rabbit in Contemporary Immunological Research*, Dubisky, S., Ed., Longman Scientific, Essex, 1987, chap. 4.
4. Klein, J., *Biology of the Mouse Histocompatibility-2 Complex: Principles of Immunogenetics Applied to a Single System*, Springer-Verlag, Berlin, 1975.
5. Medawar, P. B., A second study of the behaviour and fate of skin homografts in rabbits, *J. Anat.*, 79, 157, 1945.

6. Medawar, P. B., Immunity to homologous grafted skin. II. The relationship between the antigens of blood and skin, *Br. J. Exp. Pathol.*, 27, 15, 1946.
7. Tissot, R. G. and Cohen, C., Histocompatibility in the rabbit. Identification of a major locus, *Tissue Antigens*, 2, 267, 1972.
8. Chai, C. K., Effects of inbreeding in rabbits, *J. Hered.*, 60, 64, 1969.
9. Chai, C. K., Rabbit inbreeding, in *Inbred and Genetically Defined Strains of Laboratory Animals*, Altman, P. L. and Katz, D. D., Eds., Federation of American Societies for Experimental Biology, Maryland, 1979, 565.
10. Chai, C. K., Genetic studies of histocompatibility in rabbits: identification of major and minor genes, *Immunogenetics*, 1, 126, 1974.
11. Knight, K. L., Spieker-Polet, H., Kazdin, D., and Oi, V. T., Transgenic rabbits with lymphocytic leukemia induced by the c-myc oncogene fused with the immunoglobulin heavy chain enhancer, *Proc. Natl. Acad. Sci. U.S.A.,* 85, 3130, 1988.
12. Sachs, D., The major histocompatibility complex, in *Fundamental Immunology*, Paul, W. E., Ed., Raven Press, New York, 1984, chap. 13.
13. Chai, C. K., The effect of inbreeding in rabbits: skin transplantation, *Transplantation*, 6, 689, 1968.
14. Chai, C. K. and Chiang, M. S. M., A method of estimating the number of histocompatibility loci in a sib-mating mouse population, *Genetics*, 48, 1153, 1963.
15. Kapitchnikov, M. M., Ballantyne, D. L., Jr., and Stetson, C. A., Immunological reactions to skin homotransplantation in rabbits, *Ann. N.Y. Acad. Sci.*, 99, 497, 1962.
16. Cohen, C., DePalma, R. G., Colberg, J. E., Tissot, R. G., and Hubay, C. A., The relationship between blood groups and histocompatibility in the rabbit, *Ann. N.Y. Acad. Sci.*, 120, 356, 1964.
17. Tissot, R. G. and Cohen, C., Histocompatibility in the rabbit. Linkage between RL-A, MLC, and the He blood group loci, *Transplantation*, 18, 142, 1974.
18. Cohen, C. and Tissot, R. G., The effect of the RL-A locus and the MLC locus on graft survival in the rabbit, *Transplantation*, 18, 150, 1974.
19. Lancki, D. W., Tissot, R. G., and Cohen, C., Histocompatibility in the rabbit. Genetic control of rabbit mixed leukocyte culture reactivity, *Transplantation*, 27, 79, 1979.
20. Tissot, R. G., Lancki, D. W., and Blaesing, M. E., Homozygous cell typing for the rabbit RLA-D antigens in animals from commercial breeder, *Immunogenetics*, 8, 509, 1979.
21. Harris, T. N., Ogburn, C. A., and Harris, S., Soluble rabbit histocompatibility antigens: some physical properties, serologic reactions, and relation to the suppression of allogeneic lymph node cells, *Transplantation*, 6, 670, 1968.
22. Bash, J. A., Dubiski, S., and Cinader, B., A search for markers of rabbit lymphoid cells; alloantigens characterized by antisera to rabbit thymus, *Int. Arch. Allergy*, 47, 369, 1974.
23. Kimball, E. S., Coligan, J. E., and Kindt, T. J., Structural characterization of antigens encoded by rabbit RLA-11 histocompatibility genes, *Immunogenetics*, 8, 201, 1979.
24. Nairn, R., Yamaga, K., and Nathenson, S. G., Biochemistry of the gene products from the murine Mhc mutants, *Annu. Rev. Genet.*, 14, 241, 1980.
25. Grey, H. M., Kubo, R. T., Colon, S. M., Poulik, M. D., Cresswell, P., Springer, T., Turner, M., and Strominger, J. L., The small subunit of HLA antigens is β_2-microglobulin, *J. Exp. Med.*, 138, 1608, 1973.
26. Gates, F. T., III, Coligan, J. E., and Kindt, T. J., Complete amino acid sequence of rabbit β_2-microglobulin, *Biochemistry*, 8, 2267, 1979.
27. Wilkinson, J. M., Tykocinski, M. L., Coligan, J. E., Kimball, E. S., and Kindt, T. J., Rabbit Mhc antigens: occurrence of non-β_2-microglobulin-associated class I molecules, *Mol. Immunol.*, 19, 1441, 1982.

28. Tykocinski, M. L., Marche, P. N., Max, E. E., and Kindt, T. J., Rabbit class I Mhc genes: cDNA clones defined full-length transcripts of an expressed gene and a putative pseudogene, *J. Immunol.*, 133, 2261, 1984.
29. Rebiere, M.-C., Marche, P. N., and Kindt, T. J., A rabbit class I major histocompatibility complex gene with a T cell-specific expression pattern, *J. Immunol.*, 139, 2066, 1987.
30. Marche, P. N., Rebiere, M. C., Laverriere, A., English, D. W., LeGuern, C., and Kindt, T. J., Definition of rabbit class I and class II Mhc gene haplotypes using molecular typing procedures, *Immunogenetics*, 29, 273, 1989.
31. Kindt, T. J. and Singer, D. S., Class I major histocompatibility complex genes in vertebrate species: what is the common denominator? *Immunol. Res.*, 6, 57, 1987.
32. Marche, P. N., Tykocinski, M. L., Max, E. E., and Kindt, T. J., Structure of a functional rabbit class I Mhc gene: similarity of human class I genes, *Immunogenetics*, 21, 71, 1985.
33. Knight, K. L., Leary, A. L., and Tissot, R. G., Identification of two Ia-like alloantigens on rabbit B lymphocytes, *Immunogenetics*, 10, 443, 1980.
34. Lobel, S. A. and Knight, K. L., The role of rabbit Ia molecules in immune functions as determined with the use of an anti-Ia monoclonal antibody, *Immunology*, 51, 35, 1984.
35. Knight, K. L., Johnson, A., Coligan, J. E., and Kindt, T. J., Partial amino acid sequence analysis of rabbit Mhc class II molecules isolated from two-dimensional polyacryamide gels, *Mol. Immunol.*, 24, 449, 1987.
36. LeGuern, C., Marche, P. N., and Kindt, T. J., Molecular evidence for five distinct Mhc class II α genes in the rabbit, *Immunogenetics*, 22, 141, 1985.
37. Sittisombut, N. and Knight, K. L., Rabbit major histocompatibility complex. I. Isolation and characterization of three subregions of class II genes, *J. Immunol.*, 136, 1871, 1986.
38. Sittisombut, N., Tissot, R., and Knight, K. L., Rabbit major histocompatibility complex. III. Multiple class II DRβ genes and restriction fragment length polymorphism of the class II α and β genes, *J. Immunogenetics*, 16, 63, 1989.
39. Chouchane, L., Brown, T. J., and Kindt, T. J., Structure and expression of a nonpolymorphic rabbit class II gene with homology to HLA-DOB, *Immunogenetics*, 38, 64, 1993.
40. Sittisombut, N., Mordacq, J., and Knight, K. L., Rabbit Mhc II. Sequence analysis of the R-DPα- and β-genes, *J. Immunol.*, 140, 3237, 1988.
41. Marsh, S. G. E. and Bodmer, J. G., HLA class II nucleotide sequences, *Tissue Antigens*, 40, 229, 1992.
42. Klein, J., Satta, Y., O'hUigin, C., and Takahata, N., The molecular descent of the major histocompatibility complex, *Annu. Rev. Immunol.*, 11, 269, 1993.
43. LeGuern, C., Weissman, J. D., Marche, P. N., Jouvin-Marche, E., Laverriere, A., Bagnato, M. R., and Kindt, T. J., Sequence determination of a transcribed rabbit class II gene with homology to HLA-DQα, *Immunogenetics*, 25, 104, 1987.
44. Chouchane, L. and Kindt, T. K., Mapping of the rabbit Mhc reveals that class I genes are adjacent to the DR subregion and defines an insertion/deletion-related polymorphism in the class II region, *J. Immunol.*, 149, 1216, 1992.
45. Trowsdale, J., Genomic structure and function in the Mhc, *Trends Genet.*, 9, 117, 1993.
46. Laverriere, A., Kulaga, H., Kindt, T. J., LeGuern, C., and Marche, P. N., A rabbit class II Mhc gene with strong similarities to HLA-DRA, *Immunogenetics*, 30, 137, 1989.
47. Marche, P. N., Laverriere, A., LeGuern, C., and Kindt, T. J., A gene remnant in the rabbit Mhc related to HLA-DRα, *Res. Immunol.*, 142, 525, 1991.

48. Trowsdale, J., Ragoussis, J., and Campbell, R. D., Map of the human Mhc, *Immunology Today*, 12, 443, 1991.
49. Kindt, T. J., personal communication, 1994.
50. Han, R., Breitburd, F., Marche, P. N., and Orth, G., Linkage of regression and malignant conversion of rabbit viral papillomas to Mhc class II genes, *Nature*, 356, 66, 1992.
51. Kulaga, H., Sogn, J. A., Weissman, J. D., Marche, P. N., LeGuern, C., Long, E. O., and Kindt, T. J., Expression patterns of Mhc class II genes in rabbit tissues indicate close homology to human counterparts, *J. Immunol.*, 139, 587, 1987.
52. Spieker-Polet, H., Sittisombut, N., Yam, P.-C., and Knight, K. L., Rabbit major histocompatibility complex. IV. Expression of major histocompatibility complex class II genes, *J. Immunogenet.*, 17, 123, 1990.
53. Mengle-Gaw, L. and McDewitt, H. O., Genetics and expression of mouse Ia antigens, *Annu. Rev. Immunol.*, 3, 367, 1985.
54. Glimcher, L. H. and Kara, C. J., Sequences and factors: a guide to Mhc class-II transcription, *Annu. Rev. Immunol.*, 10, 13, 1992.
55. Benoist, C. and Mathis, D., Regulation of major histocompatibility complex class II genes: X, Y and other letters of the alphabet, *Annu. Rev. Immunol.*, 8, 681, 1990.
56. Sukiennicki, T. L., Shewey, L. M., and Nepom, G. T., Locus- and allele-specific DNA-protein interaction in the HLA-DQB1 X Box, *Immunol. Res.*, 12, 317, 1993.
57. Sittisombut, N., Two distinct nuclear factors bind the conserved regulatory sequences of a rabbit major histocompatibility complex class II gene, *Mol. Cell. Biol.*, 8, 2034, 1988.
58. Kroemer, G. A., Bernot, A., Behar, G., Chausse, A. M., Gastinel, L. N., Guillemot, F., Park, I., Thoroval, P., Zoorob, R., and Auffray, C., Molecular genetics of the chicken Mhc: current status and evolutionary aspects, *Immunol. Rev.*, 113, 119, 1990.
59. Miller, M. M., Goto, R., Bernot, A., Zoorob, R., Auffray, C., Bumstead, N., and Briles, W. E., Two Mhc class I and two Mhc class II genes map to the chicken Rfp-Y system outside the B complex, *Proc. Natl. Acad. Sci. U.S.A.*, 91, 4397, 1994.
60. Syverton, J. T., The pathogenesis of the rabbit papilloma-to-carcinoma sequence, *Ann. N.Y. Acad. Sci.*, 54, 1126, 1952.
61. Kreider, J. W. and Bartlett, G. L., The Shope papilloma-carcinoma complex of rabbits: a model system of neoplastic progression and spontaneous regression. *Adv. Cancer Res.*, 35, 81, 1981.
62. Okabayashi, M., Angell, M. G., Christensen, N. D., and Kreider, J. W., Morphometric analysis and identification of infiltrating leukocytes in regressing and progressing shope rabbit papillomas, *Int. J. Cancer*, 49, 919, 1991.
63. Okabayashi, M., Angell, M. G., Christensen, N. D., and Kreider, J. W., Shope papilloma cell and leukocyte proliferation in regressing and progressing lesions, *Am. J. Pathol.*, 142, 489, 1993.
64. Lin, Y.-L., Borenstein, L. A., Selvakumar, R., Ahmed, R., and Wettstein, F. O., Progression from papilloma to carcinoma is accompanied by changes in antibody response to papillomavirus proteins, *J. Virol.*, 67, 382, 1993.
65. Selvakumar, R., Borenstein, L. A., Lin, Y.-L., Ahmed, R., and Wettstein, F. O., T cell response to cotton tail rabbit papillomavirus structural proteins in infected rabbits, *J. Virol.*, 68, 4043, 1994.
66. Zeltner, R., Borenstein, L. A., Wettstein, F. O., and Iftner, T., Changes in RNA expression pattern during the malignant progression of cottontail rabbit papillomavirus-induced tumors in rabbits, *J. Virol.*, 68, 3620, 1994.
67. Wank, R. and Thomssen, C., High risk of squamous cell carcinoma of the cervix for women with HLA-DQw3, *Nature (London)*, 352, 723, 1991.

68. Wank, R., Schendel, D. J., and Thomssen, C., HLA antigens and cervical carcinoma (correspondence), *Nature (London)*, 356, 22, 1992.
69. Helland, A., Borresen, A. L., Kaern, J., Ronningen, K. S., and Thorsby, E., HLA antigens and cervical carcinoma (correspondence), *Nature (London)*, 356, 23, 1992.
70. Klitz, W., Viruses, cancer and the Mhc, *Nature (London)*, 356, 17, 1992.
71. Kindt, T. J., Hirsh, V. M., Johnson, P. R., and Sawasdikosol, S., Animal models for acquired immunodeficiency syndrome, *Adv. Immunol.*, 52, 425, 1992.
72. Sawasdikosol, S. and Kindt, T. J., Infection of the laboratory rabbit with HIV-1 and HTLV-I, *AIDS Res. Review*, 2, 211, 1992.

CHAPTER 12

Future Perspectives

The driving force for characterizing the major histocompatibility complex (Mhc) of domestic animals has been to develop new strategies in genetic progress in providing healthier food animals. Rising concerns for the growing widespread use of antibiotics and the impact of stress on an animal's health resulting from changing farm management practices have fueled both basic and applied animal scientists to seek additional knowledge regarding the Mhc. For as new knowledge emerges, as clearly illustrated in the chapters presented in this book, the reality of providing new tools for making genetic progress of animal health and production efficiency via integrated management of the Mhc is closer.

Clearly with reagents for each of the species, the fidelity of the locus-specific primers used for polymerase chain reaction (PCR)-based genotyping efforts will need to be verified with various diverse populations. Whether the typing reagents are allele-specific oligonucleotides or PCR-based restriction fragment length polymorphisms (RFLPs), further efforts must focus on understanding the number of loci expressed in Mhc haplotypes. As illustrated in this book, many methods have been used to characterize the polymorphisms of the Mhc genes and their products. Although serological and biochemical analysis have been very successful in characterizing these molecules, future efforts will most likely focus on the development of DNA-based methodologies because of the difficulties in producing and standardizing biological reagents such as antisera (whether allo- or monoclonal).

After further reagent development, appropriate studies associating specific Mhc haplotypes, or specific alleles, can be initiated with respect to immune responses, disease resistance, or as polymorphic anchored genetic markers linked to other economically important traits. The emergence of the mapping of farm animal genomes and the use of marker-assisted selection as a tool for improved farm animal breeding provides exciting opportunities. Moreover, the restructuring of the animal production industries, most notably of the poultry and swine producers, toward fully vertically integrated companies, provides yet another stimulus for understanding the Mhc of domestic animal species.

0-8493-8032-4/96/$0.00+$.50

Finally, to develop strategies for epitope mapping associated with recombinant subunit vaccine development, it will be imperative to have further characterization of both class I and class II encoded alleles. Recent research efforts outlined here also will permit the identification of common structural motifs for peptide vaccines. Thus, "promiscuous" peptides can be developed which bind to all Mhc molecules and increase vaccine efficacy. Such peptides can serve to induce not only protection of the animal from infectious disease but also induce signaling of cell growth or differentiation, and enhance other economically important physiological processes.

Clearly, information gleaned from investigating the Mhc of domestic animals has already proven valuable in identifying disease-resistant and -susceptible animals in ongoing production schemes. Future applications as suggested above will assist in the ever-important quest to produce safer and cheaper food in feeding the world populations.

Lawrence B. Schook
Susan J. Lamont

Index

A

B

D

E

F

G

H

J

K

L

M

N

O

P

Q

R

S

T

U

V

W

X

Z